Lecture Notes in Computer Science   10367

Commenced Publication in 1973
Founding and Former Series Editors:
Gerhard Goos, Juris Hartmanis, and Jan van Leeuwen

Lei Chen · Christian S. Jensen
Cyrus Shahabi · Xiaochun Yang
Xiang Lian (Eds.)

Web and Big Data

First International Joint Conference, APWeb-WAIM 2017
Beijing, China, July 7–9, 2017
Proceedings, Part II

 Springer

Editors
Lei Chen
Computer Science and Engineering
Hong Kong University of Science and
  Technology
Hong Kong
China

Christian S. Jensen
Computer Science
Aarhus University
Aarhus N
Denmark

Cyrus Shahabi
Computer Science
University of Southern California
Los Angeles, CA
USA

Xiaochun Yang
Northeastern University
Shenyang
China

Xiang Lian
Kent State University
Kent, OH
USA

ISSN 0302-9743   ISSN 1611-3349 (electronic)
Lecture Notes in Computer Science
ISBN 978-3-319-63563-7   ISBN 978-3-319-63564-4 (eBook)
DOI 10.1007/978-3-319-63564-4

Library of Congress Control Number: 2017947034

LNCS Sublibrary: SL3 – Information Systems and Applications, incl. Internet/Web, and HCI

Printed on acid-free paper

This Springer imprint is published by Springer Nature
The registered company is Springer International Publishing AG
The registered company address is: Gewerbestrasse 11, 6330 Cham, Switzerland

Preface

This volume (LNCS 10366) and its companion volume (LNCS 10367) contain the proceedings of the first Asia-Pacific Web (APWeb) and Web-Age Information Management (WAIM) Joint Conference on Web and Big Data, called APWeb-WAIM. This new joint conference aims to attract participants from different scientific communities as well as from industry, and not merely from the Asia Pacific region, but also from other continents. The objective is to enable the sharing and exchange of ideas, experiences, and results in the areas of World Wide Web and big data, thus covering Web technologies, database systems, information management, software engineering, and big data. The first APWeb-WAIM conference was held in Beijing during July 7–9, 2017.

As a new Asia-Pacific flagship conference focusing on research, development, and applications in relation to Web information management, APWeb-WAIM builds on the successes of APWeb and WAIM: APWeb was previously held in Beijing (1998), Hong Kong (1999), Xi'an (2000), Changsha (2001), Xi'an (2003), Hangzhou (2004), Shanghai (2005), Harbin (2006), Huangshan (2007), Shenyang (2008), Suzhou (2009), Busan (2010), Beijing (2011), Kunming (2012), Sydney (2013), Changsha (2014), Guangzhou (2015), and Suzhou (2016); and WAIM was held in Shanghai (2000), Xi'an (2001), Beijing (2002), Chengdu (2003), Dalian (2004), Hangzhou (2005), Hong Kong (2006), Huangshan (2007), Zhangjiajie (2008), Suzhou (2009), Jiuzhaigou (2010), Wuhan (2011), Harbin (2012), Beidaihe (2013), Macau (2014), Qingdao (2015), and Nanchang (2016). With the fast development of Web-related technologies, we expect that APWeb-WAIM will become an increasingly popular forum that brings together outstanding researchers and developers in the field of Web and big data from around the world.

The high-quality program documented in these proceedings would not have been possible without the authors who chose APWeb-WAIM for disseminating their findings. Out of 240 submissions to the research track and 19 to the demonstration track, the conference accepted 44 regular (18%), 32 short research papers, and ten demonstrations. The contributed papers address a wide range of topics, such as spatial data processing and data quality, graph data processing, data mining, privacy and semantic analysis, text and log data management, social networks, data streams, query processing and optimization, topic modeling, machine learning, recommender systems, and distributed data processing.

The technical program also included keynotes by Profs. Sihem Amer-Yahia (National Center for Scientific Research, CNRS, France), Masaru Kitsuregawa (National Institute of Informatics, NII, Japan), and Mohamed Mokbel (University of Minnesota, Twin Cities, USA) as well as tutorials by Prof. Reynold Cheng (The University of Hong Kong, SAR China), Prof. Guoliang Li (Tsinghua University, China), Prof. Arijit Khan (Nanyang Technological University, Singapore), and

Prof. Yu Zheng (Microsoft Research Asia, China). We are grateful to these distinguished scientists for their invaluable contributions to the conference program.

As a new joint conference, teamwork is particularly important for the success of APWeb-WAIM. We are deeply thankful to the Program Committee members and the external reviewers for lending their time and expertise to the conference. Special thanks go to the local Organizing Committee led by Jun He, Yongxin Tong, and Shimin Chen. Thanks also go to the workshop co-chairs (Matthias Renz, Shaoxu Song, and Yang-Sae Moon), demo co-chairs (Sebastian Link, Shuo Shang, and Yoshiharu Ishikawa), industry co-chairs (Chen Wang and Weining Qian), tutorial co-chairs (Andreas Züfle and Muhammad Aamir Cheema), sponsorship chair (Junjie Yao), proceedings co-chairs (Xiang Lian and Xiaochun Yang), and publicity co-chairs (Hongzhi Yin, Lei Zou, and Ce Zhang). Their efforts were essential to the success of the conference. Last but not least, we wish to express our gratitude to the Webmaster (Zhao Cao) for all the hard work and to our sponsors who generously supported the smooth running of the conference.

We hope you enjoy the exciting program of APWeb-WAIM 2017 as documented in these proceedings.

June 2017

Xiaoyong Du
Beng Chin Ooi
M. Tamer Özsu
Bin Cui
Lei Chen
Christian S. Jensen
Cyrus Shahabi

Organization

Organizing Committee

General Co-chairs

Xiaoyong Du — Renmin University of China, China
BengChin Ooi — National University of Singapore, Singapore
M. Tamer Özsu — University of Waterloo, Canada

Program Co-chairs

Lei Chen — Hong Kong University of Science and Technology, China
Christian S. Jensen — Aalborg University, Denmark
Cyrus Shahabi — The University of Southern California, USA

Workshop Co-chairs

Matthias Renz — George Mason University, USA
Shaoxu Song — Tsinghua University, China
Yang-Sae Moon — Kangwon National University, South Korea

Demo Co-chairs

Sebastian Link — The University of Auckland, New Zealand
Shuo Shang — King Abdullah University of Science and Technology, Saudi Arabia
Yoshiharu Ishikawa — Nagoya University, Japan

Industrial Co-chairs

Chen Wang — Innovation Center for Beijing Industrial Big Data, China
Weining Qian — East China Normal University, China

Proceedings Co-chairs

Xiang Lian — Kent State University, USA
Xiaochun Yang — Northeast University, China

Tutorial Co-chairs

Andreas Züfle — George Mason University, USA
Muhammad Aamir Cheema — Monash University, Australia

ACM SIGMOD China Lectures Co-chairs

Guoliang Li	Tsinghua University, China
Hongzhi Wang	Harbin Institute of Technology, China

Publicity Co-chairs

Hongzhi Yin	The University of Queensland, Australia
Lei Zou	Peking University, China
Ce Zhang	Eidgenössische Technische Hochschule ETH, Switzerland

Local Organization Co-chairs

Jun He	Renmin University of China, China
Yongxin Tong	Beihang University, China
Shimin Chen	Chinese Academy of Sciences, China

Sponsorship Chair

Junjie Yao	East China Normal University, China

Web Chair

Zhao Cao	Beijing Institute of Technology, China

Steering Committee Liaison

Yanchun Zhang	Victoria University, Australia

Senior Program Committee

Dieter Pfoser	George Mason University, USA
Ilaria Bartolini	University of Bologna, Italy
Jianliang Xu	Hong Kong Baptist University, SAR China
Mario Nascimento	University of Alberta, Canada
Matthias Renz	George Mason University, USA
Mohamed Mokbel	University of Minnesota, USA
Ralf Hartmut Güting	Fernuniversität in Hagen, Germany
Seungwon Hwang	Yongsei University, South Korea
Sourav S. Bhowmick	Nanyang Technological University, Singapore
Tingjian Ge	University of Massachusetts Lowell, USA
Vincent Oria	New Jersey Institute of Technology, USA
Walid Aref	Purdue University, USA
Wook-Shin Han	Pohang University of Science and Technology, Korea
Yoshiharu Ishikawa	Nagoya University, Japan

Program Committee

Alex Delis	University of Athens, Greece
Alex Thomo	University of Victoria, Canada

Aviv Segev	Korea Advanced Institute of Science and Technology, South Korea
Baoning Niu	Taiyuan University of Technology, China
Bin Cui	Peking University, China
Bin Yang	Aalborg University, Denmark
Carson Leung	University of Manitoba, Canada
Chih-Hua Tai	National Taipei University, China
Cuiping Li	Renmin University of China, China
Daniele Riboni	University of Cagliari, Italy
Defu Lian	University of Electronic Science and Technology of China, China
Dejing Dou	University of Oregon, USA
Demetris Zeinalipour	Max Planck Institute for Informatics, Germany and University of Cyprus, Cyprus
Dhaval Patel	Indian Institute of Technology Roorkee, India
Dimitris Sacharidis	Technische Universität Wien, Vienna, Austria
Fei Chiang	McMaster University, Canada
Ganzhao Yuan	South China University of Technology, China
Giovanna Guerrini	Universita di Genova, Italy
Guoliang Li	Tsinghua University, China
Guoqiong Liao	Jiangxi University of Finance and Economics, China
Hailong Sun	Beihang University, China
Han Su	University of Southern California, USA
Hiroaki Ohshima	Kyoto University, Japan
Hong Chen	Renmin University of China, China
Hongyan Liu	Tsinghua University, China
Hongzhi Wang	Harbin Institute of Technology, China
Hongzhi Yin	The University of Queensland, Australia
Hua Li	Aalborg University, Denmark
Hua Lu	Aalborg University, Denmark
Hua Wang	Victoria University, Melbourne, Australia
Hua Yuan	University of Electronic Science and Technology of China, China
Iulian Sandu Popa	Inria and PRiSM Lab, University of Versailles Saint-Quentin, France
James Cheng	Chinese University of Hong Kong, SAR China
Jeffrey Xu Yu	Chinese University of Hong Kong, SAR China
Jiaheng Lu	University of Helsinki, Finland
Jiajun Liu	Renmin University of China, China
Jialong Han	Nanyang Technological University, Singapore
Jian Yin	Zhongshan University, China
Jianliang Xu	Hong Kong Baptist University, SAR China
Jianmin Wang	Tsinghua University, China
Jiannan Wang	Simon Fraser University, Canada
Jianting Zhang	City College of New York, USA
Jianzhong Qi	University of Melbourne, Australia

Jinchuan Chen	Renmin University of China, China
Ju Fan	National University of Singapore, Singapore
Jun Gao	Peking University, China
Junfeng Zhou	Yanshan University, China
Junhu Wang	Griffith University, Australia
Kai Zeng	University of California, Berkeley, USA
Karine Zeitouni	PRISM University of Versailles St-Quentin, Paris, France
Kyuseok Shim	Seoul National University, Korea
Lei Zou	Peking University, China
Lei Chen	Hong Kong University of Science and Technology, SAR China
Leong Hou U.	University of Macau, SAR China
Liang Hong	Wuhan University, China
Lianghuai Yang	Zhejiang University of Technology, China
Long Guo	Peking University, China
Man Lung Yiu	Hong Kong Polytechnical University, SAR China
Markus Endres	University of Augsburg, Germany
Maria Damiani	University of Milano, Italy
Meihui Zhang	Singapore University of Technology and Design, Singapore
Mihai Lupu	Vienna University of Technology, Austria
Mirco Nanni	ISTI-CNR Pisa, Italy
Mizuho Iwaihara	Waseda University, Japan
Mohammed Eunus Ali	Bangladesh University of Engineering and Technology, Bangladesh
Peer Kroger	Ludwig-Maximilians-University of Munich, Germany
Peiquan Jin	Univerisity of Science and Technology of China
Peng Wang	Fudan University, China
Yaokai Feng	Kyushu University, Japan
Wookey Lee	Inha University, Korea
Raymond Chi-Wing Wong	Hong Kong University of Science and Technology, SAR China
Richong Zhang	Beihang University, China
Sanghyun Park	Yonsei University, Korea
Sangkeun Lee	Oak Ridge National Laboratory, USA
Sanjay Madria	Missouri University of Science and Technology, USA
Shengli Wu	Jiangsu University, China
Shi Gao	University of California, Los Angeles, USA
Shimin Chen	Chinese Academy of Sciences, China
Shuai Ma	Beihang University, China
Shuo Shang	King Abdullah University of Science and Technology, Saudi Arabia
Sourav S Bhowmick	Nanyang Technological University, Singapore
Stavros Papadopoulos	Intel Labs and MIT, USA
Takahiro Hara	Osaka University, Japan
Taketoshi Ushiama	Kyushu University, Japan

Tieyun Qian	Wuhan University, China
Ting Deng	Beihang University, China
Tru Cao	Ho Chi Minh City University of Technology, Vietnam
Vicent Zheng	Advanced Digital Sciences Center, Singapore
Vinay Setty	Aalborg University, Denmark
Wee Ng	Institute for Infocomm Research, Singapore
Wei Wang	University of New South Wales, Australia
Weining Qian	East China Normal University, China
Weiwei Sun	Fudan University, China
Wei-Shinn Ku	Auburn University, USA
Wenjia Li	New York Institute of Technology, USA
Wen Zhang	Wuhan University, China
Wolf-Tilo Balke	Braunschweig University of Technology, Germany
Xiang Lian	Kent State University, USA
Xiang Zhao	National University of Defence Technology, China
Xiangliang Zhang	King Abdullah University of Science and Technology, Saudi Arabia
Xiangmin Zhou	RMIT University, Australia
Xiaochun Yang	Northeast University, China
Xiaofeng He	East China Normal University, China
Xiaoyong Du	Renmin University of China, China
Xike Xie	University of Science and Technology of China, China
Xingquan Zhu	Florida Atlantic University, USA
Xuan Zhou	Renmin University of China, China
Yanghua Xiao	Fudan University, China
Yang-Sae Moon	Kangwon National University, South Korea
Yasuhiko Morimoto	Hiroshima University, Japan
Yijie Wang	National University of Defense Technology, China
Yingxia Shao	Peking University, China
Yong Zhang	Tsinghua University, China
Yongxin Tong	Beihang University, China
Yoshiharu Ishikawa	Nagoya University, Japan
Yu Gu	Northeast University, China
Yuan Fang	Institute for Infocomm Research, Singapore
Yueguo Chen	Renmin University of China, China
Yunjun Gao	Zhejiang University, China
Zakaria Maamar	Zayed University, United Arab Emirates
Zhaonian Zou	Harbin Institute of Technology, China
Zhengjia Fu	Advanced Digital Sciences Center, Singapore
Zhiguo Gong	University of Macau, SAR China
Zouhaier Brahmia	University of Sfax, Tunisia

Contents – Part II

Machine Learning

Combining Node Identifier Features and Community Priors
for Within-Network Classification 3
   Qi Ye, Changlei Zhu, Gang Li, and Feng Wang

An Active Learning Approach to Recognizing Domain-Specific Queries
From Query Log. 18
   Weijian Ni, Tong Liu, Haohao Sun, and Zhensheng Wei

Event2vec: Learning Representations of Events on Temporal Sequences 33
   Shenda Hong, Meng Wu, Hongyan Li, and Zhengwu Wu

Joint Emoji Classification and Embedding Learning 48
   Xiang Li, Rui Yan, and Ming Zhang

Target-Specific Convolutional Bi-directional LSTM Neural Network
for Political Ideology Analysis 64
   Xilian Li, Wei Chen, Tengjiao Wang, and Weijing Huang

Boost Clickbait Detection Based on User Behavior Analysis 73
   Hai-Tao Zheng, Xin Yao, Yong Jiang, Shu-Tao Xia, and Xi Xiao

Recommendation Systems

A Novel Hybrid Friends Recommendation Framework for Twitter 83
   Yan Zhao, Jia Zhu, Mengdi Jia, Wenyan Yang, and Kai Zheng

A Time and Sentiment Unification Model for Personalized
Recommendation 98
   Qinyong Wang, Hongzhi Yin, and Hao Wang

Personalized POI Groups Recommendation in Location-Based
Social Networks 114
   Fei Yu, Zhijun Li, Shouxu Jiang, and Xiaofei Yang

Learning Intermediary Category Labels for Personal Recommendation. 124
   Wenli Yu, Li Li, Jingyuan Wang, Dengbao Wang, Yong Wang,
   Zhanbo Yang, and Min Huang

Skyline-Based Recommendation Considering User Preferences 133
   Shuhei Kishida, Seiji Ueda, Atsushi Keyaki, and Jun Miyazaki

Improving Topic Diversity in Recommendation Lists: Marginally
or Proportionally? 142
   Xiaolu Xing, Chaofeng Sha, and Junyu Niu

Distributed Data Processing and Applications

Integrating Feedback-Based Semantic Evidence to Enhance Retrieval
Effectiveness for Clinical Decision Support 153
   Chenhao Yang, Ben He, and Jungang Xu

Reordering Transaction Execution to Boost High Frequency Trading
Applications 169
   Ningnan Zhou, Xuan Zhou, Xiao Zhang, Xiaoyong Du, and Shan Wang

Bus-OLAP: A Bus Journey Data Management Model for Non-on-time
Events Query 185
   Tinghai Pang, Lei Duan, Jyrki Nummenmaa, Jie Zuo, and Peng Zhang

Distributed Data Mining for Root Causes of KPI Faults
in Wireless Networks. 201
   Shiliang Fan, Yubin Yang, Wenyang Lu, and Ping Song

Precise Data Access on Distributed Log-Structured Merge-Tree 210
   *Tao Zhu, Huiqi Hu, Weining Qian, Aoying Zhou, Mengzhan Liu,
   and Qiong Zhao*

Cuttle: Enabling Cross-Column Compression in Distributed Column Stores ... 219
   *Hao Liu, Jiang Xiao, Xianjun Guo, Haoyu Tan, Qiong Luo,
   and Lionel M. Ni*

Machine Learning and Optimization

Optimizing Window Aggregate Functions via Random Sampling 229
   Guangxuan Song, Wenwen Qu, Yilin Wang, and Xiaoling Wang

Fast Log Replication in Highly Available Data Store. 245
   *Donghui Wang, Peng Cai, Weining Qian, Aoying Zhou, Tianze Pang,
   and Jing Jiang*

New Word Detection in Ancient Chinese Literature. 260
   Tao Xie, Bin Wu, and Bai Wang

Identifying Evolutionary Topic Temporal Patterns Based on Bursty
Phrase Clustering 276
   Yixuan Liu, Zihao Gao, and Mizuho Iwaihara

Personalized Citation Recommendation via Convolutional
Neural Networks... 285
    Jun Yin and Xiaoming Li

A Streaming Data Prediction Method Based on Evolving
Bayesian Network.. 294
    Yongheng Wang, Guidan Chen, and Zengwang Wang

A Learning Approach to Hierarchical Search Result Diversification........ 303
    Hai-Tao Zheng, Zhuren Wang, and Xi Xiao

Demo Papers

TeslaML: Steering Machine Learning Automatically in Tencent 313
    Jiawei Jiang, Ming Huang, Jie Jiang, and Bin Cui

DPHSim: A Flexible Simulator for DRAM/PCM-Based Hybrid Memory.... 319
    *Dezhi Zhang, Peiquan Jin, Xiaoliang Wang, Chengcheng Yang,
    and Lihua Yue*

CrowdIQ: A Declarative Crowdsourcing Platform for Improving the
Quality of Web Tables... 324
    Yihai Xi, Ning Wang, Xiaoyu Wu, Yuqing Bao, and Wutong Zhou

OICPM: An Interactive System to Find Interesting Co-location Patterns
Using Ontologies 329
    Xuguang Bao, Lizhen Wang, and Qing Xiao

BioPW: An Interactive Tool for Biological Pathway Visualization
on Linked Data... 333
    Yuan Liu, Xin Wang, and Qiang Xu

ChargeMap: An Electric Vehicle Charging Station Planning System 337
    *Longlong Xu, Wutao Lin, Xiaorong Wang, Zhenhui Xu, Wei Chen,
    and Tengjiao Wang*

Topic Browsing System for Research Papers Based on Hierarchical
Latent Tree Analysis 341
    *Leonard K.M. Poon, Chun Fai Leung, Peixian Chen,
    and Nevin L. Zhang*

A Tool of Benchmarking Realtime Analysis for Massive Behavior Data 345
    Mingyan Teng, Qiao Sun, Buqiao Deng, Lei Sun, and Xiongpai Qin

Interactive Entity Centric Analysis of Log Data 349
    Qiao Sun, Xiongpai Qin, Buqiao Deng, and Wei Cui

A Tool for 3D Visualizing Moving Objects   353
Weiwei Wang and Jianqiu Xu

Author Index   359

Contents – Part I

Tutorials

Meta Paths and Meta Structures: Analysing Large Heterogeneous
Information Networks   3
   Reynold Cheng, Zhipeng Huang, Yudian Zheng, Jing Yan, Ka Yu Wong,
   and Eddie Ng

Spatial Data Processing and Data Quality

TrajSpark: A Scalable and Efficient In-Memory Management System
for Big Trajectory Data   11
   Zhigang Zhang, Cheqing Jin, Jiali Mao, Xiaolin Yang, and Aoying Zhou

A Local-Global LDA Model for Discovering Geographical Topics
from Social Media   27
   Siwei Qiang, Yongkun Wang, and Yaohui Jin

Team-Oriented Task Planning in Spatial Crowdsourcing   41
   Dawei Gao, Yongxin Tong, Yudian Ji, and Ke Xu

Negative Survey with Manual Selection: A Case Study
in Chinese Universities   57
   Jianguo Wu, Jianwen Xiang, Dongdong Zhao, Huanhuan Li, Qing Xie,
   and Xiaoyi Hu

Element-Oriented Method of Assessing Landscape of Sightseeing Spots
by Using Social Images   66
   Yizhu Shen, Chenyi Zhuang, and Qiang Ma

Sifting Truths from Multiple Low-Quality Data Sources   74
   Zizhe Xie, Qizhi Liu, and Zhifeng Bao

Graph Data Processing

A Community-Aware Approach to Minimizing Dissemination in Graphs   85
   Chuxu Zhang, Lu Yu, Chuang Liu, Zi-Ke Zhang, and Tao Zhou

Time-Constrained Graph Pattern Matching in a Large Temporal Graph   100
   Yanxia Xu, Jinjing Huang, An Liu, Zhixu Li, Hongzhi Yin, and Lei Zhao

Efficient Compression on Real World Directed Graphs   116
   Guohua Li, Weixiong Rao, and Zhongxiao Jin

Keyphrase Extraction Using Knowledge Graphs   132
  Wei Shi, Weiguo Zheng, Jeffrey Xu Yu, Hong Cheng, and Lei Zou

Semantic-Aware Partitioning on RDF Graphs   149
  Qiang Xu, Xin Wang, Junhu Wang, Yajun Yang, and Zhiyong Feng

An Incremental Algorithm for Estimating Average Clustering Coefficient
Based on Random Walk   158
  Qun Liao, Lei Sun, He Du, and Yulu Yang

Data Mining, Privacy and Semantic Analysis

Deep Multi-label Hashing for Large-Scale Visual Search Based
on Semantic Graph   169
  Chunlin Zhong, Yi Yu, Suhua Tang, Shin'ichi Satoh, and Kai Xing

An Ontology-Based Latent Semantic Indexing Approach
Using Long Short-Term Memory Networks   185
  Ningning Ma, Hai-Tao Zheng, and Xi Xiao

Privacy-Preserving Collaborative Web Services QoS Prediction
via Differential Privacy   200
  *Shushu Liu, An Liu, Zhixu Li, Guanfeng Liu, Jiajie Xu, Lei Zhao,
  and Kai Zheng*

High-Utility Sequential Pattern Mining with Multiple Minimum
Utility Thresholds   215
  Jerry Chun-Wei Lin, Jiexiong Zhang, and Philippe Fournier-Viger

Extracting Various Types of Informative Web Content via Fuzzy
Sequential Pattern Mining   230
  Ting Huang, Ruizhang Huang, Bowei Liu, and Yingying Yan

Exploiting High Utility Occupancy Patterns   239
  *Wensheng Gan, Jerry Chun-Wei Lin, Philippe Fournier-Viger,
  and Han-Chieh Chao*

Text and Log Data Management

Translation Language Model Enhancement for Community Question
Retrieval Using User Adoption Answer   251
  Ming Chen, Lin Li, and Qing Xie

Holographic Lexical Chain and Its Application in Chinese
Text Summarization   266
  *Shengluan Hou, Yu Huang, Chaoqun Fei, Shuhan Zhang,
  and Ruqian Lu*

Authorship Identification of Source Codes   282
   Chunxia Zhang, Sen Wang, Jiayu Wu, and Zhendong Niu

DFDS: A Domain-Independent Framework for Document-Level
Sentiment Analysis Based on RST   297
   Zhenyu Zhao, Guozheng Rao, and Zhiyong Feng

Fast Follower Recovery for State Machine Replication   311
   Jinwei Guo, Jiahao Wang, Peng Cai, Weining Qian, Aoying Zhou,
   and Xiaohang Zhu

Laser: Load-Adaptive Group Commit in Lock-Free Transaction Logging   320
   Huan Zhou, Huiqi Hu, Tao Zhu, Weining Qian, Aoying Zhou,
   and Yukun He

Social Networks

Detecting User Occupations on Microblogging Platforms:
An Experimental Study   331
   Xia Lv, Peiquan Jin, Lin Mu, Shouhong Wan, and Lihua Yue

Counting Edges and Triangles in Online Social Networks
via Random Walk...   346
   Yang Wu, Cheng Long, Ada Wai-Chee Fu, and Zitong Chen

Fair Reviewer Assignment Considering Academic Social Network   362
   Kaixia Li, Zhao Cao, and Dacheng Qu

Viral Marketing for Digital Goods in Social Networks.   377
   Yu Qiao, Jun Wu, Lei Zhang, and Chongjun Wang

Change Detection from Media Sharing Community.   391
   Naoki Kito, Xiangmin Zhou, Dong Qin, Yongli Ren, Xiuzhen Zhang,
   and James Thom

Measuring the Similarity of Nodes in Signed Social Networks with Positive
and Negative Links   399
   Tianchen Zhu, Zhaohui Peng, Xinghua Wang, and Xiaoguang Hong

Data Mining and Data Streams

Elastic Resource Provisioning for Batched Stream Processing System
in Container Cloud   411
   Song Wu, Xingjun Wang, Hai Jin, and Haibao Chen

An Adaptive Framework for RDF Stream Processing   427
   Qiong Li, Xiaowang Zhang, and Zhiyong Feng

Investigating Microstructure Patterns of Enterprise Network
in Perspective of Ego Network   444
    Xiutao Shi, Liqiang Wang, Shijun Liu, Yafang Wang, Li Pan, and Lei Wu

Neural Architecture for Negative Opinion Expressions Extraction   460
    Hui Wen, Minglan Li, and Zhili Ye

Identifying the Academic Rising Stars via Pairwise Citation
Increment Ranking   475
    Chuxu Zhang, Chuang Liu, Lu Yu, Zi-Ke Zhang, and Tao Zhou

Fuzzy Rough Incremental Attribute Reduction Applying
Dependency Measures   484
    Yangming Liu, Suyun Zhao, Hong Chen, Cuiping Li, and Yanmin Lu

Query Processing

SET: Secure and Efficient Top-k Query in Two-Tiered Wireless
Sensor Networks   495
    Xiaoying Zhang, Hui Peng, Lei Dong, Hong Chen, and Hui Sun

Top-k Pattern Matching Using an Information-Theoretic Criterion
over Probabilistic Data Streams   511
    Kento Sugiura and Yoshiharu Ishikawa

Sliding Window Top-K Monitoring over Distributed Data Streams   527
    Zhijin Lv, Ben Chen, and Xiaohui Yu

Diversified Top-k Keyword Query Interpretation on Knowledge Graphs   541
    Ying Wang, Ming Zhong, Yuanyuan Zhu, Xuhui Li, and Tieyun Qian

Group Preference Queries for Location-Based Social Networks   556
    Yuan Tian, Peiquan Jin, Shouhong Wan, and Lihua Yue

A Formal Product Search Model with Ensembled Proximity   565
    Zepeng Fang, Chen Lin, and Yun Liang

Topic Modeling

Incorporating User Preferences Across Multiple Topics into Collaborative
Filtering for Personalized Merchant Recommendation   575
    Yunfeng Chen, Lei Zhang, Xin Li, Yu Zong, Guiquan Liu,
    and Enhong Chen

Joint Factorizational Topic Models for Cross-City Recommendation   591
    Lin Xiao, Zhang Min, and Zhang Yongfeng

Aligning Gaussian-Topic with Embedding Network
for Summarization Ranking   610
    Linjing Wei, Heyan Huang, Yang Gao, Xiaochi Wei, and Chong Feng

Improving Document Clustering for Short Texts by Long Documents
via a Dirichlet Multinomial Allocation Model   626
    *Yingying Yan, Ruizhang Huang, Can Ma, Liyang Xu, Zhiyuan Ding,
    Rui Wang, Ting Huang, and Bowei Liu*

Intensity of Relationship Between Words: Using Word Triangles
in Topic Discovery for Short Texts   642
    Ming Xu, Yang Cai, Hesheng Wu, Chongjun Wang, and Ning Li

Context-Aware Topic Modeling for Content Tracking in Social Media   650
    Jinjing Zhang, Jing Wang, and Li Li

Author Index   659

Machine Learning

Combining Node Identifier Features
and Community Priors
for Within-Network Classification

Qi Ye[✉], Changlei Zhu, Gang Li, and Feng Wang

Sogou Inc., Beijing, China
{yeqi,zhuchanglei,ligang,wangfeng}@sogou-inc.com

Abstract. With widely available large-scale network data, one hot topic is how to adopt traditional classification algorithms to predict the most probable labels of nodes in a partially labeled network. In this paper, we propose a new algorithm called identifier based relational neighbor classifier (IDRN) to solve the within-network multi-label classification problem. We use the node identifiers in the egocentric networks as features and propose a within-network classification model by incorporating community structure information to predict the most probable classes for unlabeled nodes. We demonstrate the effectiveness of our approach on several publicly available datasets. On average, our approach can provide Hamming score, Micro-F_1 score and Macro-F_1 score up to 14%, 21% and 14% higher than competing methods respectively in sparsely labeled networks. The experiment results show that our approach is quite efficient and suitable for large-scale real-world classification tasks.

Keywords: Within-network classification · Node classification · Collective classification · Relational learning

1 Introduction

Massive networks exist in various real-world applications. These networks may be only partially labeled due to their large size, and manual labeling can be highly cost in real-world tasks. A critical problem is how to use the network structure and other extra information to build better classifiers to predict labels for the unlabelled nodes. Recently, much attention has been paid to this problem, and various prediction algorithms over nodes have been proposed [19,22,25].

In this paper, we propose a within-network classifier which makes use of the first-order Markov assumption that labels of each node are only dependent on its neighbors and itself. Traditional relational classification algorithms, such as WvRn [13] and SCRN [27] classifier, make statistical estimations of the labels through statistics, class label propagation or relaxation labeling. From a different viewpoint, many real-world networks display some useful phenomena, such as clustering phenomenon [9] and scale-free phenomenon [2]. Most real-world networks show high clustering property or community structure, i.e., their nodes are

© Springer International Publishing AG 2017
L. Chen et al. (Eds.): APWeb-WAIM 2017, Part II, LNCS 10367, pp. 3–17, 2017.
DOI: 10.1007/978-3-319-63564-4_1

organized into clusters which are also called communities [8,9]. The clustering phenomenon indicates that the network can be divided into communities with dense connections internally and sparse connections between them. In the dense connected communities, the identifiers of neighbors may capture link patterns between nodes. The scale-free phenomenon indicates the existence of nodes with high degrees [2], and we regard that the identifiers of these high degree nodes can also be useful to capture local patterns. By introducing the node identifiers as fine-grained features, we propose identifier based relational neighbor classifier (IDRN) by incorporating the first Markov assumption and community priors. As well, we demonstrate the effectiveness of our algorithm on 10 public datasets. In the experiments, our approach outperforms some recently proposed baseline methods.

Our contributions are as follows. First, to the best of our knowledge, this is the first time that node identifiers in the egocentric networks are used as features to solve network based classification problem. Second, we utilize the community priors to improve its performance in sparsely labeled networks. Finally, our approach is very effective and easily to implement, which makes it quite applicable for different real-world within-network classification tasks. The rest of the paper is organized as follows. In the next section, we first review related work. Section 3 describes our methods in detail. In Sect. 4, we show the experiment results in different publicly available datasets. Section 5 gives the conclusion and discussion.

2   Related Work

One of the recent focus in machine learning research is how to extend traditional classification methods to classify nodes in network data, and a body of work for this purpose has been proposed. Bhagat et al. [3] give a survey on the node classification problem in networks. They divide the methods into two categories: one uses the graph information as features and the other one propagate existing labels via random walks. The relational neighbor (RN) classifier provides a simple but effective way to solve the node classification problems. Macskassy and Provost [13] propose the weighted-vote relational neighbor (WvRN) classifier by making predictions based on the class distribution of a certain node's neighbors. It works reasonably well for within-network classification and is recommended as a baseline method for comparison. Wang and Sukthankar [27] propose a multi-label relational neighbor classification algorithm by incorporating a class propagated probability obtained from edge clustering. Macskassy et al. [14] also believe that the very high cardinality categorical features of identifiers may cause the obvious difficulty for classifier modeling. Thus there is very little work that has incorporated node identifiers [14]. As we regard that node identifiers are also useful features for node classification, our algorithm does not solely depend on neighbors' class labels but also incorporating local node identifiers as features and community structure as priors.

For within-network classification problem, a large number of algorithms for generating node features have been proposed. Unsupervised feature learning approaches typically exploit the spectral properties of various matrix representations of graphs. To capture different affiliations of nodes in a network, Tang and Liu [23] propose the SocioDim algorithm framework to extract latent social dimensions based on the top-d eigenvectors of the modularity matrix, and then utilize these features for discriminative learning. Using the same feature learning framework, Tang and Liu [24] also propose an algorithm to learn dense features from the d-smallest eigenvectors of the normalized graph Laplacian. Ahmed et al. [1] propose an algorithm to find low-dimensional embeddings of a large graph through matrix factorization. However, the objective of the matrix factorization may not capture the global network structure information. To overcome this problem, Tang et al. [22] propose the LINE model to preserve the first-order and the second-order proximities of nodes in networks. Perozzi et al. [20] present DeepWalk which uses the SkipGram language model [12] for learning latent representations of nodes in a network by considering a set of short truncated random walks. Grover and Leskovec [10] define a flexible notion of a node's neighborhood by random walk sampling, and they propose node2vec algorithm by maximizing the likelihood of preserving network neighborhoods of nodes. Nandanwar and Murty [19] also propose a novel structural neighborhood-based classifier by random walks, while emphasizing the role of medium degree nodes in classification. As the algorithms based on the features generated by heuristic methods such as random walks or matrix factorization often have high time complexity, thus they may not easily be applied to large-scale real-world networks. To be more effective in node classification, in both training and prediction phrases we extract community prior and identifier features of each node in linear time, which makes our algorithm much faster.

Several real-world network based applications boost their performances by obtaining extra data. McDowell and Aha [16] find that accuracy of node classification may be increased by including extra attributes of neighboring nodes as features for each node. In their algorithms, the neighbors must contains extra attributes such as textual contents of web pages. Rayana and Akoglu [21] propose a framework to detect suspicious users and reviews in a user-product bipartite review network which accepts prior knowledge on the class distribution estimated from metadata. To address the problem of query classification, Bian and Chang [4] propose a label propagation method to automatically generate query class labels for unlabeled queries from click-based search logs. With the help of the large amount of automatically labeled queries, the performance of the classifiers has been greatly improved. To predict the relevance issue between queries and documents, Jiang et al. [11] and Yin et al. [28] propose a vector propagation algorithm on the click graph to learn vector representations for both queries and documents in the same term space. Experiments on search logs demonstrate the effectiveness and scalability of the proposed method. As it is hard to find useful extra attributes in many real-world networks, our approach only depends on the structural information in partially labeled networks.

3   Methodology

In this section, as a within-network classification task, we focus on performing multi-label node classification in networks, where each node can be assigned to multiple labels and only a few nodes have already been labeled. We first present our problem formulation, and then show our algorithm in details.

3.1   Problem Formulation

The multi-label node classification we addressed here is related to the within-network classification problem: estimating labels for the unlabeled nodes in partially labeled networks. Given a partially labeled undirected network $G = \{\mathcal{V}, \mathcal{E}\}$, in which a set of nodes $\mathcal{V} = \{1, \cdots, n_{max}\}$ are connected with edge $e(i,j) \in \mathcal{E}$, and $\mathcal{L} = \{l_1, \cdots, l_{max}\}$ is the label set for nodes.

3.2   Objective Formulation

In a within-network single-label classification scenario, let Y_i be the class label variable of node i, which can be assigned to one categorical value $c \in \mathcal{L}$. Let G_i denote the information node i known about the whole graph, and let $P(Y_i = c|G_i)$ be the probability that node i is assigned to the class label c. The relational neighbor (RN) classifier is first proposed by Macskassy and Provost [13], and in the relational learning context we can get the probability $P(Y_i = c|G_i)$ by making the first order Markov assumption [13]:

$$P(Y_i = c|G_i) = P(Y_i = c|\mathcal{N}_i), \qquad (1)$$

where $\mathcal{N}_i$ is the set of nodes that are adjacent to node i. Taking advantage of the Markov assumption, Macskassy and Provost [13] proposed the weighted-vote relational neighbor (WvRN) classifier whose class membership probability can be defined as follows:

$$P(Y_i = c|G_i) = P(Y_i = c|\mathcal{N}_i) = \frac{1}{Z} \sum_{j \in \mathcal{N}_i} w_{i,j} \times P(Y_j = c|\mathcal{N}_j), \qquad (2)$$

where Z is a normalizer and $w_{i,j}$ represents the weight between i and j.

IDRN Classifier. As shown in Eq. 2, traditional relational neighbor classifiers, such as WvRN [13], only use the class labels in neighborhood as features. However, as we will show, by taking the identifiers in each node's egocentric network as features, the classifier often performs much better than most baseline algorithms.

In our algorithm, the node identifiers, i.e., unique symbols for individual nodes, are extracted as features for learning and inference. With the first order Markov assumption, we can simplify $G_i = G_{\mathcal{N}_i} = \mathbf{X}_{\mathcal{N}_i} = \{x|x \in \mathcal{N}_i\} \cup \{i\}$

as a feature vector of all identifiers in node i's egocentric graph $G_{\mathcal{N}_i}$. The ego-centric network $G_{\mathcal{N}_i}$ of node i is the subgraph of node i's first-order zone [15]. Aside from just considering neighbors' identifiers, our approach also includes the identifier of node i itself, with the assumption that both the identifiers of node i's neighbors and itself can provide meaningful representations for its class label. For example, if node i $(ID = 1)$ connects with three other nodes where $ID = 2, 3, 5$ respectively, then its feature vector $\mathbf{X}_{\mathcal{N}_i}$ of node i will be $[1, 2, 3, 5]$. Eq. 2 can be simplified as follows:

$$P(Y_i = c|G_i) = P(Y_i = c|G_{\mathcal{N}_i}) = P(Y_i = c|\mathbf{X}_{\mathcal{N}_i}). \tag{3}$$

By taking the strong independent assumption of naive Bayes, we can simplify $P(Y_i = c|\mathbf{X}_{\mathcal{N}_i})$ in Eq. 3 as the following equation:

$$P(Y_i = c|\mathbf{X}_{\mathcal{N}_i}) = \frac{P(Y_i = c)P(\mathbf{X}_{\mathcal{N}_i}|Y_i = c)}{P(\mathbf{X}_{\mathcal{N}_i})}$$

$$\propto P(Y_i = c)P(\mathbf{X}_{\mathcal{N}_i}|Y_i = c) \tag{4}$$

$$\propto P(Y_i = c) \prod_{k \in \mathbf{X}_{\mathcal{N}_i}} P(k|Y_i = c),$$

where the last step drops all values independent of Y_i.

Multi-label Classification. Traditional ways of addressing multi-label classification problem is to transform it into a one-vs-rest learning problem [23, 27]. When training IDRN classifier, for each node i with a set of true labels T_i, we transform it into a set of single-label data points, i.e., $\{\langle \mathbf{X}_{\mathcal{N}_i}, c \rangle | c \in T_i\}$. After that, we use naive Bayes training framework to estimate the class prior $P(Y_i = c)$ and the conditional probability $P(k|Y_i = c)$ in Eq. 4.

Algorithm 1 shows how to train IDRN to get the maximal likelihood estimations (MLE) for the class prior $P(Y_i = c)$ and conditional probability $P(k|Y_i = c)$, i.e., $\hat{\theta}_c = P(Y_i = c)$ and $\hat{\theta}_{kc} = P(k|Y_i = c)$. As it has been suggested that multinomial naive Bayes classifier usually performs better than Bernoulli naive Bayes model in various real-world practices [26], we take the multinomial approach here. Suppose we observe N data points in the training dataset. Let N_c be the number of occurrences in class c and let N_{kc} be the number of occurrences of feature k and class c. In the first 2 lines, we initialize the counting values of N, N_c and N_{kc}. After that, we transform each node i with a multi-label set T_i into a set of single-label data points and use the multinomial naive Bayes framework to count the values of N, N_c and N_{kc} as shown from line 3 to line 12 in Algorithm 1. After that, we can get the estimated probabilities, i.e., $\hat{\theta}_c = P(Y_i = c)$ and $\hat{\theta}_{kc} = P(k|Y_i = c)$, for all classes and features.

In multi-label prediction phrase, the goal is to find the most probable classes for each unlabeled node. Since most methods yield a ranking of labels rather than an exact assignment, a threshold is often required. To avoid the affection of introducing a threshold, we assign s most probable classes to a node, where

s is the number of labels assigned to the node originally. Unfortunately a naive implementation of Eq. 4 may fail due to numerical underflow, the value of $P(Y_i = c|\mathbf{X}_{\mathcal{N}_i})$ is proportional to the following equation:

$$P(Y_i = c|\mathbf{X}_{\mathcal{N}_i}) \propto \log P(Y_i = c) + \sum_{k \in \mathbf{X}_{\mathcal{N}_i}} \log P(k|Y_i = c). \tag{5}$$

Defining $b_c = \log P(Y_i = c) + \sum_{k \in \mathbf{X}_{\mathcal{N}_i}} \log P(k|Y_i = c)$ and using **log-sum-exp** trick [18], we get the precise probability $P(Y_i = c|\mathbf{X}_{\mathcal{N}_i})$ for each class label c as follows:

$$P(Y_i = c|\mathbf{X}_{\mathcal{N}_i}) = \frac{e^{(b_c - B)}}{\sum_{c \in \mathcal{L}} e^{(b_c - B)}}, \tag{6}$$

where $B = \max_c b_c$. Finally, to classify unlabeled nodes i, we can use the Eq. 6 to assign s most probable classes to it.

Algorithm 1. Training the **Id**entifier based **r**elational **n**eighbor classifier.

Input: Graph $G = \{\mathcal{V}, \mathcal{E}\}$, the labeled nodes $\mathcal{V}'$ and the class label set $\mathcal{L}$.
Output: The MLE for each class c's prior $\hat{\theta}_c$ and the MLE for conditional
probability $\hat{\theta}_{kc}$.

1 $N := 0$;
2 $N_c := 0$ and $N_{kc} := 0$, $\forall c \in \mathcal{L}$ and $\forall k \in \mathcal{V}$.
3 **for** $i \in \mathcal{V}'$ **do**
4     $C = T_i$; // Get the true label set C of node i.
5   **for** $c \in C$ **do**
6   **for** $k \in X_{\mathcal{N}_i}$ **do**
7           $N := N + 1$;
8           $N_c := N_c + 1$;
9           $N_{kc} := N_{kc} + 1$;
10   **end**
11   **end**
12 **end**
13 **for** $c \in \mathcal{L}$ **do**
14    $\hat{\theta}_c := \frac{N_c}{N}$;
15   **for** $k \in \mathcal{V}$ **do**
16       $\hat{\theta}_{kc} := \frac{N_{kc}+1}{N+|\mathcal{V}|}$; // Corresponding to Laplace adding-one smoothing.
17   **end**
18 **end**
19 **return** $\hat{\theta}_c$ and $\hat{\theta}_{kc}$, $\forall c \in \mathcal{L}$ and $\forall k \in V$.

Community Priors. Community detection is one of the most popular topics of network science, and a large number of algorithms have been proposed recently [7,8]. It is believed that nodes in communities share common properties or play similar roles. Grover and Leskovec [10] also regard that nodes from the same community should share similar representations. The availability of such pre-detected community structure allows us to classify nodes more precisely especially with insufficient training data. Given the community partition of a certain network, we can estimate the probability $P(Y_i = c|C_i)$ for each class c through the empirical counts and adding-one smoothing technique, where C_i indicates the community that node i belongs to. Then, we can define the probability $P(Y_i = c|\mathbf{X}_{\mathcal{N}_i})$ in Eq. 3 as follows:

$$P(Y_i = c|\mathbf{X}_{\mathcal{N}_i}, C_i) = \frac{P(Y_i = c|C_i)P(\mathbf{X}_{\mathcal{N}_i}|Y_i = c, C_i)}{P(\mathbf{X}_{\mathcal{N}_i}|C_i)}, \tag{7}$$

where $P(\mathbf{X}_{\mathcal{N}_i}|C_i)$ refers to the conditional probability of the event $\mathbf{X}_{\mathcal{N}_i}$ occurring given that node i belongs to community C_i. Obviously, given the knowledge of C_i will not influence the probability of the event $X_{\mathcal{N}_i}$ occurring, thus we can assume that $P(\mathbf{X}_{\mathcal{N}_i}|C_i) = P(\mathbf{X}_{\mathcal{N}_i})$ and $P(\mathbf{X}_{\mathcal{N}_i}|Y = c, C_i) = P(\mathbf{X}_{\mathcal{N}_i}|Y = c)$. So Eq. 7 can be simplified as follows:

$$\begin{aligned}
&P(Y_i = c|\mathbf{X}_{\mathcal{N}_i}, C_i) \\
&= \frac{P(Y_i = c|C_i)P(\mathbf{X}_{\mathcal{N}_i}|Y_i = c)}{P(\mathbf{X}_{\mathcal{N}_i})} \\
&\propto P(Y_i = c|C_i)P(\mathbf{X}_{\mathcal{N}_i}|Y_i = c) \\
&\propto \log P(Y_i = c|C_i) + \sum_{k \in \mathbf{X}_{\mathcal{N}_i}} \log P(k|Y_i = c).
\end{aligned} \tag{8}$$

As shown in Eq. 8, we assume that different communities have different priors rather than sharing the same global prior $P(Y_i = c)$. To extract communities in networks, we choose the Louvain algorithm [5] in this paper which has been shown as one of the best performing algorithms.

3.3 Efficiency

Suppose that the largest node degree of the given network $G = \{\mathcal{V}, \mathcal{E}\}$ is K. In the training phrase, as shown in Algorithm 1, the time complexity from line 1 to line 12 is about $O(K \times |\mathcal{L}| \times |\mathcal{V}|)$, and the time complexity from line 13 to line 18 is $O(|\mathcal{L}| \times |\mathcal{V}|)$. So the total time complexity of the training phrase is $O(K \times |\mathcal{L}| \times |\mathcal{V}|)$. Obviously, it is quite simple to implement this training procedure. In the training phrase, the time complexity of each node is linear with respect to the product of the number of its degree and the size of class label set $|\mathcal{L}|$.

In the prediction phrase, suppose node i contains n neighbors. It takes $O(n + 1)$ time to find its identifier vector $\mathbf{X}_{\mathcal{N}_i}$. Given the knowledge of i's community membership C_i, in Eqs. 5 and 8, it only takes $O(1)$ time to get the values of $P(Y_i = c|C_i)$ and $P(Y_i = c)$, respectively. As it takes $O(1)$ time to get the value of $P(k|Y_i = c)$, for a given class label c the time complexities of Eqs. 5 and 8 both are $O(n)$. Thus for a given node, the total complexity of predicting the probability scores on all labels $\mathcal{L}$ is $O(|\mathcal{L}| \times n)$ even we consider predicting the precise probabilities in Eq. 6. For each class label prediction, it takes $O(n)$ time which is linear to its neighbor size. Furthermore, the prediction process can be greatly sped-up by building an inverted index of node identifiers, as the identifier features of each class label can be sparse.

4   Experiments

In this section, we first introduce the dataset and the evaluation metrics. After that, we conduct several experiments to show the effectiveness of our algorithm. Code to reproduce our results will be available at the authors' website[1].

4.1   Dataset

The task is to predict the labels for the remaining nodes. We use the following publicly available datasets described below.

Amazon. The dataset contains a subset of books from the amazon co-purchasing network data extracted by Nandanwar and Murty [19]. For each book, the dataset provides a list of other similar books, which is used to build a network. Genre of the books gives a natural categorization, and the categories are used as class labels in our experiment.

CoRA. It contains a collection of research articles in computer science domain with predefined research topic labels which are used as the ground-truth labels for each node.

IMDb. The graph contains a subset of English movies from **IMDb**[2], and the links indicate the relevant movie pairs based on the top 5 billed stars [19]. Genre of the movies gives a natural class categorization, and the categories are used as class labels.

PubMed. The dataset contains publications from **PubMed** database, and each publication is assigned to one of three diabetes classes. So it is a single-label dataset in our learning problem.

Wikipedia. The network data is a dump of Wikipedia pages from different areas of computer science. After crawling, Nandanwar and Murty [19] choose 16 top level category pages, and recursively crawled subcategories up to a depth of 3. The top level categories are used as class labels.

Youtube. A subset of Youtube users with interest grouping information is used in our experiment. The graph contains the relationships between users and the user nodes are assigned to multiple interest groups.

Blogcatalog and Flickr. These datasets are social networks, and each node is labeled by at least one category. The categories can be used as the ground-truth of each node for evaluation in multi-label classification task.

---

[1] https://github.com/yeqi-adrs/IDRN.
[2] http://www.imdb.com/interfaces.

PPI. It is a protein-protein interaction (PPI) network for Homo Sapiens. The labels of nodes represent the bilolgical states.

POS. This is a co-occurrence network of words appearing in the Wikipedia dump. The node labels represent Part-of-Speech (POS) tags of each word.

The **Amazon**, **CoRA**, **IMDb**, **PubMed**, **Wikipedia** and **Youtube** datasets are made available by Nandanwar and Murty [19]. The **Blogcatalog** and **Flickr** datasets are provided by Tang and Liu [23], and the **PPI** and **POS** datasets are provided by Grover and Leskovec [10]. The statistics of the datasets are summarized in Table 1.

Table 1. Summary of undirected networks used for multi-label classification.

Dataset	#Nodes	#Edges	#Classes	Average Category	$\frac{\text{\#Edges}}{\text{\#Nodes}}$
Amazon	83742	190097	30	1.546	2.270
CoRA	24519	92207	10	1.004	3.782
IMDb	19359	362079	21	2.301	18.703
PubMed	19717	44324	3	1.000	2.248
Wikipedia	35633	495388	16	1.312	13.903
Youtube	22693	96361	47	1.707	4.246
Blogcatalog	10312	333983	39	1.404	32.387
Flickr	80513	5899882	195	1.338	73.278
PPI	3890	37845	50	1.707	9.804
POS	4777	92295	40	1.417	19.320

4.2 Evaluation Metrics

In this part, we explain the details of the evaluation metrics: Hamming score, Micro-F_1 score and Macro-F_1 score which have also widely been used in many other multi-label within-network classification tasks [19,23,27]. Given node i, let T_i be the true label set and P_i be the predicted label set, then we have the following scores:

Definition 1. *Hamming Score* $= \sum_{i=1}^{|\mathcal{V}|} \frac{|T_i \cap P_i|}{|T_i \cup P_i|}$,

Definition 2. *Micro-F_1 Score* $= \frac{2\sum_{i=1}^{|\mathcal{V}|} |T_i \cap P_i|}{\sum_{i=1}^{|\mathcal{V}|} |T_i| + \sum_{i=1}^{|\mathcal{V}|} |P_i|}$,

Definition 3. *Macro-F_1 Score* $= \frac{1}{|\mathcal{L}|} \sum_{j=1}^{|\mathcal{L}|} \frac{2\sum_{i \in \mathcal{L}_j} |T_i \cap P_i|}{\sum_{i \in \mathcal{L}_j} |T_i| + \sum_{i \in \mathcal{L}_j} |P_i|}$,

where $|\mathcal{L}|$ is the number of classes and $\mathcal{L}_j$ is the set of nodes in class j.

Baseline Methods. In this paper, we focus on comparing our work with the state-of-the-art approaches. To validate the performance of our approach, we compare our algorithms against a number of baseline algorithms. In this paper, we use IDRN to denote our approach with the global priori and use $IDRN_c$ to denote the algorithm with different community priors. All the baseline algorithms are summarized as follows:

- WvRN [13]: The **W**eighted-**v**ote **R**elational **N**eighbor is a simple but surprisingly good relational classifier. Given the neighbors $\mathcal{N}_i$ of node i, the WvRN estimates i's classification probability $P(y|i)$ of class label y with the weighted mean of its neighbors as mentioned above. As WvRN algorithm is not very complex, we implement it in Java programming language by ourselves.
- SocioDim [23]: This method is based on the SocioDim framework which generates a representation in d dimension space from the top-d eigenvectors of the modularity matrix of the network, and the eigenvectors encode the information about the community partitions of the network. The implementation of SocioDim in Matlab is available on the author's web-site[3]. As the authors preferred in their study, we set the number of social dimensions as 500.
- DeepWalk [20]: DeepWalk generalizes recent advancements in language modeling from sequences of words to nodes [17]. It uses local information obtained from truncated random walks to learn latent dense representations by treating random walks as the equivalent of sentences. The implementation of DeepWalk in Python has already been published by the authors[4].
- LINE [22]: LINE algorithm proposes an approach to embed networks into low-dimensional vector spaces by preserving both the *first-order* and *second-order* proximities in networks. The implementation of LINE in C++ has already been published by the authors[5]. To enhance the performance of this algorithm, we set embedding dimensions as 256 (i.e., 128 dimensions for the *first-order* proximities and 128 dimensions for the *second-order* proximities) in LINE algorithm as preferred in its implementation.
- SNBC [19]: To classify a node, SNBC takes a structured random walk from the given node and makes a decision based on how nodes in the respective k^{th}-level neighborhood are labeled. The implementation of SNBC in Matlab has already been published by the authors[6].
- node2vec [10]: It also takes a similar approach with DeepWalk which generalizes recent advancements in language modeling from sequences of words to nodes. With a flexible neighborhood sampling strategy, node2vec learns a mapping of nodes to a low-dimensional feature space that maximizes the likelihood of preserving network neighborhoods of nodes. The implementation of node2vec in Python is available on the authors' web-site[7].

---

[3] http://leitang.net/social_dimension.html.
[4] https://github.com/phanein/deepwalk.
[5] https://github.com/tangjianpku/LINE.
[6] https://github.com/sharadnandanwar/snbc.
[7] https://github.com/aditya-grover/node2vec.

Table 2. Experiment comparisons of baselines, IDRN and IDRN$_c$ by the metrics of Hamming score, Micro-F_1 score and Macro-F_1 score with 10% nodes labeled for training.

Metric	Network	WvRN	SocioDim	DeepWalk	LINE	SNBC	node2vec	IDRN	IDRN$_c$
Hamming Score (%)	Amazon	33.76	38.36	31.79	40.55	59.00	49.18	68.97	**72.25**
	Youtube	22.82	31.94	36.63	33.90	35.06	33.86	42.19	**44.03**
	CoRA	55.83	63.02	71.37	65.50	66.75	72.66	77.80	**77.95**
	IMDb	**33.59**	22.21	33.12	30.39	30.18	32.97	26.96	26.89
	Pubmed	50.32	65.68	77.40	68.31	79.22	79.02	80.13	**80.92**
	Wikipedia	45.10	65.29	71.10	68.812	68.78	70.69	**75.38**	73.58
	Flickr	21.37	29.67	28.73	30.96	24.20	30.65	28.22	**33.12**
	Blogcatalog	17.89	27.04	25.63	25.32	22.40	27.46	**31.87**	31.05
	PPI	6.28	8.61	8.14	9.27	7.97	8.88	19.80	**20.95**
	POS	23.05	21.06	31.40	38.24	37.73	34.59	43.92	**44.16**
	Average	31.00	37.28	41.53	41.12	43.12	43.99	49.52	**50.49**
Micro-F_1 (%)	Amazon	34.86	39.62	33.06	42.42	59.79	50.55	69.60	**73.04**
	Youtube	27.81	36.40	40.73	38.01	39.67	38.35	47.94	**49.17**
	CoRA	55.85	63.00	71.36	65.47	66.78	72.66	77.80	**77.96**
	IMDb	**42.62**	29.99	41.82	39.89	39.53	42.36	36.29	36.29
	Pubmed	50.32	65.68	77.40	68.31	79.22	79.02	80.13	**80.92**
	Wikipedia	48.51	66.95	72.19	70.21	70.68	72.07	**76.85**	75.25
	Flickr	25.40	32.91	31.66	34.03	27.60	33.76	31.86	**36.55**
	Blogcatalog	20.50	28.86	27.29	27.45	24.66	29.41	**34.25**	33.56
	PPI	18.41	12.29	11.52	13.16	11.32	12.80	24.92	**25.73**
	POS	26.04	24.42	35.98	42.70	41.99	39.09	47.70	**48.23**
	Average	35.03	40.01	44.30	44.16	46.12	47.00	52.73	**53.67**
Macro-F_1 (%)	Amazon	32.00	35.95	21.64	37.52	56.84	45.85	66.39	**70.64**
	Youtube	18.17	34.19	33.92	33.47	32.07	32.60	40.71	**42.59**
	CoRA	43.16	56.82	62.68	59.07	55.68	64.79	72.10	**72.20**
	IMDb	18.89	18.77	18.22	18.83	17.45	18.46	26.61	**27.06**
	Pubmed	41.57	64.85	75.92	66.66	77.16	77.50	71.20	**79.89**
	Wikipedia	45.58	58.93	62.29	62.17	61.99	64.90	**70.04**	69.39
	Flickr	15.54	18.28	17.13	**21.80**	7.36	18.46	13.71	21.56
	Blogcatalog	11.47	**18.88**	14.65	15.52	8.29	17.16	17.44	16.76
	PPI	7.35	10.59	9.61	10.82	8.27	11.27	21.67	**22.00**
	POS	3.91	6.05	8.26	8.93	5.92	8.61	13.49	**14.29**
	Average	23.76	32.33	32.43	33.47	33.10	35.96	41.33	**43.63**

We obtain 128 dimension embeddings for a node using DeepWalk and node2Vec as preferred in the algorithms. After getting the embedding vectors for each node, we use these embeddings further in classification. In the multi-label classification experiment, each node is assigned to one or more class labels. We assign s most probable classes to the node using these decision values, where s is equal to the number of labels assigned to the node originally. Specifically, for all vector representation models (i.e., SocioDim, DeepWalk, LINE, SNBC and node2vec), we use a one-vs-rest logistic regression implemented by LibLinear [6] to return the most probable labels as described in prior work [20, 23, 27].

4.3   Performances of Classifiers

In this part, we study the performances of within-network classifiers in different datasets respectively. As some baseline algorithms are just designed for undirected or unweighted graphs, we just transform all the graphs to undirected and unweighted ones for a fair comparison.

First, to study the performance of different algorithms on a sparsely labeled network, we show results obtained by using 10% nodes for training and the left 90% nodes for testing. The process has been repeated 10 times, and we report the average scores over different datasets.

Table 2 shows the average Hamming score, Macro-F_1 score, and Micro-F_1 score for multi-label classification results in the datasets. Numbers in bold show the best algorithms in each metric of different datasets. As shown in the table, in most of the cases, IDRN and IDRN$_c$ algorithms improve the metrics over the existing baselines. For example, in the **Amazon** network, IDRN$_c$ outperforms all baselines by at least 22.46%, 22.16% and 24.28% with respect to Hamming score, Macro-F_1 score, and Micro-F_1 score respectively. Our model with community priors, i.e., IDRN$_c$ often performs better than IDRN with global prior. For the three metrics, IDRN and IDRN$_c$ perform consistently better than other algorithms in the 10 datasets except for **IMDb, Flickr** and **Blogcatalog**. Take **IMDb** dataset for an example, we observe that Hamming score and Micro-F_1 score got by IDRN$_c$ are worse than those got by some baseline algorithm, such as node2vec and WvRN, however Macro-F_1 score got by IDRN$_c$ is the best. As Macro-F_1 score computes an average over classes while Hamming and Micro-F_1 scores get the average over all testing nodes, the result may indicate that our algorithms get more accurate results over different classes in the imbalanced **IMDb** dataset. To show the results more clearly, we also get the average validation scores for each algorithm in these datasets which are shown in the last lines of the three metrics in Table 2. On average our approach can provide Hamming score, Micro-F_1 score and Macro-F_1 score up to 14%, 21% and 14% higher than competing methods, respectively. The results indicate that our IDRN with community priors outperforms almost all baseline methods when networks are sparsely labeled.

Second, we show the performances of the classification algorithms of different training fractions. When training a classifier, we randomly sample a portion of the labeled nodes as the training data and the rest as the test. For all the datasets, we randomly sample 10% to 90% of the nodes as the training samples, and use the left nodes for testing. The process has been repeated 5 times, and we report the averaged scores. Due to limitation in space, we just summarize the results of 3 datasets for Hamming scores, Micro-F_1 scores and Macro-F_1 scores in Fig. 1. Here we can make similar observations with the conclusion given in Table 2. As shown in Fig. 1, IDRN and IDRN$_c$ perform consistently better than other algorithms in these 3 datasets in Fig. 1. In fact, nearly in all the 10 datasets, our approaches outperform all the baseline methods significantly. When the networks are sparsely labeled (i.e., with 10% or 20% labeled data), IDRN$_c$ outperforms slightly better than IDRN. However, when more nodes are labeled,

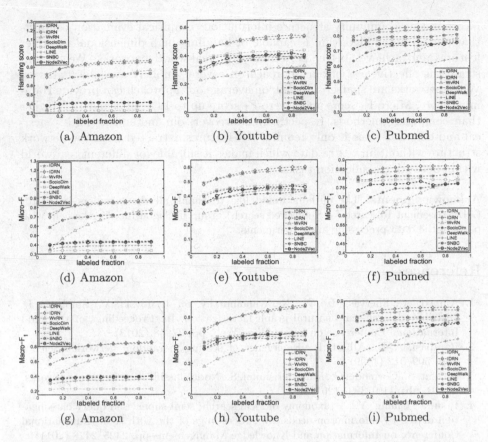

Fig. 1. Performance evaluation of Hamming scores, Micro-F_1 scores and Macro-F_1 scores on varying the amount of labeled data used for training. The x axis denotes the fraction of labeled data, and the y axis denotes the Hamming scores, Micro-F_1 scores and Macro-F_1 scores, respectively.

IDRN usually outperforms $IDRN_c$. As we see that the posterior in Eq. 3 is a combination of prior and likelihood, the results may indicate that the community prior of a given node corresponds to a strong prior, while the global prior is a weak one. The strong prior will improve the performance of IDRN when the training datasets are small, while the opposite conclusion holds for training on large datasets.

5   Conclusion and Discussion

In this paper, we propose a novel approach for node classification, which combines local node identifiers and community priors to solve the multi-label node classification problem. In the algorithm, we use the node identifiers in the egocentric networks as features and propose a within-network classification model by incorporating community structure information. Empirical evaluation confirms that our proposed algorithm is capable of handling high dimensional identifier features and achieves better performance in real-world networks. We demonstrate the effectiveness of our approach on several publicly available datasets. When networks are sparsely labeled, on average our approach can provide Hamming score, Micro-F_1 score and Macro-F_1 score up to 14%, 21% and 14% higher than competing methods, respectively. Moreover, our method is quite practical and efficient, since it only requires the features extracted from the network structure without any extra data which makes it suitable for different real-world within-network classification tasks.

Acknowledgments. The authors would like to thank all the members in ADRS (ADvertisement Research for Sponsered search) group in Sogou Inc. for the help with parts of the data processing and experiments.

References

1. Ahmed, A., Shervashidze, N., Narayanamurthy, S., Josifovski, V., Smola, A.J.: Distributed large-scale natural graph factorization. In: Proceedings of the 22nd International Conference on World Wide Web, pp. 37–48 (2013)
2. Barabási, A.-L., Albert, R.: Emergence of scaling in random networks. Science **286**, 509–512 (1999)
3. Bhagat, S., Cormode, G., Muthukrishnan, S.: Node classification in social networks. CoRR, abs/1101.3291 (2011)
4. Bian, J., Chang, Y.: A taxonomy of local search: semi-supervised query classification driven by information needs. In: Proceedings of the 20th ACM International Conference on Information and Knowledge Management, pp. 2425–2428 (2011)
5. Blondel, V.D., Guillaume, J.-L., Lambiotte, R., Lefebvre, E.: Fast unfolding of communities in large networks. J. Stat. Mech. **10**, 10008 (2008)
6. Fan, R.-E., Chang, K.-W., Hsieh, C.-J., Wang, X.-R., Lin, C.-J.: LIBLINEAR: a library for large linear classification. J. Mach. Learn. Res. **9**, 1871–1874 (2008)
7. Fortunato, S.: Community detection in graphs. Phys. Rep. **486**(3–5), 75–174 (2010)
8. Fortunato, S., Hric, D.: Community detection in networks: a user guide. Phys. Rep. **659**, 1–44 (2016)
9. Girvan, M., Newman, M.E.J.: Community structure in social and biological networks. Proc. Natl. Acad. Sci. **99**(12), 7821–7826 (2002)
10. Grover, A., Leskovec, J.: Node2vec: scalable feature learning for networks. In: Proceedings of the 22Nd ACM SIGKDD International Conference on Knowledge Discovery and Data Mining, pp. 855–864 (2016)
11. Jiang, S., Hu, Y., et al.: Learning query and document relevance from a web-scale click graph. In: Proceedings of the 39th International ACM SIGIR Conference on Research and Development in Information Retrieval, SIGIR 2016, pp. 185–194 (2016)
12. Joulin, A., Grave, E., et al.: Bag of tricks for efficient text classification. CoRR, abs/1607.01759 (2016)
13. Macskassy, S.A., Provost, F.: A simple relational classifier. In: Proceedings of the Second Workshop on Multi-Relational Data Mining (MRDM-2003) at KDD-2003, pp. 64–76 (2003)

14. Macskassy, S.A., Provost, F.: Classification in networked data: a toolkit and a univariate case study. J. Mach. Learn. Res. **8**(May), 935–983 (2007)
15. Marsden, P.V.: Egocentric and sociocentric measures of network centrality. Soc. Netw. **24**(4), 407–422 (2002)
16. McDowell, L.K., Aha, D.W.: Labels or attributes? Rethinking the neighbors for collective classification in sparsely-labeled networks. In: International Conference on Information and Knowledge Management, pp. 847–852 (2013)
17. Mikolov, T., Chen, K., Corrado, G., Dean, J.: Efficient estimation of word representations in vector space. CoRR, abs/1301.3781 (2013)
18. Murphy, K.P., Learning, M.: A Probabilistic Perspective. The MIT Press, Cambridge (2012)
19. Nandanwar, S., Murty, M.N.: Structural neighborhood based classification of nodes in a network. In: Proceedings of the 22nd ACM SIGKDD International Conference on Knowledge Discovery and Data Mining, pp. 1085–1094 (2016)
20. Perozzi, B., Al-Rfou, R., Skiena, S.: Deepwalk: online learning of social representations. In: Proceedings of the 20th ACM SIGKDD International Conference on Knowledge Discovery and Data Mining, pp. 701–710 (2014)
21. Rayana, S., Akoglu, L.: Collective opinion spam detection: bridging review networks and metadata. In: Proceedings of the 21th ACM SIGKDD International Conference on Knowledge Discovery and Data Mining, pp. 985–994 (2015)
22. Tang, J., Qu, M., Wang, M., Zhang, M., Yan, J., Mei, Q.: Line: large-scale information network embedding. In Proceedings of the 24th International Conference on World Wide Web, pp. 1067–1077 (2015)
23. Tang, L., Liu, H.: Relational learning via latent social dimensions. In: Proceedings of the 15th ACM SIGKDD International Conference on Knowledge Discovery and Data Mining, pp. 817–826 (2009)
24. Tang, L., Liu, H.: Scalable learning of collective behavior based on sparse social dimensions. In: The 18th ACM Conference on Information and Knowledge Management, pp. 1107–1116 (2009)
25. Wang, D., Cui, P., Zhu, W.: Structural deep network embedding. In: Proceedings of the 22Nd ACM SIGKDD International Conference on Knowledge Discovery and Data Mining, pp. 1225–1234 (2016)
26. Wang, S.I., Manning, C.D.: Baselines and bigrams: simple, good sentiment and topic classification. In: Proceedings of the ACL, pp. 90–94 (2012)
27. Wang, X., Sukthankar, G.: Multi-label relational neighbor classification using social context features. In: Proceedings of The 19th ACM SIGKDD Conference on Knowledge Discovery and Data Mining (KDD), pp. 464–472 (2013)
28. Yin, D., Hu, Y., et al.: Ranking relevance in yahoo search. In: Proceedings of the 22Nd ACM SIGKDD International Conference on Knowledge Discovery and Data Mining, pp. 323–332 (2016)

An Active Learning Approach to Recognizing Domain-Specific Queries From Query Log

Weijian Ni, Tong Liu[(✉)], Haohao Sun, and Zhensheng Wei

College of Computer Science and Engineering, Shandong University of Science and Technology, Qingdao 266510, Shandong, China
niweijian@gmail.com, liu_tongtong@foxmail.com, shhlat@163.com, zhensheng_wei@163.com

Abstract. In this paper, we address the problem of recognizing domain-specific queries from general search engine's query log. Unlike most previous work in query classification relying on external resources or annotated training queries, we take query log as the only resource for recognizing domain-specific queries. In the proposed approach, we represent query log as a heterogeneous graph and then formulate the task of domain-specific query recognition as graph-based transductive learning. In order to reduce the impact of noisy and insufficient of initial annotated queries, we further introduce an active learning strategy into the learning process such that the manual annotations needed are reduced and the recognition results can be continuously refined through interactive human supervision. Experimental results demonstrate that the proposed approach is capable of recognizing a certain amount of high-quality domain-specific queries with only a small number of manually annotated queries.

Keywords: Query classification · Active learning · Transfer learning · Search engine · Query log

1 Introduction

General search engines, although being an indispensable tool in people's information seeking activities, are still facing essential challenges in producing satisfactory search results. One challenge is that general search engines are always required to handle users' queries from a wide range of domains, whereas each domain often having its own preference on retrieval model. Taking two queries "steve jobs" and "steve madden" for example, the first query is for celebrity search, thus descriptive pages about Steve Jobs should be considered relevant; whereas the second one is for commodity search, thus structured items of this brand should be preferred. Therefore, if domain specificity of search query was recognized, a targeted domain-specific retrieval model can be selected to refine search results [1, 2]. In addition, with the increasing use of general search engines, search queries have become a valuable and extensive resource containing a large number of domain named entities or domain terminologies, thus domain-specific

© Springer International Publishing AG 2017
L. Chen et al. (Eds.): APWeb-WAIM 2017, Part II, LNCS 10367, pp. 18–32, 2017.
DOI: 10.1007/978-3-319-63564-4_2

query recognition can be viewed as a fundamental step in constructing large scale domain knowledge bases [3].

Domain-specific query recognition is essentially a query classification task which has been attracting much attention for decades in information retrieval (IR) community. Many traditional work views query classification as a supervised learning problem and requires a number of manually annotated queries [4,5]. However, training queries are often time-consuming and costly to obtain. In order to overcome this limitation, many studies leveraged both labeled and unlabeled queries in query classification [6,12]. The intuition behind is that queries strongly correlated in click-through graph are likely to have similar class labels.

In this paper, inspired by semi-supervised learning over click-through graph in [6,7], we propose a new query classification method that aims to recognize queries specific to a target domain, utilizing search engine's query log as the only resource. Intuitively, users' search intents mostly remain similar in short search sessions and most pages concentrate on only a small number of topics. This implies the queries frequently issued by same users or retrieve same pages are more likely to be relevant to the same domain. In other words, domain-specificity of each queries in query log follows a manifold structure. In order to exploit the intrinsic manifold structure, we represent query log as a heterogenous graph with three types of nodes, i.e., users, queries and URLs, and then formulate domain-specific query recognition as transductive learning on heterogenous graph.

The performance of graph-based transductive learning is highly rely on the set of manually pre-annotated nodes, named as seed domain-specific queries in the domain-specific query recognition task. We further introduce a novel active learning strategy in the graph-based transductive learning process that allows interactive and continuous manual adjustments of seed queries. In this way, the recognition process can be started from an insufficient or even noisy initial set of seed queries, thus alleviating the difficulty of manually specifying a complete seed set for recognizing domain-specific queries. Moreover, through introducing inter-active human supervision, the seed set generated during the recognition process tend to be more informative than the one given in advance, and is beneficial to improve the recognition performance.

We evaluate the proposed approach using query log of a Chinese commercial search engine. We provide in-depth experimental analyses on the proposed approach, and compare our approach with several state-of-the-art query classification methods. Experimental results conclude the superior performance of the proposed approach.

The rest of the paper is organized as follows. Section 2 describes the graph representation of query log. Section 3 gives a formal definition of domain-specific query recognition problem together with the details of the proposed approach. Section 4 presents the experimental results. We discuss related work in Sect. 5 and conclude the paper in Sect. 6.

2   Graph Representation of Query Log

In modern search engines, the interaction process between search users and search engine is recorded as so-called query log. Despite of the difference between search engines, query log generally contains at least four types of information: users, queries, search results w.r.t. each query and user's click behaviors on search results. Table 1 gives an example of a piece of log that is recorded for an interaction between a user and search engine.

In this work, we make use of heterogenous graph, as shown in Fig. 1, to formally represent the objects involved in the search process. More specifically, a tripartite graph composed of three types of nodes, i.e., users, queries and URLs is constructed according to the interaction process recorded in query log. There are two types of links (shown by dashed line and dotted line) in the tripartite graph that indicate query issuing behavior of search users and click-through behavior between queries and URLs, respectively. In addition, the timestamps of query issuing behaviors are attached on each links between the corresponding user and query.

Based on the graph representation, the inherent domain-specificity manifold structure in query log implies that the strongly correlated queries, through either user nodes or URL nodes, are highly likely to be relevant to the same domain. Therefore, with a set of manually annotated domain-specific queries

Table 1. Query log example

Field	Content	Description
UserId	*bc3f448598a2dbea*	The unique identifier of the search user
Query	*piglet prices sichuan*	The query issued by the user
URL	*alibole.com/57451.html*	URL of the webpage retrieved by the query
Timestamp	*20111230114640*	The time when the query was issued
ViewRank	*4*	The rank of the URL in search results
ClickRank	*1*	The rank of the URL in user's click sequence

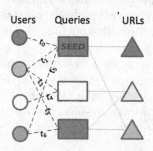

Fig. 1. User-Query-URL tripartite graph representation

(e.g., depicted as a *SEED* gray rectangle in Fig. 1), the domain-specificity of other queries can be derived through transductive learning on the tripartite graph (e.g., in Fig. 1, the gray level of each nodes indicates the mass of domain-specificity propagated from the seed node).

In the next section, we will describe the graph-based transductive learning process in more details.

3 Domain-Specific Query Recognition

3.1 Problem Definition

Formally, let $\mathcal{G}_{tri} = \langle U \cup Q \cup L, E^{(UQ)} \cup E^{(QL)} \rangle$ be the tripartite graph of query log, where $U = \{u_1, \cdots, u_{|U|}\}$, $Q = \{q_1, \cdots, q_{|Q|}\}$ and $L = \{l_1, \cdots, l_{|L|}\}$ denote the set of search users, queries and click-through URLs, respectively.

The links in $E^{(UQ)} \cup E^{(QL)}$ are weighted according to strength of the relation. Intuitively, if a user repeatedly issued the same query or only a few users issued that query, the relation between them would be strong. We thus calculate the weight of link between $u_i \in U$ and $q_j \in Q$ as follows:

$$W_{i,j}^{(UQ)} = \frac{N_{i,j}^{(UQ)}}{\sum_{i'=1}^{|U|} N_{i',j}^{(UQ)}} \cdot log \frac{|Q|}{\sum_{j'=1}^{|Q|} I_{i,j'}^{(UQ)}} \qquad (1)$$

where $N_{i,j}^{(UQ)}$ denotes the times u_i issued q_j. $I_{i,j'}^{(UQ)}$ is an indicator function that equals to 1 if these is a link between u_i and q_j in $\mathcal{G}_{tri}$, and 0 otherwise.

The weight of links in $E^{(QL)}$ dependents on frequency and rank of the URL clicked w.r.t the query and the number of the queries bringing click-through on the URL. Similarly as Eq. 1, the weight of link between $q_j \in Q$ and $l_k \in L$ can be calculated by URL frequency w.r.t. the query and inverse query frequency of the URL:

$$W_{j,k}^{(QL)} = \frac{N_{j,k}^{(QL)} / R_{j,k}}{\sum_{k'=1}^{|L|} N_{j,k'}^{(QL)} / R_{j,\bar{k'}}} \cdot log \frac{|Q|}{\sum_{j'=1}^{|Q|} I_{j',k}^{(QL)}} \qquad (2)$$

where $N_{j,k}^{(QL)}$ and $R_{j,k}$ denote the times and rank of l_k clicked w.r.t. q_j, respectively. $I_{j,k}^{(QL)}$ is an indicator function that equals to 1 if these is a link between q_j and l_k in the graph, and 0 otherwise.

In addition, Each link in $E^{(UQ)}$ is associated with a set of timestamps $T_{i,j} = \{t_{i,j}\}$, where $t_{i,j}$ is the time when user u_i issued query q_j. Note that $|T_{i,j}| = N_{i,j}^{(UQ)}$.

As for a target domain, suppose vectors $\mathbf{f} \in [0,1]^{|U|}$, $\mathbf{g} \in [0,1]^{|Q|}$ and $\mathbf{h} \in [0,1]^{|L|}$ denote the predicted domain-specificity of each user, query and URL in query log, respectively. The closer the value is to 1, the more confident the corresponding object is relevant to the target domain. Besides, vector

$\mathbf{y} \in \{0,1\}^{|Q|}$ denotes the pre-annotated domain-specificity of queries. Specifically, if query q_j is manually selected as a domain-specific one, i.e. seed query, $y_j = 1$; otherwise, $y_j = 0$. Using the above notations, the problem of learning domain-specificity is defined as follows:

Problem Definition I (DOMAIN-SPECIFIC QUERY RECOGNITION): *Given a tripartite graph $\mathcal{G}_{tri}$ with its associated link weight matrices $\mathbf{W}^{(UQ)}$ and $\mathbf{W}^{(QL)}$ for a query log, and a set of manually specified seed queries $\mathbf{y}$, the aim is to estimate $\mathbf{f}$, $\mathbf{g}$ and $\mathbf{h}$ as the prediction of domain-specificity of each objects in query log.*

3.2   Tranductive Learning on Tripartite Graph

The assumption for learning domain-specificity of objects in query log is that the domain-specificity distribution exhibits strong manifold structure, i.e., two objects tend to have similar domain-specificity if they are strongly associated with each other. Besides, the domain-specificity learned should be consistent with that pre-annotated on the seed queries. We formally design the following objective function:

$$\begin{aligned}
\mathcal{O}_{tri}(\mathbf{f},\mathbf{g},\mathbf{h}) = {} & \alpha \cdot \sum_{i=1}^{|U|}\sum_{j=1}^{|Q|} W_{ij}^{(UQ)}(f_i - g_j)^2 + \beta \cdot \sum_{j=1}^{|Q|}\sum_{k=1}^{|L|} W_{ij}^{(QL)}(g_j - h_k)^2 \\
& + \gamma \cdot \sum_{j=1}^{|Q|}\sum_{j'=1,j'\neq j}^{|Q|}\sum_{i=1}^{|U|} \frac{W_{ij}^{(UQ)} \cdot W_{ij'}^{(UQ)}}{\Delta_{i,j,j'}^{\tau}}(g_j - g_{j'})^2 \quad (3) \\
& + \delta \cdot \sum_{j=1}^{|Q|} I_j^{(seed)}(g_j - y_j)^2
\end{aligned}$$

where $I_j^{(seed)}$ is an indicator function that equals to 1 if query q_j is specified as a seed query, and 0 otherwise. $\Delta_{i,j,j'}$ is the minimum timespan of u_i issuing q_j and $q_{j'}$ and is calculated as:

$$\Delta_{i,j,j'} = \min_{t\in T_{i,j}, t'\in T_{i,j'}} |t - t'| \quad (4)$$

In Eq. 3, α, β, γ and δ ($\alpha,\beta,\gamma,\delta \geq 0$ and $\alpha + \beta + \gamma + \delta = 1$) are parameters that balance the contributions of the four items in the objective function. $\tau \geq 0$ is the parameter controlling the shrinkage effect on query weights. The larger τ gives more penalty to pair of queries with long issuing timespan, which are more likely to belong to different search sessions.

The first two items in Eq. 3 evaluate how smooth are the predictions of each objects in query log w.r.t. the manifold structured in the tripartite graph $\mathcal{G}_{tri}$. The third item evaluate the smoothness of predicted query domain-specificity w.r.t. search session. The forth item evaluates how the predicted domain-specificity of seed queries fit with that pre-annotated and it can be viewed as a soft constraint for the predictions on the seed queries.

Based on the objective function in Eq. 3, the optimal domain-specificity of each user, query and URL can be derived through the following optimization problem:

$$\min_{\mathbf{f,g,h}} \quad \frac{1}{2}\, \mathcal{O}_{tri}(\mathbf{f,g,h})$$

$$\text{s.t.} \quad \mathbf{f} \in [0,1]^{|U|}, \ \mathbf{g} \in [0,1]^{|Q|}, \ \mathbf{h} \in [0,1]^{|L|} \tag{5}$$

As for the optimization problem in Eq. 5, it is possible to derive a closed form solution. However, the closed form solution requires matrix inversion operations, which will be computationally inefficient since there are generally huge amount of objects, i.e., unique users, queries and URLs in query log. Therefore, we propose an efficient iterative algorithm to approximately solve the optimization problem in Eq. 5 based on Stochastic Gradient Descent (SGD). The basic idea is to iteratively update the domain-specificity of each objects towards the direction of negative gradient of $\mathcal{O}(\mathbf{f,g,h})$, when observing each $\langle \text{user, query, URL} \rangle$ trinity in query log. More precisely, through computing the derivatives of $\mathcal{O}(\mathbf{f,g,h})$ w.r.t. $\mathbf{f}$, $\mathbf{g}$ and $\mathbf{h}$, we derive the update formulas for $\langle u_i, q_j, l_k \rangle$ as follows:

$$f_i \leftarrow f_i - \mu \cdot \alpha \cdot W_{ij}^{(UQ)}(f_i - g_j)$$

$$g_j \leftarrow g_j - \mu \cdot \big(\alpha \cdot W_{ij}^{(UQ)}(g_j - f_i) + \beta \cdot W_{jk}^{(QL)}(g_j - h_k)$$

$$+ \gamma \cdot \sum_{j'=1, j' \neq j}^{|Q|} \sum_{i=1}^{|U|} \frac{W_{ij}^{(UQ)} \cdot W_{ij'}^{(UQ)}}{\Delta_{i,j,j'}^{\tau}}(g_j - g_{j'})$$

$$+ \delta \cdot I_j^{(seed)}(g_j - y_j) \big)$$

$$h_k \leftarrow h_k - \mu \cdot \beta \cdot W_{jk}^{(QL)}(h_k - g_j) \tag{6}$$

where μ is the learning rate.

From the update formulas in Eq. 6, it can be seen that in each iteration the domain-specificity of each object is updated by taking into account domain-specificity of its associated objects in query log; in other words, domain-specificity of each object propagates along the tripartite graph during the optimization process. This implies that the optimization process is guided by the manifold structure on domain-specificity.

3.3 Active Learning Strategy

In the above domain-specificity learning problem, the seed queries $\mathbf{y}$ have a direct effect on the prediction accuracy. If noisy queries (i.e., queries irrelevant to the target domain) are included in the seed set, the mistake will propagate along the graph during the learning process, which has negative influence on recognition precision; whereas limited or unrepresentative seed set cannot guarantee the coverage of recognized domain-specific queries. In order to construct a high-quality seed query set with the least human annotation efforts, we introduce an active learning strategy into the domain-specificity learning process.

In recent years, many active graph-based transductive learning approaches have been proposed [8–11]. Most of the existing efforts in the literature aim to develop active learning algorithms for general graph data, irrespective of the characteristics of graphs in particular applications. In this work, instead of employing existing approaches, we propose a novel graph-based active learning algorithm, specially tailored for domain-specificity learning task.

The proposed active learning algorithm works in a batch mode: a number of the informative queries are selected and annotated in each round of active learning. The possible informative queries are divided into two types: informative domain-specific queries and informative domain-irrelevant queries, which are named as DS-set and DI-set respectively. Formally, given two informativeness criterion functions $\mathcal{I}^+ : 2^Q \to \mathbb{R}$ and $\mathcal{I}^- : 2^Q \to \mathbb{R}$, the aim is to identify a DS-set $Q^{(DS)} \subseteq Q$ and a DI-set $Q^{(DI)} \subseteq Q$ ($Q^{(DS)} \cap Q^{(DI)} = \emptyset$) such that $\mathcal{I}^+(Q^{(DS)})$ and $\mathcal{I}^-(Q^{(DI)})$ are maximized, respectively.

There are two keys to the active domain-specificity learning algorithm: (1) informativeness criterion and (2) query selection algorithm. We will give the details in the following subsections.

Informativeness criterion. In order to measure the informativeness brought by annotating domain-specificity of a set of queries, three factors: prediction reliability, redundancy and authority, of each query are taken into account.

Prediction reliability. Intuitively, domain-specificity prediction accuracy can be promoted if mistakenly predicted queries were corrected and added into seed set. The informativeness criterion thus should prefer the queries that are unreliably predicted. However, it is hard to evaluate prediction reliability due to the lack of ground truth of domain-specificity for the queries beyond seed set. In this work, we make an assumption that *Queries from the same domain statistically have higher lexical similarity than that from different domains.* Thus DS-set should prefer the queries has low lexical similarity with seed queries while predicted as domain-specific; whereas DI-set should prefer the queries has high lexical similarity with seed queries while predicted as domain-irrelevant.

Redundancy. With limited annotation budget, it is better to select diverse rather than redundant queries for domain-specificity judgement because undesirable redundant queries in DS-set and DI-set will lead to unnecessary repetitive annotation efforts.

Authority. As for graph-based transductive learning, each node in the graph generally has different levels of importance. A central node generally has more influence on other part of the graph than non-central ones, because the domain-specificity of a central node can be more easily propagated along the graph during the learning process. The informativeness criterion should thus prefer the queries with high authority in the tripartite graph.

Synthesizing the above three factors, the informativeness of a query set X selected as DS-set and DI-set is calculated as:

$$\mathcal{I}^+(X; S) = \sum_{q \in X} wt(q) \left(\sum_{p \in S} (1 - sim(q, p)) - \eta \cdot \sum_{(o \in X) \wedge (o \neq q)} sim(q, o) \right) \quad (7)$$

$$\mathcal{I}^-(X;S) = \sum_{q \in X} wt(q)\Big(\sum_{p \in S} sim(q,p) - \eta \cdot \sum_{(o \in X) \wedge (o \neq q)} sim(q,o)\Big) \qquad (8)$$

where S is the current seed set, and $sim(\cdot,\cdot)$ is the lexical similarity between pair of queries. Following the prediction reliability assumption, $1 - sim(q,p)$ and $sim(q,p)$ are used as rough measures of the reliability of predicting p as domain-irrelevant or not. $sim(q,o)$ is used to measure the redundancy of the selected query set. $wt(q)$ is the function quantifying authority of each query q. We simply apply Google's PageRank on the tripartite graph and take the ranking score of each node as $wt(q)$. η is the parameter balancing the contributions of prediction reliability and redundancy in the informativeness criterion.

Query selection algorithm. Given seed set S, selecting DS-set and DI-set composed of k queries can be formulated as the following optimization problems:

$$Q^{(DS)} = \operatorname*{argmax}_{(X \subseteq P^+) \wedge (|X|=k)} \mathcal{I}^+(X;S) \qquad (9)$$

$$Q^{(DI)} = \operatorname*{argmax}_{(X \subseteq P^-) \wedge (|X|=k)} \mathcal{I}^-(X;S) \qquad (10)$$

where P^+ and P^- are the query pools that consist of queries predicted as domain-specific and domain-irrelevant using current seed set S, respectively. Since the prediction of domain-specificity learning algorithm is a rank of all the queries according to the predicted domain-specificity w.r.t. the target domain, P^+ and P^- can be practically constructed by fixed number of the queries in the top and rear of the rank list, respectively.

Essentially, the optimization problems in Eqs. 9 and 10 are knapsack packing problems and NP-hard in general. We develop a polynomial time greedy heuristic solution. The overall algorithm can be found in Algorithm 1. In what follows, for brevity we shall only consider the DI-set selection problem in Eq. 10. The same conclusions can be easily derived for the DS-set selection problem in Eq. 9.

Algorithm I. (GREEDY ALGORITHM FOR DI-SET SELECTION)

Input: Seed set S;
        Query pool P^-, in which each queries are predicted as irrelevant
        to the target domain;
        The number of queries k selected for annotation;
Output: The set of queries for annotation $Q^{(DI)}$, subject to $|Q^{(DI)}| = k$
1: Initialize $Q^{(DI)} \leftarrow \emptyset$
2: **while** $|Q^{(DI)}| \leq k$ **do**
3:     **foreach** $q \in P^- - Q^{(DI)}$ **do**
4:         $\mathcal{I}^-(q) = wt(q) \cdot \big(\sum_{p \in S} sim(q,p) - \gamma \sum_{q' \in Q^{(DI)}} sim(q,q')\big)$
5:     **end**
6:     $q^* = \operatorname{argmax}_{q \in P^- - Q^{(DI)}} \mathcal{I}^-(q)$
7: $Q^{(DI)} = Q^{(DI)} \cup \{q^*\}$
8: **end**

Simply speaking, Algorithm 1 iteratively selects the most informative query q^* out of the query pool $P^- - Q^{(DI)}$ and adds it into the result set $Q^{(DI)}$ (line 2–8). The computational complexity of the algorithm is $O(|P^-| \cdot |Q^{(DI)}|^2)$. Practically, the size of query pool and DI-set (usually varies from tens to thousands) are far less than the total number of queries in query log. Thus algorithm 1 can scale well to real-world query log with millions of objects.

Although Algorithm 1 is an approximate solution to the optimization problems in Eq. 10, it is guaranteed to have a fix error bound because of the submodular object function $\mathcal{I}^-(X; S)$. Due to space limitation, we only give the error bound of Algorithm1 in the following theorem:

Theorem I: *Let X be the query set selected using Algorithm I, and $Q^{(DI)}$ be the optimal solution of the problem in* Eq. 10, *then,*

$$\mathcal{I}^-(X; S) \geq \left(1 - \frac{1}{e}\right) \cdot \mathcal{I}^-(Q^{(DI)}; S) \qquad (11)$$

4   Experiments

4.1   Experiment Settings

Query log. We performed experimental evaluation of the proposed approach using a publicly available query log of a Chinese commercial search engine[1]. In the experiment, we selected *Science:Agriculture* in the open web directory DMOZ[2] as the target domain. Before used for domain-specific query recognition, the query log was sampled by filtering out a number of queries that are obviously irrelevant with the domain of agriculture. Table 2 shows the statistics of the query log corpus used in the experiment.

Evaluation method. For a specified target domain, it is always practically hard to find a comprehensive set of queries specific to the domain that is qualified to evaluate the coverage of the recognized results. In the experiments, we thus mainly focus on evaluating domain-specific query recognition approaches in terms of precision. In particular, we evaluated the proposed approach and baselines in terms of Precision@n (or P@n for short). P@n measures the percentage of true domain-specific queries in the top-n recognized results. Given a

Table 2. Dataset statistics

Node type	Number
User	1,608,222
Query	3,971,977
URL	7,341,534

---

[1] http://www.sogou.com/labs/dl/q.html.
[2] https://www.dmoz.org/.

recognized domain-specific query list $r = \langle t_1, \cdots, t_n \rangle$, P@$n$ is calculated as:

$$P@n = \frac{1}{n} \cdot \sum_{i=1}^{n} I_i^{(specific)} \tag{12}$$

where $I_i^{(specific)}$ is an indicator function that equals to 1 if t_i is judged as specific to the target domain, and 0 otherwise.

In the experiment, correctness of recognized domain-specific queries are manually justified by two volunteers. Given a recognized query, the volunteers are asked to answer "yes" or "no" depending on their judgments on whether the query is relevant to the target domain – DMOZ *Science:Agriculture*. If they gave different answers, domain-specificity of the query would be finally determined by an agriculture expert, i.e., a master student majored in agriculture hired for the experiment. In all the following experiments, top-500 queries recognized by each methods were manually examined for evaluation.

4.2   Parameter Sensitivity Analysis

The parameters that control the contributions of different items in the learning objective, i.e., α, β, γ and δ in Eq. 3 are the main parameters of the proposed approach. We perform sensitivity analysis of these parameters. Firstly, we evaluated the impact of the fitness constraint in Eq. 3 on domain-specific query recognition, while keeping the relative contributions of user-side, URL-side and session smoothness constraints (the first three items in Eq. 3) as constant. In particular, we fixed $\alpha = \beta$ and $\gamma = 0$ empirically, and varied δ from 0 to 1 with the step of 0.1. Secondly, we evaluated the relative impact of user-side and URL-side smoothness constraints in Eq. 3, while keeping session smoothness constraint and fitness constraint as constant. In particular, we fixed $\delta = 0.2$ and $\gamma = 0$ empirically, and varied $\frac{\alpha}{\alpha+\beta}$ from 0 to 1 with the step of 0.1. Thirdly, we evaluated the impact of session smoothness constraint, while keeping relative impact of other constraints in Eq. 3 as constant. In particular, we fixed $\alpha = \beta = \delta$ empirically, and varied γ from 0 to 1 with the step of 0.1.

In each of the three experiments, we utilized a fixed seed set with 200 manual annotated domain-specific queries. For the sake of simplicity, we didn't introduce the proposed active learning strategy in the learning process. Tables 3, 4 and 5 show domain-specific queries recognition results in terms of P@n with different parameters.

From Table 3, it can be seen that domain-specific query recognition performance is quite stable across a wide range of δ (from 0.2 to 0.7). Besides, when δ is quite small ($\delta < 0.1$), the performance drops heavily. It indicates the importance of seed queries to graph-based transductive learning.

From Table 4, it can be seen that smaller α ($\frac{\alpha}{\alpha+\beta} < 0.5$) achieves better domain-specific query recognition performance in terms of P@n ($n \leq 50$), whereas the performance in terms of P@n ($n > 100$) drops heavily when $\frac{\alpha}{\alpha+\beta} < 0.3$. Besides, larger α ($0.8 \leq \frac{\alpha}{\alpha+\beta} \leq 0.9$) achieves relatively consistent

Table 3. Domain-specific query recognition results with varying δ

δ	0	0.1	0.2	0.3	0.4	0.5	0.6	0.7	0.8	0.9	1.0
P@10	0	0.20	**1.00**	**1.00**	**1.00**	**1.00**	**1.00**	**1.00**	**1.00**	**1.00**	**1.00**
P@50	0	0.38	0.88	0.88	**0.92**	**0.92**	**0.92**	**0.92**	**0.92**	**0.92**	0.72
P@100	0.02	0.16	0.5	0.80	0.80	0.75	0.75	0.82	0.78	**0.90**	0.41
P@200	0.01	0.11	0.27	0.58	0.59	0.59	0.63	0.63	0.65	0.61	0.23
P@500	0.01	0.10	0.20	0.35	0.35	0.35	**0.38**	0.34	0.37	0.12	0.08

Table 4. Domain-specific query recognition results with varying $\frac{\alpha}{\alpha+\beta}$

$\frac{\alpha}{\alpha+\beta}$	0	0.1	0.2	0.3	0.4	0.5	0.6	0.7	0.8	0.9	1.0
P@10	**1.00**	**1.00**	**1.00**	**1.00**	**1.00**	**1.00**	0.80	0.80	0.90	0.90	0.50
P@50	**0.92**	**0.92**	**0.92**	**0.92**	**0.92**	**0.92**	0.86	0.82	0.90	0.88	0.60
P@100	0.80	0.80	0.80	0.76	0.78	0.78	0.81	**0.83**	0.73	0.76	0.61
P@200	0.74	0.74	0.69	0.67	0.61	0.65	0.65	**0.76**	0.73	0.75	0.55
P@500	0.45	0.55	0.44	0.47	0.35	0.37	0.40	0.45	0.50	**0.54**	0.33

Table 5. Domain-specific query recognition results with varying γ

γ	0	0.1	0.2	0.3	0.4	0.5	0.6	0.7	0.8	0.9	1.0
P@10	**1.00**	**1.00**	**1.00**	**1.00**	**1.00**	0.80	0.40	0.40	0.40	0	0
P@50	**0.92**	**0.92**	0.82	0.78	0.58	0.48	0.28	0.30	0.14	0.08	0
P@100	0.79	**0.87**	0.79	0.61	0.55	0.51	0.21	0.21	0.19	0.12	0
P@200	0.60	**0.67**	0.61	0.54	0.42	0.38	0.18	0.16	0.12	0.11	0
P@500	0.38	**0.43**	0.36	0.35	0.24	0.22	0.13	0.11	0.08	0.06	0

results over a large range of n. The probable reason is that for domain-specificity learning, manifold structure embedded in the relations between queries and URLs is more precious than that between queries and users, because the topic of a page generally focus on a few domains whereas a user may have a variety of interests involving a number of domains. On the other hand, the numbers of unique URLs is much larger than that of users in the query log (as shown in Table 2), making the relations between queries and URLs more sparse and helpless in discovering wider range of domain-specific queries.

From Table 5, it can be seen that the proposed approach performs not as well when $\gamma > 0.3$, so session smoothness constraint may be not so important as other constraints in optimizing the objective function in Eq. 3. However, when $\gamma = 0.1$ and 0.2, we can see improvements on P@n ($n = 100, 200, 500$) over that achieved when $\gamma = 0$. It indicates that session information still contributes in estimating domain-specificity of queries, especially when larger number of candidate queries are taken into account.

4.3 Effectiveness of Active Learning

We verified effectiveness of the proposed active learning algorithm for graph-based transductive learning. In this experiment, we started from an initial seed set consists of 50 domain-specific queries specified by an agriculture expert, and then repeatedly selected a batch of 30 queries for which the agriculture expert was inquired to annotate. Then the new queries were added into the seed set. In order to avoid the bias caused by single trial, the agriculture expert was asked to specify 200 agriculture queries as a query pool, and we conduct the active learning experiments five times, each of which was based on an initial seed set consists of 50 queries randomly sampled from the query pool. The average over the five trials was used for evaluation. Figure 2 shows the results of domain-specific query recognition based on active learning. We also show the result achieved by graph-based transductive learning without active learning strategy, which as labeled as "200 (fixed)".

We can see that domain-specific query recognition performance in terms of P@n continuously improves while new queries are selected and added into seed sets, especially when the seed set already constructed is of smaller size. We also see that the results based on actively selected 200 seed queries are roughly better than those based on 200 fixed seed queries. More accurately, the improvements are about 6.5%, 6.7% and 5.4% in terms of P@100, P@200 and P@500, respectively. This indicates that the proposed active learning algorithm is helpful in selecting informative seeds for graph-based transductive learning and thus improving the performance of domain-specific query recognition.

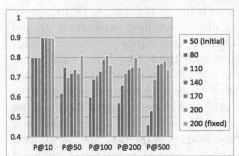
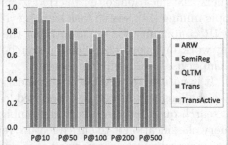

Fig. 2. Domain-specific query recognition results of active learning

Fig. 3. Comparison of domain terminology extraction results

4.4 Comparison with Baseline Methods

To demonstrate the effectiveness of the proposed approach, three state-of-the-art methods exploiting inherent structure of query log for query classification, i.e., ARW [12], SemiReg [6] and QLTM [13], were chosen as baselines for comparison. All the baselines are outlined in related work section.

To make a fair comparison, all the methods made use of the pre-specified 200 agriculture queries as the only training data. We also investigated performance of the proposed approach without/with active learning strategy (abbr. Trans and TransActive, perspectively). The comparison of the proposed approach with baselines is presented in Fig. 3.

It can be seen that the proposed approach (Trans) outperforms all the baselines in term of P@n ($n = 100, 200, 500$) and the improvements are significant especially when n gets larger: about $+24\%$ from the best baselines (QLTM) in terms of P@500. Among all the methods, QLTM performs best when n is small ($n = 10, 50$); however, the performance drops when n gets larger, even worse than SemiReg when $n = 500$. QLTM is able to exploit multi-dimensional latent relations inherent in query log and actually works under supervised learning paradigm. Although supervised learning can leverage latent query features well, but it suffers from the facts that the pre-annotated queries are rather limited and the domain-specific queries are highly sparse. In this experiment, there are only 200 positive queries and the percentage of Agriculture-specific queries in the sampled query log amounts to only about 0.1%.

Besides, among all the graph-based semi-supervised learning methods (ARW, SemiReg, Trans and TransActive), TransActive performs best in term of all the evaluation measures. This provides strong evidence that active learning is capable of guiding the semi-supervised learning process, thus beneficial to the domain-specific query recognition task.

5   Related Work

Query classification, often referred to as search intent learning [6,13,18], query topic mining [14], search task learning [15] and etc., has been extensively studied in IR community for decades. From the learning techniques perspective, existing work on query classification can be put into three categories: supervised, unsupervised and semi-supervised.

It is natural to view query classification as supervised learning; however, it is also challenging to leverage supervised learning techniques in query classification as search query is often short and ambiguous. Therefore, one key to supervised query classification is to enrich feature representation of queries. Shen et al. used the retrieved pages as expansion of the query [4]. Lee et al. found the statistics of click distribution of a query is helpful to identify the underlying search goal [5].

When there is no query category predefined, query classification will turn out to be an unsupervised learning task. Clustering techniques have been extensive utilized in query classification. For example, Hu et al. conduct clustering on the clicked URLs of queries and took each URL cluster as a subtopic of a query [14]; Li et al. proposed a clustering framework with multiple kernel to identify synonymous query intent templates [16]; Qian et al. used incremental clustering method to group clicked URLs in log stream and took each URL group as constant or bursty query intents [17].

Recently, graph-based semi-supervised learning techniques have been extensively used in query classification tasks, especially for those using query log as the resource. Fuxman et al. [12] modeled click-through graph of query log as Markov Random Fields and employed absorbing random walks to compute probability of a query to belong to a pre-defined class. Li et al. [6] formulated query classification as semi-supervised learning on click graphs, with a content-based classifier regularization to avoid erroneous propagation. In order to enhance classification accuracy, the click-through graph of query log is expanded in many existing work. For example, Jiang et al. [13] proposed a query log topic model to derive latent relations between search queries, URL, session and term. Query classifiers are learned using latent relations and further combined using several strategies such as maximum confidence and majority voting.

Our work follows semi-supervised learning theme, but differs in the introduction of active learning strategy. To our best knowledge, no previous work leverages active learning in query classification problem. Moreover, multiple types of objects, i.e., users, queries, URLs and sessions, in query log are simultaneously integrated in query classification task in a more principled way through using tripartite graph representation.

6 Conclusion and Future Work

In this paper, we propose a novel approach to recognize domain-specific queries from general search engine's query log. There are mainly two advantages of the proposed approach. Firstly, the manifold structure inherent in query log is fully exploited through using heterogenous graph representation of query log. Secondly, a novel active learning strategy is introduced into graph-based transductive leaning process to reduce the human annotation efforts and continuously refine the recognition results. Experimental results on real-world query log demonstrate the effectiveness of the proposed approach. Future work includes evaluating the proposed approach on more target domains.

Acknowledgement. This work is partially supported by Chinese Natural Science Foundation (61602278), Shandong Province Higher Educational Science and Technology Program (J14LN33) and China Postdoctoral Science Foundation (2014M561949).

References

1. Arguello, J., Diaz, F., Callan, J., Crespo, J.F.: Sources of evidence for vertical selection. In: Proceedings of the 32nd International ACM SIGIR Conference on Research and Development in Information Retrieval, pp. 315–322 (2009)
2. Giachanou, A., Salampasis, M., Paltoglou, G.: Multilayer source selection as a tool for supporting patent search and classification. Inf. Retrieval J. **18**(6), 559–585 (2015)
3. Yan, X., Liu, Y., Fand, Q., Zhang, M., Ma, S., Ru, L.: Domain-specific terms extraction based on web resource and user behavior. J. Softw. (in Chinese) **24**(9), 2089–2100 (2013)
4. Shen, D., Pan, R., Sun, J.-T., Pan, J.J., Wu, K., Yin, J., Yang, Q.: Query enrichment for web-query classification. ACM Trans. Inf. Syst. **24**, 320–352 (2006)

5. Lee, U., Liu, Z., Cho, J.: Automatic identification of user goals in Web search. In: Proceedings of the 14th International Conference on World Wide Web, pp. 391–400 (2005)
6. Li, X., Wang, Y., Acero, A.: Learning query intent from regularized click graphs. In: Proceedings of the 31st Annual International ACM SIGIR Conference on Research and Development in Information Retrieval, pp. 339–346 (2008)
7. Zhou, D., Bousquet, O., Lal, T.N., Weston, J., Schölkopf, B.: Learning with local and global consistency. Adv. NIPS **16**(16), 321–328 (2004)
8. Zhu, X., Lafferty, J., Ghahramani, Z.: Combining active learning and semi-supervised learning using gaussian fields and harmonic functions. In: ICML 2003 Workshop on the Continuum from Labeled to Unlabeled Data in Machine Learning and Data Mining (2003)
9. Gu, Q., Zhang, T., Han, J.: Batch-mode active learning via error bound minimization. In: Proceedings of the 30th Conference on Uncertainty in Artificial Intelligence, pp. 300–309 (2014)
10. Shi, L., Zhao, Y., Tang, J.: Batch mode active learning for networked data. ACM Trans. Intell. Syst. Technol. **3**(2), 1–25 (2012)
11. Ji, M., Han, J.: A variance minimization criterion to active learning on graphs. In: Proceedings of the 15th International Conference on Artificial Intelligence and Statistics, pp. 556–564 (2012)
12. Fuxman, A., Tsaparas, P., Achan, K., Agrawal, R.: Using the wisdom of the crowds for keyword generation. In: Proceeding of the 17th International World Wide Web Conference, pp. 61–70 (2008)
13. Jiang, D., Leung, K.W.T., Ng, W.: Query intent mining with multiple dimensions of web search data. World Wide Web **19**(3), 475–497 (2016)
14. Hu, Y., Qian, Y., Li, H., Jiang, D., Pei, J., Zheng, Q.: Mining query subtopics from search log data. In: Proceedings of the 35th International ACM SIGIR Conference on Research and Development in Information Retrieval, pp. 305–314 (2012)
15. Ji, M., Yan, J., Gu, S., Han, J., He, X., Zhang, W.V., Chen, Z.: Learning search tasks in queries and web pages via graph regularization. In: Proceedings of the 34th International ACM SIGIR Conference on Research and Development in Information Retrieval, pp. 55–64 (2011)
16. Li, Y., Hsu, B.J.P., Zhai, C.: Unsupervised identification of synonymous query intent templates for attribute intents. In: Proceedings of the 22nd ACM International Conference on Information & Knowledge Management, pp. 2029–2038 (2013)
17. Qian, Y., Sakai, T., Ye, J., Zheng, Q., Li, C.: Dynamic query intent mining from a search log stream. In: Proceedings of the 22nd ACM International Conference on Information & Knowledge Management, pp. 1205–1208 (2013)
18. Ren, X., Wang, Y., Yu, X., Yan, J., Chen, Z., Han, J.: Heterogeneous graph-based intent learning with queries, web pages and Wikipedia concepts. In: Proceedings of the 7th ACM International Conference on Web Search and Data Mining, pp. 23–32 (2014)

Event2vec: Learning Representations of Events on Temporal Sequences

Shenda Hong[1,2], Meng Wu[1,2], Hongyan Li[1,2(✉)], and Zhengwu Wu[3]

[1] Key Laboratory of Machine Perception, Ministry of Education, Beijing, China
[2] School of EECS, Peking University, Beijing, China
lihy@cis.pku.edu.cn
[3] Science and Technology on Information Systems Engineering Laboratory,
Beijing Institute of Control and Electronic Technology, Beijing, China

Abstract. Sequential data containing series of events with timestamps is commonly used to record status of things in all aspects of life, and is referred to as temporal event sequences. Learning vector representations is a fundamental task of temporal event sequence mining as it is inevitable for further analysis. Temporal event sequences differ from symbol sequences and numerical time series in that each entry is along with a corresponding time stamp and that the entries are usually sparse in time. Therefore, methods either on symbolic sequences such as word2vec, or on numerical time series such as pattern discovery perform unsatisfactorily. In this paper, we propose an algorithm called event2vec that solves these problems. We first present Event Connection Graph to summarize events while taking time into consideration. Then, we conducts a training Sample Generator to get clean and endless data. Finally, we feed these data to embedding neural network to get learned vectors. Experiments on real temporal event sequence data in medical area demonstrate the effectiveness and efficiency of the proposed method. The procedure is totally unsupervised without the help of expert knowledge. Thus can be used to improve the quality of health-care without any additional burden.

Keywords: Temporal event sequences · Learning representations

1   Introduction

Sequential data containing series of events with timestamps is referred to as temporal event sequences [9]. Temporal event sequences is commonly used to record status of things in all aspects of life, such as customer purchase sequences, motion gesture sequences, and hospital treatment sequences. On the left side of Fig. 1 shows the daily treatment procedures of patients in hospital [13]. Each ITEM is an event, and (ITEM, TIMESTAMP) is a temporal event. Intuitively, this temporal event sequence tells a detailed story about how patients behave in the hospital. And that story is written in the language of events.

© Springer International Publishing AG 2017
L. Chen et al. (Eds.): APWeb-WAIM 2017, Part II, LNCS 10367, pp. 33–47, 2017.
DOI: 10.1007/978-3-319-63564-4_3

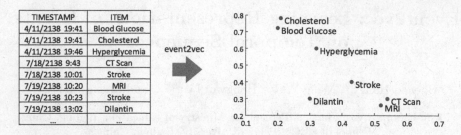

Fig. 1. (Left) An example of a temporal event sequence from EHR. The first column is TIMESTAMP, indicating the occurring time, while the second column is ITEM, indicating the event recorded during the treatment. (Right) Event2vec transforms symbolic events into numerical vectors (take two-dimension vectors for example).

Consider the task of data mining on temporal event sequence, it would be a fundamental step to transform symbolic events into numerical vectors (Fig. 1 right). Thus, learning representations has good prospect of applications such as healthcare [2,18], marketing analytics [9] and motion recognition [5]. On the one hand, the representation itself provides an insight into connection between different events. On the other hand, the representation can be used for practical tasks such as recommendation and clustering. In the above example, the records of one patient require systematic analysis from doctors to provide individualized diagnosis and treatment. It is critical in scenarios like clinical reasoning (what comes the most probability when Cholesterol appears), complication forecasting (what symptoms are Hyperglycemia complication). However, the task would be extremely difficult to fulfill manually even for experts, since the events are numerous and are usually from various medical fields.

Although some recent researches focusing on data mining from temporal event sequences have made certain achievements on learning representations of events. It is still an open challenge due to the following problems.

- **Complexity**: Symbolic methods usually use a combination of symbols as representations (referred to as patterns) [1,15]. Whereas the number of all possible patterns would be 2^n, where n is the number of distinct events. Although recent works have improved the computing efficiency dramatically, the number of patterns still explodes along with the increasing of item quantity [3], which leads to the complexity of mined representations.
- **Sparsity**: Numerical methods are based on continuous numeric sequences [7,10,18,21]. If we want to apply these methods on temporal event sequences, they should first be transformed into continuous numeric sequences (one-hot encoding for example). However, the transformation procedure will lead to high sparsity and introduce lots of noise.
- **Unbalance**: Some events rarely appear but are very important (such as "vital sign "in EHRs). Meanwhile, not all of the potential patterns have been appeared in original data. Both symbolic methods and numerical methods are unsensitive to these important but rare events. [1,10,15,18,21].

– **Time Consideration**: Recent learning representation works implemented neural network architecture like word2vec [11,12], and they did get more dense vectors and resolved some of the above problems. However, they did not take timestamp into consideration since there is no time information in text data. It is a big loss to ignore the occurring time which contains rich semantic information. For example, consider two events, one next to each other with large time interval between. We shouldn't treat them as adjacent events at all as they are irrelevant at high rates, but word2vec has no mechanism to handle that problem.

In this paper, we propose an algorithm called event2vec to embed temporal events into the vector space based on sequential data. Event2vec takes time information into consideration, while implementing neural network architecture, which makes the method benefit from both timestamp information and distributed representation. In particular, event2vec has the following properties:

– Event2vec summarizes large scale temporal event sequence to an event connection graph, with nodes representing events, edges representing relations extracted from temporal sequence. It considers time relationship of events.
– Event2vec conducts a training sample generator. We can relieve unbalance and sparsity problems with the help of the sample generator.
– Event2vec trains an embedding neural network, and represents events in a distributed way. It avoids exaggerated complexity by adjusting the number of units in hidden layer, while keeping the internal relationship among events.
– Event2vec is a totally unsupervised algorithm with no need of expert knowledge, and it can discover relations between events automatically.

We conduct comprehensive experiments on real temporal event sequence data. We unsupervisedly discover implicit relations between various kinds of events. Experiments demonstrate the effectiveness and efficiency of our proposed method. Specifically, we implement event2vec on electronic health records (EHRs) data, and map clinical events to vector space, which reveals some interesting results. The procedure is totally unsupervised. Thus can be used to improve the quality of health-care.

2   Preliminaries

In this section, we introduce our notations and some definitions. Notations are shown in Table 1.

Definition 1. *(Event Set): Event Set Eve is a collection that records all possible conditions of something that would happens. And we use N to denote the number of distinct events in Event Set Eve. For example, the event set of signal light in cross road should be $Eve = \{red, green, yellow\}$, and $N = 3$.*

Definition 2. *(Temporal Event): A temporal event is represented as a two-gram tuple (e, t), where e is the event, t is the occurring time. $e \in Eve$.*

Table 1. Notations and Meanings

Symbol	Description
S	Temporal event sequences dataset
Eve	Set of all possible events
N	Number of distinct events
G	Event connection graph
T	Time threshold (Hyper-parameter in Constructing Graph Process)
Δ	Shrink coefficient (Hyper-parameter in Constructing Graph Process)
$Corpus$	Generated data for embedding
Φ	Matrix of event representations
d	Embedding dimension (Hyper-parameter in Embedding Process)
N_w	Number of events generated in a sequence (Hyper-parameter in Embedding Process)
N_l	Number of samples in generated data (Hyper-parameter in Embedding Process)

Definition 3. *(Temporal Event Sequence): Given a set of N events Eve, $S = \{(e_i, t_i)|e_i \in Eve,\}$ denotes a temporal event sequence.*

In the following section, we will introduce the details of how event2vec learns representations of symbolic events while considering timestamp information.

3   Method

In this section, we present the detailed procedure of the event2vec algorithm. The framework of our method is shown in Algorithm 1. First of all, for the purpose of taking timestamps into consideration, we summarize temporal event sequences into a single graph called event connection graph, with nodes representing events and edges representing extracted relations. A training sample generator is then conducted with reference to the idea of probabilistic walk. It generates a minibatch of clean and endless $Corpus$ once a time. Finally, we feed these data to embedding neural network and output learned vectors (Fig. 2).

Algorithm 1. Event2vec

1: **Input:** $S, Eve, d, \Delta, T, N_w, N_l$
2: **Output:** vectors of event representations Φ
3: $G = $ ConstructingGraph(S, Eve, Δ, T)
4: **while** $iter < N_l$ **do**
5:    $Corpus = $ SampleGenerator(G, Eve, N_w)
6:    LearningRepresentations($Corpus$, Φ, d)
7:    $iter = iter + 1$
8: **end while**

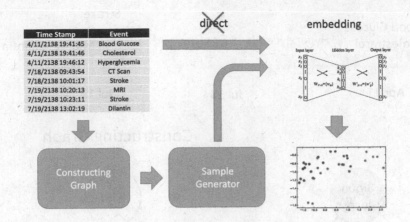

Fig. 2. Framework of event2vec

3.1 Constructing Event Connection Graph

When realizing the idea, a natural solution would be to apply the word2vec directly by treating the sequences of events as the sequences of words. However, it would be difficult to deal with the challenges introduced in Sect. 1 if using word2vec directly. Hence, for the purpose of taking timestamps into consideration, we summarize the temporal event sequence into a compact graph (Fig. 3). In this way, the following purposes would be accomplished: Firstly, extracting temporal information of events. Secondly, condensing the sparse data and settling the problems of unbalance at the same time. Lastly, simplifying the task of updating edge weight with incoming data, which can be updated incrementally instead of completing the entire re-computation.

The comparison between word2vec and event2vec in the experiment section shows the effectiveness of constructed graph. Now we introduce the process of constructing event connection graph.

Definition 4. *(Event Connection Graph): Let $G = < V, E >$ be a directed weighted graph constructed from S, where each vertex e_i in V represents an unique event from Eve and edges in E represent relations extracted from temporal event sequence. Weight of the directed edge from node e_i to node e_j is calculated by:*

$$G_{ij} = \sum_{1 \leq i < j \leq N} 1\{S(t_1) = e_i\} \wedge 1\{S(t_2) = e_j\} \delta(t_2 - t_1) \tag{1}$$

where $1\{\cdot\}$ is set to 1 if its argument is true, and $\delta(x)$ is a function mapping time interval to the relation measurement of two events.

Note that $\delta(x)$ is non-increasing, and the function satisfies $\delta(0) = 1$ and $\delta(T) = 0$. The relation measurement decreases to zero smoothly with the increase

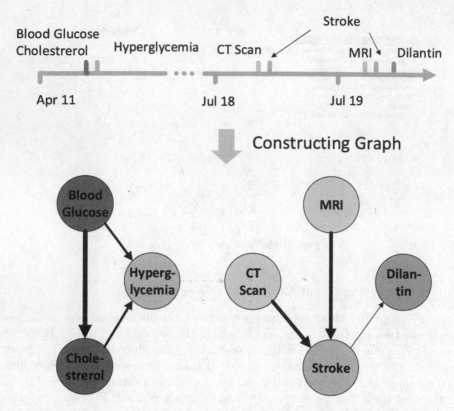

Fig. 3. The process of constructing event connection graph. Each vertex in V represents an unique event from Eve, and edges in E represent relations extracted from temporal event sequence. Weight between the node i to node j is calculated by Eq. 1

of time interval until reaching the threshold T.

$$\delta(x) = \begin{cases} exp(-x/\Delta), & 0 \leq x < T \\ 0, & otherwise \end{cases} \qquad (2)$$

3.2 Sample Generator

The next step is to get samples for the training process. It is beneficial to use Probabilistic Walk for training instead of original temporal event sequence. On the one hand, it relieves unbalance and sparsity effectively. Even if an event appears only once, it must connect with other events with a small probability. So we can resample to get more training data containing that event. On the other hand, it can generate endless training data to tackle the problem of lack of original training data and guarantee the convergence of training process. The experiment in Sect. 4.3, shows the effectiveness of sample generator.

Algorithm 2. ConstructingGraph

1: **Input:** S, Eve, Δ, T
2: **Output:** event connection graph G
3: **for** i = 1 to Length(S) **do**
4:   $(e_i, t_i) = S(i)$
5:   **for** j = i to Length(S) **do**
6:     $(e_j, t_j) = S(j)$
7:     $pos_i = Eve.\text{Index}(e_i)$
8:     $pos_j = Eve.\text{Index}(e_j)$
9:     $G[pos_i, pos_j]+ = \delta(t_i, t_j, \Delta, T)$
10:   **end for**
11: **end for**

We incorporate Probabilistic Walk into the previously constructed event graph. It is a stochastic process starting with a single randomly chosen event e_i. And one of the neighbors of e_i denoted as e_j would be added based on a probability in direct proportion to the edge weight between e_i and e_j. The probability of choosing event e_j after event e_i is

$$P(i \rightarrow j) = W_{ij}/\sum_i W_{ij} \tag{3}$$

The sample generator generates a mini-batch of $Corpus$ at a time.

Algorithm 3. SampleGenerator

1: **Input:** G, Eve, N_w
2: **Output:** a mini-batch of training data $Corpus$
3: Empty($Corpus$)
4: $e_c = \text{Random}(Eve)$
5: $Corpus.\text{Append}(e_c)$
6: $Distri_{e_c} = G[e_c]$
7: **for** i = 1 to $numBatch$ **do**
8:   **for** j = 1 to N_w **do**
9:     $e_{next} = \text{Walk}(Eve, Distri_{e_c})$
10:    $Corpus.\text{Append}(e_{next})$
11:   **end for**
12: **end for**

3.3 Learning Representations

By now, we can get any number of event sequences via Sample Generator. Although these event sequences have no timestamps, they still imply time information thanks to the event Connection Graph. The remaining challenge is to learn representations from the generated sequences.

Formally, given a generated sequence of $Corpus = \{e_1, e_2, e_3, ..., e_n\}$, we would like to predict events nearby a certain event e_i ($e_i \in V$) That is, to maximize the conditional probability of a center event within a context with width L in the training cases. The probability of nearby events $e_{c-L}, ..., e_{c+L}$ given the center event e_i is written as:

$$Pr(\{e_{c-L}, ..., e_{c+L}\}|e_c) = \prod_{k=c-L}^{c+L} Pr(e_k|\Phi(e_c)) \tag{4}$$

Where Φ is our desired representations. After transforming the expression into log-likelihood, the overall loss function for the optimization process would be:

$$J(\Phi) = -\sum_{c=1}^{n} \sum_{k=c-L}^{c+L} logPr(e_k|\Phi(e_c)) \tag{5}$$

We noticed that this loss function is exactly the same as that of SkipGram model in word2vec. And we can use the standard optimization method to solve Eq. 5, namely stochastic gradient descent (SGD). The training procedure can be endless on the basis of mini-batches generated from generator. We first randomly initialize representation vector Φ. On each iteration, we map each event e_k to its current representation vector, and then maximize the probability of its neighbors in the sampled sequence.

Algorithm 4. Learning Representations

1: **Input**: $Corpus$, Φ, d
2: **Output**: vectors of events representations Φ
3: **for** $sequence_{e_c} \in Corpus$ **do**
4:    **for** $e_k \in sequence_{e_c}$ **do**
5:       $J(\Phi) = -logPr(e_k|\Phi(e_c))$
6:       $\Phi = \Phi - \alpha * \frac{\partial J}{\partial \Phi}$
7:    **end for**
8: **end for**

---

4 Experiments

4.1 Experimental Setup

In consideration of realistic application prospect, we use MIMIC (Medical Information Mart for Intensive Care)—a real world temporal event sequence dataset, which is collected on over 58,000 ICU patients at the Beth Israel Deaconess Medical Center (BIDMC) from June 2001 to October 2012 [4]. It contains millions of records of medical events and there are 5373 unique medical events in *Eve*. Our goal is to learn distributed representation of these events.

We design the following mechanism to evaluate the quality of the learned vectors based on the fact that these medical events inherently belong to five categories: demographics, diagnosis, medications, lab tests and procedures. Thus, aiming to estimate whether the learned vectors of events from different categories are well distinguishable, we use the same cluster method on the learned vectors and compare the clustering results. We use the typical evaluation index of clustering, namely Rand Index [16]. It is defined as:

$$RI = \frac{TP + TN}{TP + TN + FP + FN} \tag{6}$$

where, T stands for true, represents that two events belong to the same cluster, P stands for positive, represents that two events are assigned to the same cluster. Then, TP is the number of event pairs in which two events belong to the same cluster and are assigned to the same cluster as well. TN, FP, FN can be explained as well. Intuitively, RI goes higher if the cluster method correctly assign more pairs of events. Similarly, we also define Precision ($P = \frac{TP}{TP+FP}$) and F_1 score($F_1 = \frac{2TP}{2TP+FP+FN}$) to evaluate the quality of representation.

All tasks are executed on a machine equipped with Intel Core i5, 16 GB RAM using Python 2.7.3. The clustering algorithm is K-means, which is realized by scikit-learn 0.18.

4.2 Event2vec Hyper-parameter Analysis

Firstly, we evaluate different groups of Δ, T, N_l, d, and see the influence of these hyper-parameters.

We evaluate Δ, T in graph construction process, and fix other hyper-parameters. As shown in Fig. 4, we can see that the method achieves highest RI, Precision and F_1 when Δ is about 300. As Δ increases, RI, Precision and F_1 will increase as first, but decrease after Δ exceeds a certain threshold. There are two main reasons. On the one hand, higher T means that events with longer time interval are considered relevant, which will bring more noise in the connection graph. On the other hand, higher Δ means events with shorter time interval will add more weight to the corresponding edge of the connection graph. As a result, there exists a balance between T and Δ.

Secondly, we evaluate N_l, d with other hyper-parameters fixed. These two parameters are from sample generate and embedding process. N_l controls sample scale input to embedding neural network, or iteration number in other words. d is the number of nodes in hidden layer of embedding neural network. Result is shown in Fig. 5. As we can see, RI, Precision and F_1 all increase when N_l goes large, and get steady when $N_l > 50,000$. Besides, RI, Precision and F_1 remains steady when d is larger than 6. It demonstrates that our model achieves good performance even when the vectors are in six-dimension space, which also shows the robustness of our model.

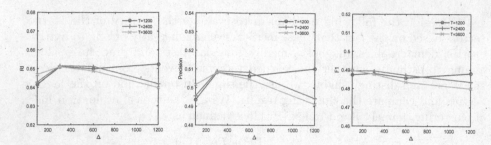

Fig. 4. The effect of hyper-parameters Δ and T. Measured by RI (left), Precision (middle) and F_1 (right).

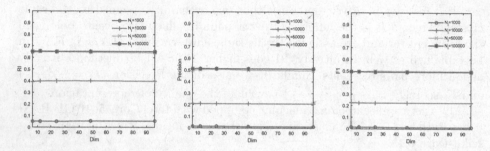

Fig. 5. The effect of hyper-parameters N_l and d measured by RI (left), Precision (middle) and F_1 (right).

4.3   Comparison with Other Methods

As far as we know, there does not exist a general framework for unsupervised learning representation while considering timestamps. We compare our model with the following state-of-the-art methods mentioned before.

- **Count-hot:** For each event, Count-hot constructs a vector using the occurrence number of the event in temporal event sequence of each patient. It means that Count-hot uses each patient to represent one dimension. So the number of dimensions of Count-hot vectors is equal to the number of patients.
- **word2vec:** As introduced in Sect. 3, word2vec applied the skip-gram model directly by treating the sequences of events as the sequences of words. It can be regard as event2vec without the graph construction step.
- **DirectGraph:** After constructing the event graph, DirectGraph regards each column (or row) from the adjacent matrix as a vector. It can be regard as event2vec without the skip-gram embedding step.
- **KNN:** KNN constructs event graph based on their k nearest neighbors. It means that the weight between e_i and e_j is the occurrence number of (e_i, e_j) pair being k nearest neighbors. The remaining procedures of sample generating and embedding process are the same as event2vec. It can be regard as event2vec with a modified graph construction step.

Table 2. Comparison with Different Methods. The result shows that event2vec achieves better effect than other methods on metrics of RI, Precision and F_1.

	Count-hot	word2vec	DirectGraph	KNN (k=3)	KNN (k=5)	KNN (k=7)	Event2vec
RI	0.3938	0.5486	0.3932	0.6202	0.6157	0.6210	**0.6516**
Precision	0.3390	0.3863	0.3372	0.4462	0.4458	0.4475	**0.5088**
F_1	0.4667	0.4238	0.4629	0.3586	0.3913	0.3572	**0.4894**

Table 3. Relativity

	Demographics	Diagnosis	Medication	Lab tests	Procedures
Relativity	1.42	1.50	1.33	1.24	1.25

The event2vec configuration is $\Delta = 300, T = 2400, L = 5$. To make it a fair comparison, we also: (1) Introduce PCA to reduce the dimension of Count-hot and DirectGraph to 48; (2) Set the hidden layer node numbers to 48 in all three methods, namely event2vec, word2vec and KNN; (3) Set N_l to 100,000 to generate the same scale of corpus for word2vec and event2vec. The experiment results are shown in Table 2.

The result shows that on all three metrics, event2vec achieves the highest performance. The only difference between word2vec and event2vec is that, word2vec does not have the procedures of graph construction and sample generation. It shows the effectiveness of these two steps, explanations of which can be found in Sect. 3.1. Besides, the differences between event2vec and DirectGraph is that, DirectGraph does not have the procedures of random walk and skip-gram, explanations of which can be found in Sect. 3.2. As for KNN, the whole process also contains graph construction, random walk and skip-gram, but a different schema is used for computing weight of edges in the graph. It demonstrates that our method of graph constructing is better than KNN, after taking timestamps into account.

4.4 Interpretation

Consider the importance of interpretability in healthcare, we conduct the following experiments in collaboration with medical experts, and they confirm the explanation of our learned representations.

We first perform a relativity assessment by randomly selecting 74 events along with their top 10 most similar events. This shows whether the learned representations effectively capture the latent relationships among them. As shown in Table 3, two medical experts checked the result and decided whether an event pair is related, possible or unrelated. Obviously, we can come to the conclusion that all of the relativity scores are greater than 1, which means that event2vec successfully finds the high related events of the given ones.

Table 4. Five selected events and their top 10 most related events

Seizure	Apnea	ECG	Cholesterol	Intra Cranial Pressure
Ataxia	high min volume	BP Right Leg Diastolic	HDL	SVV Arterial
CIWA Sum Total	high mv	ECG	Fibrinogen	Intra Cranial Pressure
Tactile Disturbances	apnea	BP Right Arm Mean	Triglyceride	Central Venous Pressure
Auditory Disturbance	Inspired Gas Temp	BP Right Leg Systolic	C Reactive Protein CRP	Cerebral Perfusion Pressure
Orient/Clouding Sens	ETT Location	BP Right Leg Mean	Amylase	CO Arterial
Visual Disturbances	ETT Mark	BP Right Arm Systolic	AST	Bladder Pressure
Aggitation	Apnea Time Interval	BP Left Leg Mean	Cholesterol	Transpulmonary Pressure Exp. Hold
Seizure	Ventilator Type	BP Left Arm Diastolic	LDL calculated	Glucose whole blood
Headache	Airway Size	BP Left Arm Mean	Alkaline Phosphate	Arterial Blood Pressure mean
Anxiety	Airway Type	BP Left Arm Systolic	Lipase	Arterial Blood Pressure systolic

Table 5. Four selected dimensions along with the top 10 events in the corresponding dimension

Dimension 1	Dimension 2	Dimension 3	Dimension 4
Religion	Sodium ApacheIV	WhiteBloodC 4.0–11.0	Ventilator Mode
Highest Level	RRScore ApacheIV	TCO2 (other)	Tracheostomy Cuff
Height (cm)	RR ApacheIV	Red Blood C(3.6–6.2)	Tidal Volume (Set)
Height	OxygenScore ApacheIV	RBC(3.6–6.2)	Return Pressure
Gestational Age	MAP ApacheIV	ph (other)	Plateau Off
Code Status	CreatScore ApacheIV	pH (cap)	Minute Volume Alarm - Low
Birthweight (kg)	BiliScore ApacheIV	pCO2 (other)	Low Exhaled Min Vol
Age	Bilirubin ApacheIV	pCO2 (cap)	Flow By (lpm)
Admission Weight (lbs.)	Apache IV Age	HGB (10.8–15.8)	Cuff Leak
Admission Weight (Kg)	AgeScore ApacheIV	Base Excess (other)	Access Pressure

Furthermore, we list five selected events along with their top 10 most related events in Table 4 A comprehensive example would be that ECG, the measurement of cardiac electrical activity, can be reflected by BP (Blood Pressure).

In addition, we randomly choose 4 dimensions, and list events with the largest 10 values in the chosen 4 dimensions in Table 5. We can see that the events in Dimension 1 are mostly related to demographics of patients; Dimension 2 are mostly related to Apache score (a commonly used health measurement); Dimension 3 are mostly related to chemical index of blood lab test reports; Dimension 4 are mostly related to clinical procedures by medical devices.

In conclusion, event2vec successfully finds meaningful events. These results demonstrate that event2vec can be used for clinical reasoning, medication recommendation and complication forecasting.

5   Related Work

For a comprehensive review, we introduce learning representation from both continuous sequential data and symbolic sequential data.

Learning representation from continuous sequential data also refers to pattern discovery on sequential data. The methods in this category usually extract a small fraction of sequences such as motif [10], shapelet [21], signature [18], and PLR [17]. [18] proposed a convolutional method that learns temporal event signatures and uses these signatures to learn representations. In addition, [19] using pattern-based hidden Markov model to find dynamics for semantic representation from time series data. We should first transform temporal event sequence to continuous numeric sequence (one hot for example), and then apply numerical methods. However, the data would have high sparsity. Besides, Some events rarely appear but are very important. It is difficult to capture the micro relationship while keeping the macro information [9] follows [8], propose skeleton of the graph constructed by temporal event sequence, and discover higher granularity representation from original data.

Learning representation from symbolic sequential data usually use a combination of symbols as representations (referred to as patterns) methods usually using a combination of symbols as representations (referred to as) [1,6,15]. Although some efforts to reduce the complexity of pattern sets, they still get the initial patterns and then reduce them by approximation [14] or clustering [20] and it might lead to a complexity of mined representations. Recently, neural network architecture was proposed for distributed representations for word embedding [11,12], does get more dense vectors and resolve above problems. However, they did not take timestamps into consideration since no time information in text data. It is a big loss to ignore the occur time which provide rich semantic information. [22] use transitive distance to transform categorical data to numerical representations. However, the works mentioned above also suffer from complexity and unbalance.

6   Conclusion and Future Work

In this paper, we propose the event2vec algorithm that transforms large scale symbolic events from sequential data into numerical vectors. Event2vec takes temporal information into consideration, and solves the problem of high complexity, sparsity and unbalance. Concretely, our Event Connection Graph takes time into consideration; Sample Generator solves sparsity and unbalance in raw data, final embedding neural network controls complexity of learned vectors. Experiments show that event2vec outperforms other learning representation methods, while providing good interpretation of events. The procedure is totally unsupervised without the help of expert knowledge. Thus, the algorithm can be used in real-world applications, such as to improve the quality of health-care without any additional burden.

In the future, we plan to modify the embedding neural network, and take the structure information extracted from constructed graph into consideration.

References

1. Agrawal, R., Srikant, R.: Mining sequential patterns. In: Proceedings of the 11th International Conference on Data Engineering (ICDE 1995), March 6–10, 1995, Taipei, Taiwan, pp. 3–14 (1995). http://dx.doi.org/10.1109/ICDE.1995.380415
2. Choi, E., Bahadori, M.T., Searles, E., Coffey, C., Thompson, M., Bost, J., Tejedor-Sojo, J., Sun, J.: Multi-layer representation learning for medical concepts. In: Proceedings of the 22nd ACM SIGKDD International Conference on Knowledge Discovery and Data Mining, San Francisco, CA, USA, August 13–17, 2016, pp. 1495–1504 (2016). http://doi.acm.org/10.1145/2939672.2939823
3. Fournier-Viger, P., Gomariz, A., Campos, M., Thomas, R.: Fast vertical mining of sequential patterns using co-occurrence information. In: Tseng, V.S., Ho, T.B., Zhou, Z.-H., Chen, A.L.P., Kao, H.-Y. (eds.) PAKDD 2014. LNCS (LNAI), vol. 8443, pp. 40–52. Springer, Cham (2014). doi:10.1007/978-3-319-06608-0_4
4. Goldberger, A.L., Amaral, L.A., Glass, L., Hausdorff, J.M., Ivanov, P.C., Mark, R.G., Mietus, J.E.: Moody: physiobank, physiotoolkit, and physionet components of a new research resource for complex physiologic signals. Circulation **101**(23), e215–e220 (2000)
5. Jaimes, A., Sebe, N., Boujemaa, N., Gatica-Perez, D., Shamma, D.A., Worring, M., Zimmermann, R. (eds.): ACM Multimedia Conference, MM 2013, Barcelona, Spain, October 21–25, 2013. ACM (2013). http://dl.acm.org/citation.cfm?id=2502081
6. Kim, Y., Han, J., Yuan, C.: TOPTRAC: topical trajectory pattern mining. In: Proceedings of the 21th ACM SIGKDD International Conference on Knowledge Discovery and Data Mining, Sydney, NSW, Australia, August 10–13, 2015. pp. 587–596 (2015),. http://doi.acm.org/10.1145/2783258.2783342
7. Lin, J., Keogh, E.J., Lonardi, S., Chiu, B.Y.: A symbolic representation of time series, with implications for streaming algorithms. In: Proceedings of the 8th ACM SIGMOD Workshop on Research Issues in Data Mining and Knowledge Discovery, DMKD 2003, San Diego, California, USA, June 13, 2003, pp. 2–11 (2003). http://doi.acm.org/10.1145/882082.882086
8. Liu, C., Wang, F., Hu, J., Xiong, H.: Temporal phenotyping from longitudinal electronic health records: A graph based framework. In: Proceedings of the 21th ACM SIGKDD International Conference on Knowledge Discovery and Data Mining, Sydney, NSW, Australia, August 10–13, 2015, pp. 705–714 (2015). http://doi.acm.org/10.1145/2783258.2783352

9. Liu, C., Zhang, K., Xiong, H., Jiang, G., Yang, Q.: Temporal skeletonization on sequential data: patterns, categorization, and visualization. IEEE Trans. Knowl. Data Eng. **28**(1), 211–223 (2016). http://dx.doi.org/10.1109/TKDE.2015.2468715
10. McGovern, A., Rosendahl, D.H., Brown, R.A., Droegemeier, K.: Identifying predictive multi-dimensional time series motifs: an application to severe weather prediction. DMKD **22**(1–2), 232–258 (2011). http://dx.doi.org/10.1007/s10618-010-0193-7
11. Mikolov, T., Chen, K., Corrado, G., Dean, J.: Efficient estimation of word representations in vector space. CoRR abs/1301.3781 (2013). http://arxiv.org/abs/1301.3781
12. Mikolov, T., Sutskever, I., Chen, K., Corrado, G.S., Dean, J.: Distributed representations of words and phrases and their compositionality. In: Advances in Neural Information Processing Systems 26: 27th Annual Conference on Neural Information Processing Systems 2013 (NIPS 2013). Proceedings of a meeting held December 5–8, 2013, Lake Tahoe, Nevada, United States, pp. 3111–3119 (2013). http://papers.nips.cc/paper/5021-distributed-representations-of-words-and-phrases-and/-their-compositionality
13. Pathak, J., Kho, A.N., Denny, J.C.: Electronic Health Records-driven Phenotyping: challenges, Recent Advances, and Perspectives. AMIA **20**(December) (2013)
14. Pei, J., Dong, G., Zou, W., Han, J.: On computing condensed frequent pattern bases. In: Proceedings of the 2002 IEEE International Conference on Data Mining (ICDM 2002), 9–12, Maebashi City, Japan, pp. 378–385 (2002). http://dx.doi.org/10.1109/ICDM.2002.1183928
15. Pei, J., Han, J., Mortazavi-Asl, B., Pinto, H., Chen, Q., Dayal, U., Hsu, M.: Prefixspan: mining sequential patterns by prefix-projected growth. In: Proceedings of the 17th International Conference on Data Engineering (ICDE 2001), April 2–6, 2001, Heidelberg, Germany, pp. 215–224 (2001). http://dx.doi.org/10.1109/ICDE.2001.914830
16. Rand, W.M.: Objective criteria for the evaluation of clustering methods. J. Am. Stat. Assoc. **66**(336), 846–850 (1971). http://www.jstor.org/stable/2284239
17. Tang, L., Cui, B., Li, H., Miao, G., Yang, D., Zhou, X.: Effective variation management for pseudo periodical streams. In: Proceedings of the ACM SIGMOD International Conference on Management of Data, Beijing, China, June 12–14, 2007, pp. 257–268 (2007). http://doi.acm.org/10.1145/1247480.1247511
18. Wang, F., Lee, N., Hu, J., Sun, J., Ebadollahi, S., Laine, A.F.: A framework for mining signatures from event sequences and its applications in healthcare data. IEEE Trans. Pattern Anal. Mach. Intell. **35**(2), 272–285 (2013). http://dx.doi.org/10.1109/TPAMI.2012.111
19. Wang, P., Wang, H., Wang, W.: Finding semantics in time series. In: Proceedings of the ACM SIGMOD International Conference on Management of Data, SIGMOD 2011, Athens, Greece, June 12–16, 2011, pp. 385–396 (2011). http://doi.acm.org/10.1145/1989323.1989364
20. Xin, D., Han, J., Yan, X., Cheng, H.: Mining compressed frequent-pattern sets. In: Proceedings of the 31st International Conference on Very Large Data Bases (VLDB 2005), Trondheim, Norway, August 30 - September 2, 2005, pp. 709–720 (2005). http://www.vldb2005.org/program/paper/thu/p709-xin.pdf
21. Ye, L., Keogh, E.J.: Time series shapelets: a novel technique that allows accurate, interpretable and fast classification. DMKD **22**(1–2), 149–182 (2011)
22. Zhang, K., Wang, Q., Chen, Z., Marsic, I., Kumar, V., Jiang, G., Zhang, J.: From categorical to numerical: multiple transitive distance learning and embedding, pp. 46–54 (2015). http://dx.doi.org/10.1137/1.9781611974010.6

Joint Emoji Classification and Embedding Learning

Xiang Li[1], Rui Yan[2,3], and Ming Zhang[1(✉)]

[1] School of EECS, Peking University, Beijing, China
{lixiang.eecs,mzhang_cs}@pku.edu.cn
[2] Institute of Computer Science and Technology, Peking University, Beijing, China
ruiyan@pku.edu.cn
[3] Beijing Institute of Big Data Research, Beijing, China

Abstract. Under conversation scenarios, emoji is widely used to express humans' feelings, which greatly enriches the representation of plain text. Plentiful utterances with emoji are produced by humans manually in social media platforms every day, which make emoji great influence on the human life. For the academic community, researchers are always with the help of utterances including emoji as annotated data to work on sentiment analysis, yet lack of adequate attention to emoji itself. The challenges lie in how to discriminate so many different kinds of emoji, especially for those with similar meanings, which make this problem quite different from traditional sentiment analysis. In this paper, in order to gain an insight into emoji, we propose a matching architecture using deep neural networks to jointly learn emoji embeddings and make classification. In particular, we use a convolutional neural network to get the embedding of the utterance and match it with the embedding of the corresponding emoji, to obtain its best classification, and otherwise also train the emoji embeddings. Experiments based on a massive dataset demonstrate the effectiveness of our proposed approach better than traditional softmax methods in terms of p@1, p@5 and MRR evaluation metrics. Then a test of human experience shows the performance could meet the requirement of practice systems.

Keywords: Emoji classification · Embedding learning · Deep learning · Neural networks

1 Introduction

Conversation is one of the most important activities for humans, which could communicate their thought and feelings. For face-to-face conversation, humans use expression to indicate their emotion. Recently with the prosperity of Web 2.0, more and more conversation occurs on web platforms like Facebook and Twitter, or using chat tools, which could make humans communicate with each other overcoming distance. For those scenarios, plain text is used instead of face-to-face talking, and emoji is used as the expression on human's face.

© Springer International Publishing AG 2017
L. Chen et al. (Eds.): APWeb-WAIM 2017, Part II, LNCS 10367, pp. 48–63, 2017.
DOI: 10.1007/978-3-319-63564-4_4

Emoji is a kind of symbols to present one's expression, for instance, 😐 and 😟 , which express happy and sad respectively. Emoji is important in humans' daily interaction, like social networks and conversation, through uttering their feelings, which could make the expression more interesting and lively. Thus, platforms like Facebook and Twitter are adding increasing numbers of emoji sets to improve user experience, and Unicode, as an international encoding standard, extends its emoji set continuously, including more than 7000 kinds in its recent 8.0 version.

The importance of emoji is increasingly realized by the academic community of artificial intelligence. For the task of sentiment analysis, researchers often use emoji as a kind of distant supervision [4]. Since large amount of annotation data is needed, they regard emoji as the ground truth, which means that, some emoji expressing happy feelings such as 😊 indicates a positive polarity, and sad expressions like 😟 indicates a passive polarity. However, emoji is just used as an assistant way, and research work for emoji itself is really lack.

Researches for emoji itself is very significative, and could support many applications. Besides as a kind of human annotation for sentiment analysis, it is convenient for humans to know when and what emoji he may use during inputting the utterance, if there is a function of recommending emoji for an utterance on the social media platform. Moreover, in automatic human-computer conversation systems, it could make the computer side more proactive through adding emoji behind the reply, just like a human expressing his feeling, considering that in human-human conversation on social network platforms, emoji is often used as a part of humans' utterances to indicate their current emotion.

In practice utterances generated by humans, the use of emoji is widely. The phenomenon is supported by the statistics of utterances collected from an online forum[1], which is shown in Fig. 1, consisting of over 5 billion items of data manually input by humans. Our observation is that, people often use emoji in their everyday life, in order to make the communication more vivid and make it easier to deliver their feelings. Therefore, many applications like automatic human-computer conversation systems, which need to make the computer side be more like a human, especially for commercial products, should add the function of automatic emoji expression in order to attract more people.

Emoji is quite different to traditional sentiment or emotion analysis, which only need to discriminate polarities or several pre-defined kinds of emotion, since there are many kinds of emoji, and meanings of some emoji are so similar that it is difficult to distinguish them well for traditional classification methods using softmax. For example, 😄 and 😊, which means laugh and smile respectively, both express emotion of happiness, with only a little difference on degree. Therefore, we propose a joint architecture to learn emoji embeddings and classification through matching them in multi-modal vector space. To be specific, based on the layer of word embeddings, we get the embedding of the whole utterance by a convolutional neural network (CNN), and then a HingeLoss function is used to match it with the emoji embeddings, which should be also trained. Comparing

---

[1] http://www.weibo.com.

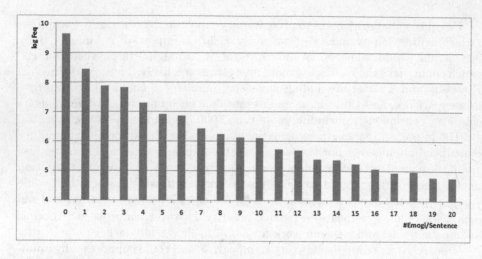

Fig. 1. For utterances manually input by humans, we plot the frequency of utterances including different amount of emoji. The range of emoji amount is from 0 to 20, where 0 means no emoji in the utterance, and the frequency has been in the logarithm. Generally speaking, 9.18% utterances include at least one kind of emoji, and some proportion of them consist of different kinds, which indicates that people often use emoji in their everyday life.

to the traditional softmax function, our approach could better distinguish emoji with similar meanings.

To sum up, the main contributions of this paper are as follows.

- We conduct scientific experiments to analyze the problem of emoji classification and embedding learning in conversation scenarios[2].
- We propose a matching approach using deep neural networks by utilizing emoji embeddings and observe that the performance of emoji classification is better than traditional softmax methods.
- Empirical experiments demonstrate the effectiveness of both our embedding learning and emoji classification, and the analysis shows a good human experience in practice.

2   Related Work

Traditional sentiment or emotion analysis is a significant research task which has attracted many researchers in the domain of natural language processing. Research work on sentiment analysis often focuses on classifying the polarities

---

[2] We notice a piece of parallel work [2], which is an application named Dango on Android platform, and also suggest emoji for conversation between humans. However, we are the first to conduct scientific experiments, showing the effectiveness of matching.

of positive and negative [5,6], or extends to the third polarity of neutral [7,8], or sometimes adds fine-grained classes like a spectrum such as very positive and very negative [9–11]. Pre-defined kinds of emotion are also involved into some work on sentiment analysis, such as happy, sad, and so on [12–14], while sometimes the emotion classification could be multi-label [45].

With the development of natural language processing, many theories and technologies have been used to deal with traditional sentiment analysis. Lexicon-based models using sentiment dictionaries are an effective series of approaches to deal with sentiment analysis [15,16], since some words have clear trends of sentiment polarities. Feature-based models using traditional classifiers are another kind of methods with high performance, which is called distant supervision by leverage utterances with emoticons as annotated data [3,17]. Other theories like statistical machine translation [18], graph-based approach [19] and topic model [20] are also used to analyze sentiment.

Recent years, with the development of word embeddings and neural networks, research work appears continuously using these technologies to improve the performance of sentiment analysis. Since word embeddings could well represent its semantic features and also latent information [35,36], it is natural to add sentiment-specific information into the word embeddings while training by neural networks [8,21,23]. Another series of approaches is to propose novel structures of neural networks [1,22,24,46], which means adapting the theory of deep learning to sentiment analysis. Furthermore, under some specific scenarios, especially on social networks, context of human interaction are considered to improve sentiment analysis [25–27].

Besides using emoji as a kind of distant supervision, emoji or emoticons themselves are also related to sentiment expression. Emoticons could indicate sentiment polarities in plain-text computer-mediated communication [44] and a sentiment map for several hundred kinds of most frequently used emoji is established [37], both in order to improve the performance of sentiment analysis.

Although there has been research work which proposes a multi-modal approach to generate emoji labels for an image [28], it is still lack of effort to match emoji with plain text. Thus, we adjust the problem of emoji classification and embedding learning, which is more complicated than traditional sentiment or emotion analysis, and then propose a match approach to obtain better performance than softmax classifiers.

3 Approach

3.1 Task Definition

Given an utterance set $Y = \{y_1, y_2, ..., y_n\}$ and a emoji set $X = \{x_1, x_2, ..., x_k\}$, our aim is to train a classification model which could predict the correct emoji $g(y) \in X$ for an utterance y, meanwhile get a vector set $E = \{e_1, e_2, ..., e_k\}$ after training, and each vector e_i is the embedding of emoji x_i in X, as a distributed representation indicating its latent semantic information.

3.2 Structure Overview

Our proposed approach is shown in Fig. 2, which is a matching structure based on neural networks, and consists of two parts. The left component is a sub convolutional neural network to get a sentence embedding which could represent the utterance, while the right one is the embeddings of emoji that should also be trained, and finally joint the two parts through a matching score. Our intuition to use a matching structure is that the embeddings in continuous vector space could well represent emoji, and perform better than discrete softmax classifiers, since meanings of some emoji are amphibolous and difficult to distinguish.

3.3 Layers

Word Embedding Layer. This word embedding layer is at the bottom of the left CNN part, which aims to get the distributed representation of each word. The embedding of a word is a vector, and each element in the vector is a real number to represent one dimension feature of the word. Thus, the embedding vector could be regarded as a feature set of the word in a low-dimension space, and could indicate latent information of the word both semantically and syntactically, with wonderful performance [36] for some tasks in the domain of natural language processing. Instead of manually designed as feature engineering methods, embeddings of words are often trained by neural networks or calculated by matrix decomposition.

For a plain text utterance, it could be regarded as a sequence of words, and for each word w_i, we first represent it as a one-hot vector of dictionary dimension, with one 1 on the corresponding word bit and other bits 0. Then using an embedding matrix $\mathbf{E}_1 \in R^{D \times V}$, where D is the dimension of word embeddings and V is the dimension of the word dictionary, we could obtain the embedding of the word $\mathbf{e}_1(w_i)$. Thus, we get each word embedding after this layer, and the matrix $\mathbf{E}_1$ consists of all the embeddings $\mathbf{e}_1(w_i)$ for words in the dictionary, which is randomly initialized and trained during the training process. In practice, we have over 30 thousand words in the dictionary and choose the parameter D equal to 128.

Convolutional Layer. The convolutional structure of neural networks is believed suitable to synthesize lexicon n-gram information of a sequence, especially for short text [29]. Different from full connection layers, CNN uses the concept of sliding windows, which is like a local feature extractor, to get information from word embeddings. If the window size is t, and the corresponding words embeddings are $\mathbf{e}_1(w_1), \mathbf{e}_1(w_2), \cdots, \mathbf{e}_1(w_t)$, then we have:

$$\mathbf{y}_1 = f(\mathbf{W}_1[\mathbf{e}_1(w_1); \mathbf{e}_1(w_2); \cdots; \mathbf{e}_1(w_t)] + \mathbf{b}_1) \tag{1}$$

where $\mathbf{b}_1$ is the bias vector, f is the non-linear activation function, and $\mathbf{W}_1$ is the parameter matrix which needs trained. We choose a window size of 3 in practice.

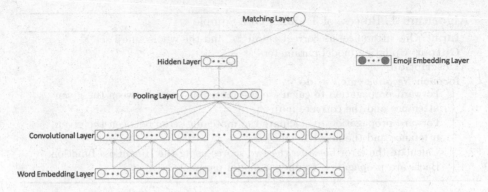

Fig. 2. The whole structure of our matching approach based on neural networks.

Pooling Layer. After the convolutional layer, we get a series continuous representations of local features. Since we need to synthesize these local embeddings into one vector as the distributed representation of the whole utterance, we use the theory of dynamic pooling [31]. To be specific, let $\mathbf{y}_1^1, \mathbf{y}_1^2, \cdots, \mathbf{y}_1^l$ be the output vectors from the convolutional layer, and then we have:

$$\mathbf{y}_2[i] = max\{\mathbf{y}_1^1[i], \mathbf{y}_1^2[i], \cdots, \mathbf{y}_1^l[i]\} \tag{2}$$

which means we use max pooling actually, and then obtain $\mathbf{y}_2$ as the sentence embedding of the utterance, which is also 128-dimension in practice. Obviously, a sentence embedding is also a vector to represent the utterance in low-dimension space, which could indicate latent information, and generally be used in the domain of natural language processing, just like word embeddings.

Hidden Layer. Then we use a hidden layer of full connection to exert a non-linear transformation to the sentence embedding, which could be calculated by:

$$\mathbf{y}_3 = f(\mathbf{W}_2\mathbf{y}_2 + \mathbf{b}_2) \tag{3}$$

where $\mathbf{b}_2$ is the bias vector, f is the non-linear activation function, and finally we get another 128-dimension vector to represent the utterance.

Emoji Embedding Layer. Since we have dealt with the plain text side through the CNN which is shown as the left part of our proposed matching structure, next we need to embed the emoji, aiming to learn continuous representations of emoji in vector space, just like word embeddings. Thus, in a similar way like the word embedding layer, we could also represent each emoji x_i as a one-hot vector of K-dimension, where K is equal to the amount of emoji, and then use a matrix $\mathbf{E}_2 \in R^{D \times K}$ to obtain its embedding $\mathbf{e}_2(x_i)$. The matrix $\mathbf{E}_2$ includes all the emoji embeddings, and each element is one parameter of the neural network, which is randomly initialized and trained during the training process.

Algorithm 1. Process of Training One Sample

Input: One utterance, its correct emoji x_t, and the whole emoji set X
Output: Updating model parameters
Description:
foreach *Negative emoji x_j* **do**

> Forward propagation to calculate the matching score between the given utterance and the correct emoji x_t
> Forward propagation to calculate the matching score between the given utterance and the negative emoji x_j
> Calculate the error between the two scores using the HingeLoss function
> Backward propagation to update model parameters

Matching Layer. In this layer, we plan to match the embedding of the plain text $\mathbf{y}_3$ from the non-linear hidden layer with the emoji embedding $\mathbf{e}_2(x_i)$, and then obtain a score as the final result which could indicate their matching degree. We choose the cosine similarity as the measurement of matching and then have:

$$score(\mathbf{y}_3, \mathbf{e}_2(x_i)) = \frac{< \mathbf{y}_3, \mathbf{e}_2(x_i) >}{||\mathbf{y}_3|| \cdot ||\mathbf{e}_2(x_i)||} \tag{4}$$

where $< \cdot, \cdot >$ means inner product of two vectors and $||\cdot||$ means the length of a vector. Since a higher score indicates more matching for the utterance embedding and the emoji embedding, we choose $argmax_{x_i} \, score(\mathbf{y}_3, \mathbf{e}_2(x_i))$ as the final emoji which should be added into the plain text.

3.4 Training

In the training process, we use the HingeLoss function, which is convex with wonderful properties, through punishing negative matching to optimize parameters. Different from the softmax function, the HingeLoss function exerts pairwise comparisons between the positive matching and each negative one, in order to distinguish them, especially for similar kinds of emoji. For each plain text $\mathbf{y}^i$ in the training set, let $\mathbf{y}_3^i$ denote the sentence embedding of $\mathbf{y}^i$ from the non-linear hidden layer, and our optimal objective is

$$Obj(\mathbf{y}^i) = min \sum_{j \neq t} max(0, \alpha + score(\mathbf{y}_3^i, \mathbf{e}_2(x_j)) - score(\mathbf{y}_3^i, \mathbf{e}_2(x_t))) \tag{5}$$

where x_t is the correct emoji for training samples $\mathbf{y}^i$ according to the ground truth, and α is the margin of the HingeLoss function. The process of training one sample is depicted in Algorithm 1.

Before the training process, we calculate the frequency of each emoji, and select top 20, 50 and 100 respectively, as three emoji sets. We use the theory of stochastic gradient descent to train our proposed neural networks, and adjust the learning rate on the set consisting of 20 kinds of emoji. Finally we get a learning rate of 10^{-6} and adapt it to the other two emoji sets.

4 Experiments

4.1 Datasets

We collect massive conversation resources of human interaction from microblog websites including Sina Weibo[3], which contains over 170 thousand kinds of emoji in all. Then we extract tens of millions of utterances with top 20, 50 and 100 most frequent kinds of emoji respectively, as our three datasets. The datasets are randomly divided into training sets, validation sets and test sets, and the details are summarized in Table 1.

Table 1. Statistics of the datasets

#Emoji	#Training	#Validation	#Test
20classes	11.6M	50K	56K
50classes	14.4M	50K	75K
100classes	15.9M	50K	83K

To be specific, each item of data is a pair of $\langle utterance, emoji \rangle$, where emoji comes from the utterance itself and is regarded as the ground truth produced by humans. Since we only need to put one utterance into one class, any expression including more than one kind of emoji is filtered, aiming to produce a clear training set. Moreover, different people may give different emoji to the same plain text, so in order to avoid this kind of confusion, we also filter any expression having different emoji for all its appearance. Besides, since most expression has no emoji, we should not give emoji to all utterances, so finally, we filter the utterances with a max-majority of non-emoji for all its appearance.

4.2 Qualitative Analysis of Embeddings

Since we regard the emoji embedding layer as a continuous representation of emoji, we firstly have a qualitative analysis of these embeddings. As mentioned before, we aim to get a meaningful set of emoji embeddings, so similar kinds of emoji should have shorter distances than others. Thus, a hierarchical clustering is applied on the embeddings of the most commonly used 20 kinds of emoji (Fig. 3), with a cosine similarity as the metric of distance and using nearest neighbourhood. Our observation is that, the embeddings could represent semantic information of the emoji in some degree, due to some phenomenon like "laugh" close to "smile" and the six kinds of emoji on the left denoting the positive polarity while the two on the right representing the negative polarity. Yet there are also confusing aspects like unclear clustering especially for clusters with low similarity.

---

[3] http://www.weibo.com.

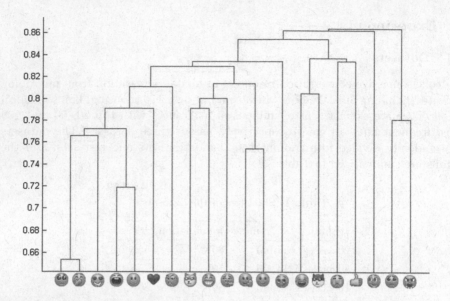

Fig. 3. The result of the hierarchical clustering of top 20 kinds of emoji.

4.3   Quantitative Analysis of Emoji Classification

In this section, we have a quantitative analysis of emoji classification directly, with lots of other experimental analysis such as parameter sensitivity and case studies.

Baseline Algorithms. We include the following methods as baselines to compare with our proposed approach. Since our approach is a matching architecture based on neural networks, besides the method based on textural similarity, we mainly use some basic neural network structures to make the comparison, with a traditional softmax function as the objective. For fairness, we conduct the same data cleaning and layer dimensions in neural networks for all algorithms. Besides, we also adjust the learning rate for baseline neural networks just the same as the process of our proposed approach, and finally get a learning rate of 10^{-3}.

Textural Similarity. This method ranks candidates according to textual similarity, which is a basic way to calculate relevance between queries and documents in the domain of information retrieve (IR). Here we regards each emoji as a document consisting of all utterances including the particular emoji, while the utterance as a query, and each word is weighed by tf-idf.

CBOW. Since an utterance consists of words, bag-of-words is a natural way to model the utterance and widely used [36]. The basic thought is to regard the utterance as a set of words, so from the word embedding layer, a summation or average operation is simply done to get the embedding of the whole utterance.

The bag-of-words method could get the utterance representation quickly and concisely, however lose the sequentiality of natural languages.

CNN. Convolutional neural networks is a kind of structure that could extract local features, and is believed to get better performance on images [38] or short text [29,30]. Instead of the way of full-connection, a convolutional layer has a sliding window which means that a neuron could only have particular numbers of connections from the last layer. After the convolutional operation, there is often a pooling layer to integrate the information, and here we also use max pooling for fairness.

RNN. Due to the sequentiality of natural languages, recurrent neural networks and its variants are also widely used to represent an utterance, especially for the generation process [39,40,42,43]. For each hidden layer, the inputs are the current word embedding as well as the last hidden layer, until the end of the utterance, and the final hidden layer is regarded as the embedding of the whole utterance, which could represent all the sequential information. In practice, due to the increasing sparsity with the propagation going on, the Long Short-term Memory (LSTM) [41] or the Gated Recurrent Unit (GRU) [32] is often used to improve its performance. Here we use the GRU version.

Table 2. Performance of the emoji classification

#Emoji	20classes			50classes			100classes		
Metrics	p@1	p@5	MRR	p@1	p@5	MRR	p@1	p@5	MRR
Textural similarity	15.61%	57.69%	32.38%	12.37%	47.91%	24.83%	11.60%	43.85%	20.93%
CBOW	22.22%	60.03%	39.74%	18.10%	49.62%	33.24%	16.41%	44.43%	30.12%
CNN	22.42%	60.31%	39.96%	18.62%	50.06%	33.67%	16.74%	44.78%	30.50%
RNN	22.21%	59.64%	39.61%	17.90%	48.96%	32.86%	16.23%	43.78%	29.75%
Matching	**24.30%**	**63.01%**	**41.39%**	**20.16%**	**51.29%**	**34.74%**	**18.69%**	**46.74%**	**31.94%**

Evaluation Metrics. We first evaluate the performance using p@1 metric, which could reflect the accuracy of algorithm results, and is believed to be the most direct judgment for classification tasks. Besides, for most applications possibly developed based on emoji classification, p@1 is also appropriate since the utterance should match at least one emoji when we want to add emoji behind the plain text.

Next, since the results we returned based on our approach or baseline algorithm are matching scores between the sentence and emoji, or distributions on emoji, they could be regarded as ranking lists of emoji given the plain text. So we could also evaluate the performance in terms of p@k, and here we choose k equal to 5.

Another evaluation metric is the Mean Reciprocal Rank (MRR)[4], which is also able to evaluate a ranking list:

$$MRR = \frac{1}{|T|} \sum_{i=1}^{|T|} \frac{1}{rank_i} \tag{6}$$

Here $rank_i$ refers to the rank position of the correct emoji according to the ground truth for the i-th plain text, and T is the set of plain text.

Performance. In this section, we show the performance of the proposed matching approach against other baselines. The results are summarized in Table 2, in which we report the performance of emoji classification in all the mentioned evaluation metrics. It is obvious that methods based on neural networks perform better than textual similarity, and our matching approach is the best of all. Although a softmax objective function could make a good classification, the matching structure is more appropriate to distinguish little difference between emoji especially for those kinds with similar or confusing meanings. The improvement is small yet consistent, which is similar to the conclusion of a prior work [33], and our observation is that in some degree, our matching approach seems similar to use a cross-entropy objective function instead of the traditional one-hot vector in the softmax structure.

Analysis. In this section, we have some analysis on emoji classification with grouping similar kinds, parameter sensitivity of our matching approach, and human experience for possible applications.

Grouped Classification. Since some kinds of emoji are similar and could express the same meaning, they could be merged together as one class. In our experiment, we group emoji in the data set of 100 kinds into 20 classes manually according to prior knowledge, and then use our matching approach as well as all the baselines to make the classification. The results comparing to the original classification of 20 classes are summarized in Table 3, which shows that, all the methods perform better than the original result, with a small increasing on p@1 yet a large improvement on p@5. The main reason of these phenomenon is that after grouping, the meaning of each class is clearer so that they could be distinguish better, especially for the situation of 5 candidates.

Parameter Sensitivity. For our matching approach, we regard the correct emoji according to the ground truth as positive samples while others as negative ones. So if the task is k classification, then the amount of negative samples is k-1. In the situation of 20 classes, we investigate the influence to the performance with changing the amount of negative samples. To be specific, besides the whole negative set of 19 samples, we randomly sample 1, 5, 10, 15 negative samples using uniform distribution. Figure 4 shows that on all the three metrics, with

---

[4] https://en.wikipedia.org/wiki/Mean_reciprocal_rank.

Table 3. Analysis of the grouped emoji classification

#Emoji	20classes			Grouped20classes		
Metrics	p@1	p@5	MRR	p@1	p@5	MRR
Textural Similarity	15.61%	57.69%	32.38%	19.89%	67.74%	38.50%
CBOW	22.22%	60.03%	39.74%	25.20%	70.47%	44.85%
CNN	22.42%	60.31%	39.96%	25.45%	70.91%	45.09%
RNN	22.21%	59.64%	39.61%	25.09%	70.41%	44.70%
Matching	24.30%	63.01%	41.39%	27.62%	72.77%	46.48%

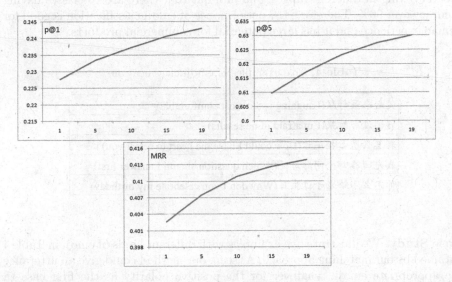

Fig. 4. The illustration of parameter influence to the results of 20 classification. The top left chart presents results of p@1 with the increasing of the amount of negative samples, the top right one is for p@5, and the bottom one is for MRR.

the negative samples increasing, the performance become better, and the rising trend becomes slower.

Human Experience. The potential of being adapted into industrial applications could obviously demonstrate the meaning and importance of research work. Yet it seems that the performance of emoji classification is far away from practice applications, because the results of the automatic testing listed before generally do not meet the high accuracy required by commercial products, such as an automatic human-computer conversation system. However, The experience of users is very different from the automatic testing, since some kinds of emoji are similar. Even if we put an utterance into a wrong class, it may not hurt the user experience, which means that in practice applications, more than one kind of emoji could be appropriate for a given utterance.

Therefore, to investigate the performance in user experience, we use human annotation with **1 Point** meaning appropriate or **0 Point** meaning inappropriate. One utterance and the emoji given by our approach are annotated by 3 individuals in an independent and blind fashion. We regard the majority voting as the "ground truth" indicating whether the emoji is appropriate for the utterance from the vision of users. We also evaluate the kappa score: $\kappa = 0.413$, showing moderate inner-annotator agreement [34].

The test data for human annotation consists of 694 items randomly selected from the 20 classification. The performance of our matching approach shows accuracy of 73.05% for the annotation dataset, which obviously is much higher than it of the automatic testing. Moreover, we also investigate the amount of classes with accuracy lying in different ranges, and find out that there are 10 classes having accuracy over 90%. Therefore, the performance of emoji classification is good for human experience and it has large potential for commercial products.

Table 4. Five examples of emoji classification

笑起来最舒服(Laughing is the most comfortable) 😆
好怕没有结果(I'm afraid of no results) 😭
我怎么忍心拒绝你(How could I have the heart to reject you) 😊
我先回答哪个问题好呢(Which question should I answer first) 😵
你怎么不给老子过生日(Why don't you celebrate my birthday) 😡

Case Study. We illustrate five examples with different kinds of emoji in Table 4 obtained by our matching approach. As seen, our method could give an utterance the appropriate emoji, whatever for the positive polarity as the first case (a laughing face), the passive polarity as the second (a crying face) and fifth (an angry face) cases, or just neutral without clear emotional tendency as the third (a shy face) and fourth (a thinking face) cases. Besides, we have the ability to express much more emotion other than polarities. With emoji, the form of expression become more liberal and indeed more vivid when communicating with others.

5   Conclusion

In this paper, for the problem of emoji classification and embedding learning in conversation scenarios, we propose a matching approach and deeply analyze its performance through both qualitative and quantitative experiments. Empirical results demonstrate our approach better than traditional softmax classifiers in terms of different metrics, and the embeddings trained from our neural networks could also represent the emoji well. For the future work, one direction is to consider contextual information in the conversation process and propose more progressive models to analyze emoji better, another one is to explore the relation between emoji and sentiments in order to improve the performance of sentiment analysis.

Acknowledgements. This paper is partially supported by the National Natural Science Foundation of China (NSFC Grant Nos. 61472006 and 91646202) as well as the National Basic Research Program (973 Program No. 2014CB340405).

References

1. Tang, D., Wei, F., Qin, B., Zhou, M., Liu, T.: Building large-scale twitter-specific sentiment lexicon: a representation learning approach. In: Proceedings of the 25th International Conference on Computational Linguistics, pp. 172–182 (2014)
2. Snelgrove, X.: http://getdango.com/emoji-and-deep-learning.html
3. Pak, A., Paroubek, P.: Twitter as a corpus for sentiment analysis and opinion mining. In: International Conference on Language Resources and Evaluation, pp. 1320–1326 (2010)
4. Yan, J.L.S., Turtle, H.R.: Exploring fine-grained emotion detection in tweets. In: Proceedings of NAACL-HLT, pp. 73–80 (2016)
5. Wu, F., Song, Y., Huang, Y.: Microblog sentiment classification with contextual knowledge regularization. In: Proceedings of the 29th AAAI Conference on Artificial Intelligence, pp. 2332–2338 (2015)
6. Deng, L., Wiebe, J., Choi, Y.: Joint inference and disambiguation of implicit sentiments via implicature constraints. In: Proceedings of the 25th International Conference on Computational Linguistics, pp. 79–88 (2014)
7. Wilson, T., Wiebe, J., Hoffmann, P.: Recognizing contextual polarity in phrase-level sentiment analysis. In: Proceedings of the Conference on Human Language Technology and Empirical Methods in Natural Language Processing, pp. 347–354 (2005)
8. Tang, D., Wei, F., Yang, N., Zhou, M., Liu, T., Qin, B.: Learning sentiment-specific word embedding for twitter sentiment classification. In: Proceedings of the 52nd Annual Meeting of the Association for Computational Linguistics, pp. 1555–1565 (2014)
9. Socher, R., Perelygin, A., Wu, J.Y., Chuang, J., Manning, C.D., Ng, A.Y., Potts, C.: Recursive deep models for semantic compositionality over a sentiment treebank. In: Proceedings of the Conference on Human Language Technology and Empirical Methods in Natural Language Processing, pp. 1631–1642 (2013)
10. dos Santos, C.N., Gatti, M.: Deep convolutional neural networks for sentiment analysis of short texts. In: Proceedings of the 25th International Conference on Computational Linguistics, pp. 69–78 (2014)
11. Mou, L., Peng, H., Li, G., Xu, Y., Zhang, L., Jin, Z. Discriminative neural sentence modeling by tree-based convolution. In: Proceedings of the Conference on Human Language Technology and Empirical Methods in Natural Language Processing, pp. 2315–2325 (2015)
12. Beck, D., Cohn, T., Specia, L.: Joint emotion analysis via multi-task Gaussian processes. In: Proceedings of the Conference on Human Language Technology and Empirical Methods in Natural Language Processing, pp. 1798–1803 (2014)
13. Wang, Z., Lee, S.Y.M., Li, S., Zhou, G.: Emotion detection in code-switching texts via bilingual and sentimental information. In: Proceedings of the 53rd Annual Meeting of the Association for Computational Linguistics, pp. 763–768 (2015)

14. Chang, Y.C., Chen, C.C., Hsieh, Y.L., Chen, C.C., Hsu, W.L.: Linguistic template extraction for recognizing reader-emotion and emotional resonance writing assistance. In: Proceedings of the 53rd Annual Meeting of the Association for Computational Linguistics, pp. 775–780 (2015)

15. Turney, P.D.: Thumbs up or thumbs down?: semantic orientation applied to unsupervised classification of reviews. In: Proceedings of the 40th Annual Meeting of the Association for Computational Linguistics, pp. 417–424 (2002)

16. Taboada, M., Brooke, J., Tofiloski, M., Voll, K., Stede, M.: Lexicon-based methods for sentiment analysis. Comput. Linguist. **37**(2), 267–307 (2011)

17. Kouloumpis, E., Wilson, T., Moore, J.D.: Twitter sentiment analysis: the good the bad and the omg!. In: Proceedings of the 15th International AAAI Conference on Web and Social Media, pp. 538–541 (2011)

18. Lambert, P.: Aspect-level cross-lingual sentiment classification with constrained SMT. In: Proceedings of the 53rd Annual Meeting of the Association for Computational Linguistics, pp. 781–787 (2015)

19. Wang, X., Wei, F., Liu, X., Zhou, M., Zhang, M.: Topic sentiment analysis in twitter: a graph-based hashtag sentiment classification approach. Proceedings of the 20th ACM International Conference on Information and Knowledge Management, pp. 1031–1040 (2011)

20. Yang, M., Peng, B., Chen, Z., Zhu, D., Chow, K.P.: A topic model for building fine-grained domain-specific emotion lexicon. Proceedings of the 52nd Annual Meeting of the Association for Computational Linguistics, pp. 421–426 (2014)

21. Zhou, H., Chen, L., Shi, F., Huang, D.: Learning bilingual sentiment word embeddings for cross-language sentiment classification. In: Proceedings of the 53rd Annual Meeting of the Association for Computational Linguistics, pp. 430–440 (2015)

22. Dong, L., Wei, F., Zhou, M., Xu, K.: Adaptive multi-compositionality for recursive neural models with applications to sentiment analysis. In: Proceedings of the 28th AAAI Conference on Artificial Intelligence, pp. 1537–1543 (2014)

23. Ren, Y., Zhang, Y., Zhang, M., Ji, D.: Improving twitter sentiment classification using topic-enriched multi-prototype word embeddings. In: Proceedings of the 30th AAAI Conference on Artificial Intelligence, pp. 3038–3044 (2016)

24. Zhang, M., Zhang, Y., Vo, D.T.: Gated neural networks for targeted sentiment analysis. In: Proceedings of the 30th AAAI Conference on Artificial Intelligence, pp. 3087–3093 (2016)

25. Vanzo, A., Croce, D., Basili, R.: A context-based model for sentiment analysis in twitter. In: Proceedings of the 25th International Conference on Computational Linguistics, pp. 2345–2354 (2014)

26. Ren, Y., Zhang, Y., Zhang, M., Ji, D.: Context-sensitive twitter sentiment classification using neural network. In: Proceedings of the 30th AAAI Conference on Artificial Intelligence, pp. 215–221 (2016)

27. Li, S., Huang, L., Wang, R., Zhou, G.: Sentence-level emotion classification with label and context dependence. In: Proceedings of the 53rd Annual Meeting of the Association for Computational Linguistics, pp. 1045–1053 (2015)

28. Cappallo, S., Mensink, T., Snoek, C.G.: Image2emoji: Zero-shot emoji prediction for visual media. In: Proceedings of the 23rd ACM International Conference on Multimedia, pp. 1311–1314 (2015)

29. Hu, B., Lu, Z., Li, H., Chen, Q.: Convolutional neural network architectures for matching natural language sentences. In: Annual Conference on Neural Information Processing Systems, pp. 2042–2050 (2014)

30. Yan, R., Song, Y., Wu, H.: Learning to Respond with Deep Neural Networks for Retrieval based Human-Computer Conversation System. In: Proceedings of SIGIR, pp. 55–64 (2016)
31. Socher, R., Huang, E.H., Pennin, J., Ng, A.Y., Manning, C.D.: Dynamic pooling and unfolding recursive autoencoders for paraphrase detection. In: Annual Conference on Neural Information Processing Systems, pp. 801–809 (2011)
32. Cho, K., Van Merriënboer, B., Gulcehre, C., Bahdanau, D., Bougares, F., Schwenk, H., Bengio, Y.: Learning phrase representations using RNN encoder-decoder for statistical machine translation. In: Proceedings of the Conference on Human Language Technology and Empirical Methods in Natural Language Processing, pp. 1724–1734 (2014)
33. Tang, Y.: Deep learning using linear support vector machines. arXiv preprint arXiv:1306.0239 (2013)
34. Fleiss, J.L.: Measuring nominal scale agreement among many raters. Psychol. Bull. **76**(5), 378 (1971)
35. Bengio, Y., Ducharme, R., Vincent, P., Jauvin, C.: A neural probabilistic language model. J. Mach. Learn. Res. **3**(Feb), 1137–1155 (2003)
36. Mikolov, T., Sutskever, I., Chen, K., Corrado, G.S., Dean, J.: Distributed representations of words and phrases and their compositionality. In: Advances in Neural Information Processing Systems, pp. 3111–3119 (2013)
37. Novak, P.K., Smailović, J., Sluban, B., Mozetič, I.: Sentiment of emojis. PloS One **10**(12), e0144296 (2015)
38. Babu, G.S., Zhao, P., Li, X.L.: Deep convolutional neural network based regression approach for estimation of remaining useful life. In: Proceedings of the 21st International Conference on Database Systems for Advanced Applications, pp. 214–228 (2016)
39. Sutskever, I., Vinyals, O., Le, Q.V.: Sequence to sequence learning with neural networks. In: Advances in Neural Information Processing Systems, pp. 3104–3112 (2014)
40. Shang, L., Lu, Z., Li, H.: Neural responding machine for short-text conversation. In: Proceedings of the 53rd Annual Meeting of the Association for Computational Linguistics, pp. 1577–1586 (2015)
41. Hochreiter, S., Schmidhuber, J.: Long short-term memory. Neural Comput. **9**(8), 1735–1780 (1997)
42. Sordoni, A., Galley, M., Auli, M., Brockett, C., Ji, Y., Mitchell, M., Nie, J.Y., Gao, J., Dolan, B.: A neural network approach to context-sensitive generation of conversational responses. In: Proceedings of NAACL-HLT, pp. 196–205 (2015)
43. Mou, L., Song, Y., Yan, R., Li, G., Zhang, L., Jin, Z.: Sequence to backward and forward sequences: a content-introducing approach to generative short-text conversation. In: Proceedings of the 26th International Conference on Computational Linguistics, pp. 3349–3358 (2016)
44. Hogenboom, A., Bal, D., Frasincar, F., Bal, M., de Jong, F., Kaymak, U.: Exploiting emoticons in polarity classification of text. J. Web Eng. **14**, 22–40 (2015)
45. Wang, Y., Feng, S., Wang, D., Yu, G., Zhang, Y.: Multi-label chinese microblog emotion classification via convolutional neural network. In: Li, F., Shim, K., Zheng, K., Liu, G. (eds.) APWeb 2016. LNCS, vol. 9931, pp. 567–580. Springer, Cham (2016). doi:10.1007/978-3-319-45814-4_46
46. Zhao, Z., Liu, T., Hou, X., Li, B., Du, X.: Distributed text representation with weighting scheme guidance for sentiment analysis. In: Li, F., Shim, K., Zheng, K., Liu, G. (eds.) APWeb 2016. LNCS, vol. 9931, pp. 41–52. Springer, Cham (2016). doi:10.1007/978-3-319-45814-4_4

Target-Specific Convolutional Bi-directional LSTM Neural Network for Political Ideology Analysis

Xilian Li, Wei Chen[✉], Tengjiao Wang, and Weijing Huang

Key Lab of High Confidence Software Technologies (MOE), School of EECS,
Peking University, Beijing 100871, China
{xilianli,pekingchenwei,tjwang,huangweijing}@pku.edu.cn

Abstract. Ideology detection from text plays an important role in identifying the political ideology of politicians who have expressed their beliefs on many issues. Most existing approaches based on bag-of-words features fail to capture semantic information. And other sentence modeling methods are inefficient to extract ideological target context which is significant for identifying the political ideology. In this paper, we propose a target-specific Convolutional and Bi-directional Long Short Term Memory neural network (CB-LSTM) which is suitable in intensifying ideological target-related context and learning semantic representations of the text at the same time. We conduct experiments on two commonly used datasets and a well-designed dataset extracted from tweets. The experimental results show that the proposed method outperforms the state-of-the-art methods.

Keywords: Ideology detection · Ideological target · Convolutional neural network · Recurrent neural network

1 Introduction

Recently, more and more politicians turn to social networks such as Twitter and Facebook to express their beliefs instead of relying on traditional interviews or magazines. Ideology detection from these user generated texts play an important role in identifying the political ideology (conservative or liberal) of politicians on many issues, such as predicting poll ratings [8] or evaluating political leadership abilities, and has attracted increasing research interests [1].

Text-based ideology detection remains two significant challenges. One is the semantic information capturing. For example, two phrases *"tax less"* and *"increase taxes"* about *"taxation"* from opposite alignments are considered highly similar if ignoring the syntax and word orders. Another crucial challenge is how to detect the target or target-related semantic context which makes it more difficult than sentiment analysis [6]. One sentence *"Abortion is murder of a human being."* from conservatives opposes *"abortion"*, while another sentence *"We can't deprive the rights that a woman can decide what happens with her baby."* from liberals supports

© Springer International Publishing AG 2017
L. Chen et al. (Eds.): APWeb-WAIM 2017, Part II, LNCS 10367, pp. 64–72, 2017.
DOI: 10.1007/978-3-319-63564-4_5

"*abortion*", even it does not contain keyword "*abortion*". Both texts express negative sentiment towards the same target, but the ideologies are just opposite which proved the importance of recognizing key target.

However, existing studies for ideology detection often fail to capture semantic information and extract ideological target context at the same time. Traditional methods exploit machine learning algorithms [7,10] to build ideology classifiers based on bag-of-words representations with various hand-crafted features [4,9], but neglect the intrinsic contextual relations between sub-sentences. Recently, dominant neural networks which focus on sentence representations can grasp the semantic information while inefficient to extract ideological target context. For instance, some recurrent neural networks [11,24] are capable of capturing abundant semantic correlations for sentiment analysis [3], but are not dedicated for extracting local features. Other frameworks represented by recursive neural networks fetch lexical characteristics to identify ideological influence [13], while depend on complex syntactic tree parsing [15,22]. Convolutional models [14] which toward stance classification [23] can extract local features, yet can not learn sequential correlations. These methods alone are insufficient for purifying target-related context or learning contextual information to the greatest extent.

To address these limitations, we propose a target-specific Convolutional Bi-directional LSTM neural network (CB-LSTM) which is suitable in capturing target-related context and learning semantic representations simultaneously. We apply a modified convolutional structure which is designed to obtain target-related context and reserve position information between n-grams. Then we integrate the output of convolutional layer into BLSTM, which helps to avoid missing long text information, to learn the sequential correlations. By combining the convolutional and LSTM structure, our model takes advantage of both convolutional neural models and recurrent neural networks.

We evaluate the performance on Convote, IBC [13] and a well-refined Twitter dataset with the work of our team [5]. Experimental results show that the CB-LSTM model has superior performances over other existing methods. The main contributions of this work can be summarized as follows:

- We present a unified neural network CB-LSTM which can extract core local features and encode sequential correlations between sub-sentences.
- We adopt the CB-LSTM model to enhance target-related context and learn semantic representations simultaneously for ideology detection task. The empirical results show that it outperforms other state-of-the-art approaches.
- We introduce an innovative and generic method to obtain ideology-related dataset from massive data.

2   CB-LSTM Model

We propose a target-specific convolutional bi-directional long-short term neural network to capture the target context and semantics of the text which are further used as features for ideology detection. First the representations of target

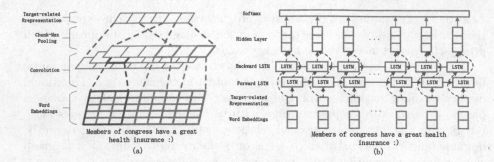

Fig. 1. (a) is the CNN architecture for learning target representations. (b) is the main structure of BLSTM. Dashed arrows indicate dropout layers are applied.

semantic or n-grams for each sentence are learned through convolutional structure in Fig. 1(a) with word embeddings as inputs. Next, the target-level vectors are concatenated with the word embeddings and are fed into the BLSTM network. Finally, the output is passed to a hidden layer and the softmax layer determines the ideology score with maximum probability of the word as shown in Fig. 1(b).

2.1 Target Context Representation Through CNN

In this section, we will present a convolutional network with multiple filters to learn semantic representations of the implied target n-grams. In our network, the CNN only has one layer of convolution on top of pre-trained word vectors. In addition, there is no fully connected layer at the end.

A convolution operation involves multiple filters. They are applied to a window of h words in a tweet t expressed as $x_{i:i+h-1}$ which is the concatenation of words $w_1, w_2, ..., w_n$. As a result, some feature maps are produced. A local feature $t_i \in \mathbb{R}^{1 \times (|t|-h+1)}$ related to the ideological target is computed by

$$t_i = f(W_t \cdot x_{i:i+h-1} + b_t) \tag{1}$$

where $W_t \in \mathbb{R}^{h_f \times h \cdot k}$ is the weight matrix between input and convolution, b_t is a bias term for each feature map f is a hyperbolic tangent ($tanh$) function and k is the dimension of word vectors. We use three filters whose window sizes are 2, 3 and 4.

Then a pooling layer to get a fixed length vector is applied. We refer to a Chunk-Max Pooling [18] operation which has been proved improvement to capture necessary position information between target features.

2.2 Sentence Representation with BLSTM

In this section, the target-level representation vectors learned through the convolutional layer are concatenated with the original word embeddings. Then we feed the concatenated vectors with intensified target semantic information into

BLSTM to learn rich contextual correlations of the sentence. The detailed structure is shown in Fig. 1(b).

As a special RNN structure, LSTM has shown strong capability for modeling long-range dependencies over standard RNN in various previous studies [12, 16]. The key element of LSTM is cell state c_t. It is able to remove or add information to the cellular state through a carefully designed structure called "gates": the *input gate* i_t, *forget gate* f_t and *output gate* o_t, which are used to protect and control cell status and avoid gradient vanishing or exploding [11]. We refer to the structure of Graves [20] and the main steps are as follows.

$$
\begin{aligned}
i_t &= f(W_i \cdot [c_{t-1}, h_{t-1}, x_t] + b_i) \\
f_t &= f(W_f \cdot [c_{t-1}, h_{t-1}, x_t] + b_f) \\
c_t &= f_t \circ c_{t-1} + i_t \circ tanh(W_c[h_{t-1}, x_t] + b_c) \\
o_t &= f(W_o \cdot [c_{t-1}, h_{t-1}, x_t] + b_o) \\
h_t &= o_t \circ tanh(c_t)
\end{aligned}
\tag{2}
$$

where $\circ$ denotes element-wise multiplication, f is the sigmoid function and $b_i, i \in [i, f, c, o]$, is the different bias term belonging to different gates.

BLSTM has both a forward and a backward LSTM in the hidden layer. The forward LSTM captures the past feature context, while the backward LSTM captures the future feature information. Dropout is used to avoid overfitting.

2.3 Ideology Detection

In the last layer of the CB-LSTM, we intend to predict the ideology polarity of one sentence based on learned semantic sentence representation h_o with several hidden layers and a *softmax* function. The output is defined as

$$
\widetilde{y}_i = \frac{exp((W_o \cdot h_o + b)^i)}{\sum_{j=1}^{k} exp((W_o \cdot h_o + b)^j)}
\tag{3}
$$

where k is the number of known labels.

The loss function applied is cross-entropy error between predicted label $\widetilde{y}_i$ and true label y_i.

$$
L(X, y) = \sum_{j=1}^{N} \sum_{i=1}^{k} \widetilde{y}_i^j \cdot log(y_i^j) + \lambda ||\theta||^2
\tag{4}
$$

where N is the number of train data, λ is the regularization coefficient and θ stands for the model parameters. We apply RMSprop algorithm to update the parameters by minimizing the loss function on training dataset.

3 Experiments

3.1 Datasets

To perform the effectiveness of our model, we focus on Convote [19], IBC [13] and Twitter data. Convote dataset is the Congressional debates data that has

Table 1. The overall information of the datasets. #class is the number of classes, #sen represents the number of sentences in per dataset, #word is the vocabulary size of words, #avgLen is the average number of words contained in each sentence.

Dataset	#class	#sen	#word	#avgLen
Convote	2	7,816	15,024	25.5
IBC	2	3,412	13454	41.4
Twt4w1kw	3	40,000	23246	26.5
Twt2w1kw	3	20,000	19958	23.4
Twt2w2kw	3	20,000	19958	23.4

annotation on the author level. IBC is the Ideological Books Corpus which contains a million sentences written by authors with well-known political leanings. It was after annotated on the sentence and phrase level by crowdsourcing [13].

Twitter Datasets: In order to get enough ideology-related datasets, we collected 88 liberal and 131 conservative user accounts from American political forums and crawled 668763 tweets posted by them until June 2015 as our Twitter corpus. Next we apply an innovative transfer learning model for accurate event detection [5] to extract 16 category-level ideology keyword lists from Wikipedia, ranked by Chi-square score which is the degree of prominence used to differentiate from other keywords in the same categories. After correction by political domain experts, we get the final outstanding ideology keyword lists. With these comprehensive keyword lists, we filter the tweets corpus. Table 1 provides detailed information about each final dataset. Suffix $i$kw represents that the dataset is filtered by i keywords.

3.2 Experimental Setting

We preprocess the datasets as follows. For all datasets, we use Standford CoreNLP[1] tool to get tokenization of the sentences. We split each dataset into training and testing sets with 80/20, with 10% of the training set as development set. The evaluation metric of all these datasets is accuracy.

All of the parameters are initialized from a uniform distribution *uniform* $(-0.05, 0.05)$. We pre-train a 100-dimension word embedding matrix on a 2 billion tweets set crawled in the year of 2015 and 2016 by skip-gram model of Google word2vec[2]. Other hyperparameter settings of our neural networks are depending on which dataset is being used. For convolutional layers, we use some of the hyper-parameters following previous work [14]. A mini-batch size is 50. We set the size of memory dimension as 50 and the learning rate of RMSprop to 0.01. Dropout rate is set to be 0.5 in convolutional and BLSTM layer. L2 regularization of le-6 is used to the last softmax layer.

---

[1] http://stanfordnlp.github.io/CoreNLP/tokenize.html.
[2] http://code.google.com/p/word2vec/.

4 Results and Discussion

4.1 Ideology Detection

The experimental results are shown in Table 2. We compare the traditional baselines SVM and Logistic Regression to other neural network approaches. The
experiment results show that the latter models outperform the formers on all
datasets. It proves the deep learning methods are better at capturing semantic
information. It's worth noting that, as a rapid approach, FastText [17] has better performance far more than the basic methods and also obtains competitive
results compared to several neural networks.

When comparing the CB-LSTM to the RecursiveNNs which include RNN1-
w2v and RNN2-(w2v) [13], we find it outperforms the latter on all datasets
except IBC. We believe the reason is that RNN2-(w2v) significantly benefit from
phrase-level annotations with a syntactic parsing tree. But most of the datasets
lack phrase annotations, RNN2-(w2v) is only applied to IBC dataset. Despite the
competitive results, our model does not require complicated syntactic parsing
process, which means that it is more simple to train with lower complexity.

By comparing many variants of LSTM and CNN [14] implemented by us,
we discover that (1) BLSTM has better performance than single directional
LSTM. Because the BLSTM is able to learn contextual information at every
time step from both previous and future text. (2) CNNs achieve better results
than RecursiveNNs on most of the datasets which illustrates that CNNs are good
at capturing the context information through the convolution layer and extract
key features by pooling layer. (3) CNN-LSTM outperforms LSTM on all datasets
which strongly proves that the importance of target-context representations.

Table 2. The accuracy of ideology detection on five datasets. The top block includes the
baselines. The models in the second block are the state-of-the-art neural networks. The
penultimate is a fast-trained method. The best method in each settings is in bold.

Model	Convote	IBC	Twt4w1kw	Twt2w1kw	Twt2w2kw
SVM + BoW	68.1%	64.5%	62.6%	64.2%	64.4%
LR + BoW	67.7%	65.3%	63.3%	63.7%	64.3%
SVM + word2vec	69.8%	62.7%	65.9%	66.2%	65.5%
LR + word2vec	67.8%	62.8%	66.5%	67.0%	66.7%
RNN1-(w2v)	70.2%	67.1%	61.4%	62.9%	67.5%
RNN2-(w2v)	-	**69.3%**	-	-	-
CNN-nonstatic-rand	69.9%	62.7%	63.4%	65.3%	67.4%
CNN-static-word2vec	71.3%	63.3%	65.1%	67.0%	68.0%
CNN-nonstatic-word2vec	73.5%	66.1%	67.7%	68.9%	69.4%
LSTM	68.1%	64.7%	67.1%	66.4%	65.9%
BLSTM	69.7%	65.5%	67.5%	67.9%	67.4%
CNN-LSTM	73.1%	68.1%	64.1%	65.0%	70.5%
FastText (Mikolov et al. 2016)	71.8%	64%	64.7%	66.8%	67.0%
CB-LSTM	**75.2%**	68.9%	**69.1%**	**69.8%**	**71.8%**

With the ability of intensifying target-related semantic context and representing contextual correlations with lower time complexity, the CB-LSTM model outperforms or achieves competitive results with respect to other methods. Furthermore, we find that all methods perform better on Twt2w2kw dataset than Twt2w1kw which reveals that our methods can obtain better results with more accurate data.

4.2 Model Analysis

We do extensive experiments to check how the performance changes with various parameters. Figure 2 shows the stability of CB-LSTM, CNN (CNN-nonstatic-word2vec), LSTM and RNN1-(w2v) for different learning rates, hidden sizes, sentence lengths, word embeddings initialization and filter lengths. In the fourth subgraph, r100 represents random 100-dimension word embeddings. w300 and g100 represent the available Google word2vec and Glove word embeddings[3]. Specifically, s100 stands for our pre-trained word vectors with 100-dimension.

As shown in Fig. 2, all models are relatively smooth when learning rate is greater than 0.05. The hidden size and sentence length do not have a significant effect on the accuracy. Moreover, we find that CB-LSTM always performs better than others which proved it is good at capturing contextual information. For word embeddings initialization, the accuracy is higher with Glove initial vectors. In addition, our pre-trained word vectors based on unlabeled tweets perform best. The last but one shows that multiple filter lengths of 2, 3, 4 performs best among all filter configurations. The last sub-graph shows that the fine-tuned learning rates deviate about 0.45 from initial values.

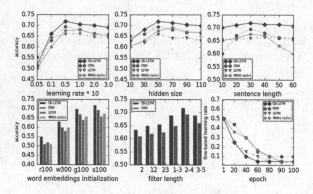

Fig. 2. Accuracy for ideology detection as a function of five hyperparameters: learning rate, hidden size, sentence length, word embeddings initialization and filter length.

---

[3] http://nlp.standford.edu/projects/glove/.

5   Conclusion

In this paper we proposed a novel, integrated model, CB-LSTM for political ideology detection. Our model is ideal for intensifying ideological target-related context and learning comprehensive contextual information at the same time without depending on complicated tree parsing. The experiment results demonstrate that our model outperforms several baselines on real datasets. Furthermore, we publish an ideology-related dataset which can be used to many ideology analysis tasks. Our model can be also applied into general sentiment analysis, question classification and many other tasks.

In the future, we plan to integrate other public profile information of users, such as marriage status and income, into the current model to better understand user's ideology.

Acknowledgments. This research is supported by the Natural Science Foundation of China (Grant No. 61572043) and National Key Research and Development Program (Project Number: 2016YFB1000704).

References

1. Conover, M.D., Gonçalves, B., Ratkiewicz, J., Flammini, A.: Predicting the political alignment of twitter users. In: IEEE PASSAT and SocialCom, pp. 192–199 (2011)
2. Conover, M., Ratkiewicz, J., Francisco, M.R., Gonçalves, B., Menczer, F., Flammini, A.: Political polarization on twitter. In: ICWSM, vol. 133, pp. 89–96 (2011)
3. Sarlan, A., Nadam, C., Basri, S.: Twitter sentiment analysis. In: IEEE ICIMU, pp. 212–216 (2014)
4. O'Connor, B., Balasubramanyan, R., Routledge, B.R.: From tweets to polls: linking text sentiment to public opinion time series. ICWSM **11**(122–129), 1–2 (2010)
5. Huang, W., Wang, T., Chen, W., Wang, Y.: Category-level transfer learning from knowledge base to microblog stream for accurate event detection. In: Candan, S., Chen, L., Pedersen, T.B., Chang, L., Hua, W. (eds.) DASFAA 2017. LNCS, vol. 10177, pp. 50–67. Springer, Cham (2017). doi:10.1007/978-3-319-55753-3_4
6. Menini, S., Tonelli, S.: Agreement and disagreement: comparison of points of view in the political domain. In: ACL, pp. 2461–2470 (2016)
7. Pennacchiotti, M., Popescu, A.M.: Democrats, republicans and starbucks afficionados: user classification in twitter. In: Proceedings of KDD, pp. 430–438 (2011)
8. Gruzd, A., Roy, J.: Investigating political polarization on twitter: a Canadian perspective. Policy Internet **6**(1), 28–45 (2014)
9. Gerrish, S., Blei, D.M.: Predicting legislative roll calls from text. In: ICML, pp. 489–496 (2011)
10. Volkova, S., Coppersmith, G., Van Durme, B.: Inferring user political preferences from streaming communications. In: ACL, pp. 186–196 (2014)
11. Hochreiter, S., Schmidhuber, J.: Long short-term memory. Neural Comput. **9**(8), 1735–1780 (1997)
12. LeCun, Y., Bengio, Y., Hinton, G.: Deep learning. Nature **521**(7553), 436–444 (2015)

13. Iyyer, M., Enns, P., Boyd-Graber, J., Resnik, P.: Political ideology detection using recursive neural networks. In: ACL, pp. 1113–1122 (2014)
14. Kim, Y.: Convolutional neural networks for sentence classification. In: EMNLP, pp. 1746–1751 (2014)
15. Lai, S., Xu, L., Liu, K., Zhao, J.: Recurrent convolutional neural networks for text classification. In: AAAI, vol. 333, pp. 2267–2273 (2015)
16. Tai, K.S., Socher, R., Manning, C.D.: Improved semantic representations from tree-structured long short-term memory networks. In: ACL (2015)
17. Joulin, A., Grave, E., Bojanowski, P., Mikolov, T.: Bag of tricks for efficient text classification. arXiv preprint (2016). arXiv:1607.01759
18. Chen, Y., Xu, L., Liu, K., Zeng, D., Zhao, J.: Event extraction via dynamic multi-pooling convolutional neural networks. In: ACL, vol. 1, pp. 167–176 (2015)
19. Homas, M., Pang, B., Lee, L.: Get out the vote: determining support or opposition from congressional floor-debate transcripts. In: EMNLP, pp. 327–335 (2006)
20. Graves, A., Jaitly, N., Mohamed, A.R.: Hybrid speech recognition with deep bidirectional lstm. In: ASRU, IEEE Workshop, pp. 273–278 (2013)
21. Huang, E.H., Socher, R., Manning, C.D., Ng, A.Y.: Improving word representations via global context and multiple word prototypes. In: ACL, pp. 873–882 (2012)
22. Li, D., Furu, W., Chuanqi, T., Duyu, T.: Adaptive recursive neural network for target-dependent twitter sentiment classification. In: ACL, vol. 2, pp. 49–54 (2015)
23. Yu, N., Pan, D., Zhang, M., Fu, G.: Stance detection in Chinese microblogs with neural networks. In: Lin, C.-Y., Xue, N., Zhao, D., Huang, X., Feng, Y. (eds.) ICCPOL/NLPCC -2016. LNCS (LNAI), vol. 10102, pp. 893–900. Springer, Cham (2016). doi:10.1007/978-3-319-50496-4_83
24. Augenstein, I., Rocktäschel, T., Vlachos, A., Bontcheva, K.: Stance detection with bidirectional conditional encoding. arXiv preprint (2016). arXiv:1606.05464

Boost Clickbait Detection Based on User Behavior Analysis

Hai-Tao Zheng[✉], Xin Yao, Yong Jiang, Shu-Tao Xia, and Xi Xiao

Tsinghua-Southampton Web Science Laboratory, Graduate School at Shenzhen,
Tsinghua University, Shenzhen, China
{zheng.haitao,yaox14,jiangy,xiast,xiaox}@sz.tsinghua.edu.cn

Abstract. Article in the web is usually titled with a misleading title to attract the users click for gaining click-through rate (CTR). A clickbait title may increase click-through rate, but decrease user experience. Thus, it is important to identify the articles with a misleading title and block them for specific users. Existing methods just consider text features, which hardly produce a satisfactory result. User behavior is useful in clickbait detection. Users have different tendencies for the articles with a clickbait title. User actions in an article usually indicate whether an article is with a clickbait title. In this paper, we design an algorithm to model user behavior in order to improve the impact of clickbait detection. Specifically, we use a classifier to produce an initial clickbait-score for articles. Then, we define a loss function on the user behavior and tune the clickbait score toward decreasing the loss function. Experiment shows that we improve precision and recall after using user behavior.

1 Introduction

Nowadays, headline of news or online article is inclined to be titled more attractively in order to attract more click. Various strategies such as building suspense, sensation, luring and teasing are used to make the title attract users' click. Users have different tendencies for the articles with such a clickbait title. Some of users hate these kinds of articles and think they are fooled. Others are willing to browse these kinds of articles. Hence, it is important to identify whether an article has a clickbait title and block them in the information stream for the specific users.

Existing methods only take text feature and meta-data (source, url feature etc.) into account. There is not a great difference between clickbait and non-clickbait in the body. Hence, traditional text classification methods have little effect in clickbait detection. Although the title text looks helpful for our task, length of the title is too short to make results reliable. As for meta-data, it just offers little help in our experiment. We do not have enough cues to finish our task well from the information of article itself. However, user behavior provides us with extra evidence. Users have different tendencies for articles with clickbait title. The effect of the classifier is enhanced by analyzing user behavior based on the above assumption.

© Springer International Publishing AG 2017
L. Chen et al. (Eds.): APWeb-WAIM 2017, Part II, LNCS 10367, pp. 73–80, 2017.
DOI: 10.1007/978-3-319-63564-4_6

In this paper, we firstly use a classifier using a serial of features to produce an initial clickbait score for every article. We considers the effect of a clickbait title as the residual between click-through rate (CTR) predictor and real value. Then we learn the residual from real data. We define a loss function based on user behavior. At last, we minimize the loss function by tuning user interest and article clickbait-score. Experiments show that our strategy is effective. Our main contributions are:

1. we present a model combined with user behavior for clickbait detection.
2. we define a loss function on user behavior and tune clickbait score of article toward decreasing the loss function.
3. we conduct a series of experiments to verify effectiveness of our method.

The rest of the paper is organized as follows. Section 2 gives an overview of the related areas. Section 3 introduces our method Sect. 4 introduces our experiments. At last, we summarize our work in Sect. 5.

2   Related Work

Recently, as the development of information stream application, clickbait detection has attracted some interest. Potthast et al. [13] studies clickbait detection in Twitter. They make use of three kinds of features: teaser message, linked web page and Twitter meta information. Teaser message and content of linked web page in their work are similar to title and body of article in our work. They use three kinds of classifier respectively (logistic regression, naive Bayes and random forest) with these features.

Chakraborty et al. [4] compare clickbaits and non-clickbaits from different angles. They detect clickbait mainly based on topical similarity and linguistic patterns.

Biyani et al. [2] also purpose a serial of features similar to the above paper. None of them considers user behavior. The above researches mainly focus on textual features.

Chen et al. [5] discus what kinds of cue would help clickbait detection. They also believe lexical and syntactic features are helpful. Besides, they propose that image analysis and user behavior analysis, which we mainly discuss in this paper, may be useful. However, there are not any experiment to prove their idea in the paper.

3   Clickbait Detection Based on User Behavior

In this section, we will introduce our method. Firstly, we establish model to estimate whether or not an article is clickbait. The model gives an article an initial click-bait score. We introduce this part in Sect. 3.1. Secondly, we model the effect of clickbait title on the users click as the residual between the real action and Click-Through Rate of our CTR predictor. We introduce this part in Sect. 3.2. At last, we define a loss function and tune the score by minimizing our defined loss function to product a final clickbait score. We introduce this part in Sect. 3.3.

Table 1. Description of features. Source T, B, M imply that the feature is extracted from the title, the body and the mixture of them respectively. Feature categories N, B, VEC denote numeric, binary and vector respectively.

Feature name	Description	Source	Type
{Has, Num}exclm	Presence/number of exclamation mark	T	B/N
{Has, Num}ques	Presence/number of question mark	T	B/N
Haspron	Presence of pronoun	T	B
Hasint	Presence of interrogative	T	B
Numwords	Number of words	T	N
Numdots	Number of dots	T	N
Stoprate	Stop words ratio	T	N
Unigram	tf-idf weight of words	T,B	VEC{N}
Bigram	tf-idf weight of bigrams	T,B	VEC{N}

3.1 Initializing Clickbait Score for Articles

Firstly, we train a classifier to produce an initial clickbait score for each article. Intuitively, the more reliable initial clickbait score is, the better result will be. Thus, we train a classifier as well as possible. In order to select suitable features, we refer to the related work [1,10–12]. A complete description of all features is shown in Table 1.

We compare with the five well-known machine learning algorithms including Logistic Regression (LR) [3], Naive Bayes (NB) [9], Random Forest (RF) [7], Support Vector Machine (SVM) [6], and Gradient Boost Decision Tree (GBDT) [8]. We choose GBDT to produce initial clickbait score based on the experiment due to the best performance. After training, we initial the clickbait score of an article as output of the classifier.

3.2 Fitting Residual Error

The click-through rate prediction should take thousands of, even millions of features into account. Many factors influence whether a user clicks an article. We have had a ready-made predictor, which use 20 thousands-dimension features based on the personas and attributes of articles. We model the effect of a clickbait title as the residual between the real click and the predicted value given by original CTR predictor. We call CTR predicted by original CTR predictor original CTR (a given value in our dataset). We define real CTR as the modified CTR after considering clickbait factor. The residual is equal to real CTR minus original CTR as Eq. 1.

$$residual = realCTR - originalCTR \tag{1}$$

We build a logistic regression model with a linear scaling to fit the residual between the real click and the predicted value. Two factors could help us fit the residual: the user interest in clickbait title and the clickbait-score of article. We choose three combinations of them as features:

1. The Product of User Interest and Clickbait-score. It is easy to assume that users tend to click the article with a clickbait title if he/she is interested in clickbait. This value will be large on the above condition.
2. Difference between Clickbait-score and User Interest through Rectified Linear Function. A user with a low interest is almost impossible to click an article with a clickbait title. This feature will be large, when this happens. But a user with a high interest can still click an article with a normal title. Therefore, we add a rectified linear function to handle this situation.
3. The Clickbait-score itself. Our statistics shows that the articles with a clickbait title gain more traffic. Therefore, we choose it as one of our features.

The input of the model is the three features above. The output indicates the impact of features on click (1 for positive impact and 0 for negative impact). Logistic regression model is as Eq. 2.

$$p_r(click|pair_i) = (1 + e^{-w^T x})^{-1} \qquad (2)$$

where $pair_i$ is a pair of an article and a user, w is a four-dimensional vector including weights of the above three features and a bias term.

The output value of logistic regression model is between 0 and 1. It does not meet our demand. Thus, we take the probability through the linear scaling to be the residual as Eq. 3.

$$residual = k(p_r(click|pair_i) - 0.5) \qquad (3)$$

where k is scaling parameter, which indicate how important is the effect of a clickbait title.

Now the real click-through rate is represented as the standard click-through rate plus residual as Eq. 4.

$$p(click|pair_i) = p_b(click|pair_i) + residual \qquad (4)$$

where $p_b(click|pair_i)$ is the standard CTR of $pair_i$.

3.3   Tuning Clickbait Score Based on User Behavior

We define the loss function as Eq. 5.

$$L(I, C; Pair) = \sum_i^m (y_i - p(click|pair_i))^2 + \alpha(\sum_{x \in C}(x - x^2) + \sum_{y \in I}(1 - y)^2) \qquad (5)$$

where I is the user interest vector, C is the clickbait-score of articles vector, is a weighting parameter, m is the size of training set and Pair is a serial of records of user action. Each record is represented as a quadruple {user_id, article_id, standard CTR, action (1 for click and 0 for only-view)}.

The loss function consists of two terms. The first term is the first sigma symbol. It is an empirical risk term and is represented as a square loss. It define the loss of user behaviors. The second term is used to control the distributions of clickbait score and user interest. The role of first term is easy to understand. We will explicate the role of second term in experiment section.

In order to get more reliable clickbait-score, we minimize the loss function by gradient descent. Firstly, we fix the clickbait-score vector and calculate gradient of interest vector and clickbait score vector respectively as Eqs. 6 and 7.

$$\frac{\partial L}{\partial I} = 2k \sum_i^m (y_i - p(click|pair_i))p_r(click|pair_i)'_I + 2\alpha \sum_{y \in I}(1 - y) \qquad (6)$$

$$\frac{\partial L}{\partial C} = 2k \sum_i^m (y_i - p(click|pair_i))p_r(click|pair_i)'_C + \alpha \sum_{x \in C}(1 - 2x) \qquad (7)$$

Then we update interest vectors and clickbait score by gradient descent algorithm.

4 Experiments

4.1 Dataset

We have two kinds of datasets: the article sets and the user behavior record sets. The article set are crawled from four major website portals (Sohu, Tencent, Netease and Sina), WeChat public accounts and some professional BBS.

There are two article sets. The first dataset contains 32,037 articles (7,461 clickbait and 24,576 non-clickbait) which are manually annotated if the article with a clickbait title. It is used for training and testing the basic model. The second contains 11,193 annotated articles (2,638 clickbait and 8,555 non-clickbait). It is user for testing the entire model. The two article sets are crawled in different periods.

The user behavior record set is the behavior records of 12,451 relatively active users. We also have two user behavior sets, which is the corresponding user record in two article sets. The first one has 5,688,369 records. It is used for training and testing the residual predictor. The second one has 6,254,448 records. It is used for testing the entire model.

4.2 Classifier Model Selection

To address the issues of imbalances between the positive examples and the negative examples in first dataset, we randomly select 7,461 examples from all non-clickbait examples as negative examples. We randomly select 80% examples as training set. The remaining 20% is used for testing. We compare the five well-known learning algorithms as implemented in sklearn using default parameters.

The experiment results can be found in Table 2. From the perspective of comparing classifier, GBDT gets the best performance in general. GBDT also

Table 2. AUC of CTR predictor of different k.

	LR	NB	RF	GBDT	SVM
Precision	**0.75**	0.71	0.72	**0.75**	0.71
Recall	0.76	0.77	0.74	**0.81**	0.80

Table 3. AUC of CTR predictor of different k.

k	0.00	0.02	0.04	0.06	0.08
AUC	0.7258	0.7484	0.7581	**0.7603**	0.7581

performs well in recent kaggle competition. One of the reason is that GBDT has a better anti-noise capability than the other methods above. Our datasets come from real world and inevitably include noise.

4.3 Training the Residual Predictor

We divide the first record set into three parts. The first part is used for initializing user interest. We set user interest to the proportion of clickbait in articles that user click recently. The second part is used for training residual predictor. We train the residual predictor based on Sect. 4.1. The third part is used for evaluating and choosing an appropriate parameter k. We compare the effects of different k for $p(clickpair_i)$. Table 3 shows the ROC curve and area under roc curve (AUC) value with different k.

Experiments demonstrate that clickbait title has a nonnegligible impact on users click. Although the increase of AUC is not very significant, it is enough for us. Our goal is not to improve AUC, but to model the effect of clickbait title on users click. Thus, such a performance is enough for us. There is only a little difference between different parameter k. AUC reaches a maximum, when k is equal to 0.06. Therefore, we set k equal to 0.06 in the following experiments.

4.4 The Effect of User Behavior

We choose the result of the GBDT classifier with all features as the initial clickbait-score due to the best performance. The basic classifier gets 75.85% precision and 80.76% recall in article set 2. We also initialize user interest with the proportion of clickbait in articles that user click recently.

From the Table 4, Initial precision and initial recall are evaluated by the basic classifier in article set 2. Terminate precision and terminate recall are the value that precision and recall reach a stable point in iteration. We can see that precision gradually reaches 1.0 and recall falls dramatically if we do not add the second term into the loss function. Users usually click many non-click articles whether they are interested in clickbait or not. User interest decline because of

Table 4. Performance with different alpha.

Alpha	Initial precision	Initial recall	Terminate precision	Terminate recall
0	0.7585	0.8076	/	0.0000
2	0.7585	0.8076	0.9535	0.6846
4	0.7585	0.8076	0.8127	0.8203
6	0.7585	0.8076	0.7680	0.8097

this behavior. Then all clickbait scores decline. Clickbait scores and user interest will have a further decline in next iteration. Thus, precision increases and recall decreases gradually at the beginning. At last, all articles are classified as non-clickbait. Therefore, we need to add a term into the loss function to control the distributions of user interest and clickbait score. The term $(1-y)^2$ make it easy to increase and hard to decline for user interest. Besides, the term $(x - x^2)$ is close to 0, when clickbait score is close to 0 or 1. It is close to maximum, when clickbait score is close to 0.5. This term make it hard to change their result of classification for articles with explicit result and stop the trend of overall deviation toward high clickbait score and user interest.

We can see that the distribution control term play a part role when alpha is equal to 2. However, it is not enough. In this condition, a large proportion of articles are classified as non-clickbait. It results in a high precision and low recall. When alpha is equal to 4, we reach the best performance. Precision increase 0.05 to 0.8127 and recall increase 0.01 to 0.8203 comparing with initial precision and recall. The performance indicate that our model confirm to the real situation. As alpha increases, precision and recall are almost constant and keep the original level. This is because the weight of distribution control term is too large, which make the cost too much to change categorized result. Thus, articles tend to keep the original result.

5 Conclusion and Future Work

In this paper, we look for a new way to help us achieve the higher precision and recall. We propose a method that models user behavior and conduct relative experiments. The main innovation of our algorithm is the following two ideas:

1. we model the effect of a clickbait title as the residual between the real click and the predicted value from standard click-through rate predictor.
2. we tune the clickbait score based on user behavior.

Experiments show that we increase 0.05 precision and 0.01 recall comparing with the methods not considering user behavior by utilizing user behavior. It demonstrates that user behavior indeed is useful for clickbait detection. Besides, our method can integrate with any other clickbait detection method. The only need is to put the result of other methods as the initial clickbait score in our method.

Acknowledgements. This research is supported by National Natural Science Foundation of China (Grant No. 61375054), Natural Science Foundation of Guangdong Province Grant No. 2014A030313745, Basic Scientific Research Program of Shenzhen City (Grant No. JCYJ20160331184440545), and Cross fund of Graduate School at Shenzhen, Tsinghua University (Grant No. JC20140001).

References

1. Abbasi, A., Chen, H.: A comparison of fraud cues and classification methods for fake escrow website detection. Inf. Technol. Manag. **10**(2), 83–101 (2009)
2. Biyani, P., Tsioutsiouliklis, K., Blackmer, J.: "8 amazing secrets for getting more clicks": detecting clickbaits in news streams using article informality (2016)
3. Cessie, S.L., Houwelingen, J.C.V.: Ridge estimators in logistic regression. J. R. Stat. Soc. **41**(1),*191–201 (1992)
4. Chakraborty, A., Paranjape, B., Kakarla, S., Ganguly, N.: Stop clickbait: detecting and preventing clickbaits in online news media. In: IEEE/ACM International Conference on Advances in Social Networks Analysis and Mining, pp. 9–16 (2016)
5. Chen, Y., Conroy, N.J., Rubin, V.L.: Misleading online content: recognizing clickbait as "false news". In: ACM on Workshop on Multimodal Deception Detection, pp. 15–19 (2015)
6. Cortes, C., Vapnik, V.: Support-Vector Networks. Kluwer Academic Publishers, Boston (1995)
7. Cutler, A., Cutler, D.R., Stevens, J.R.: Random forests. Mach. Learn. **45**(1), 157–176 (2004)
8. Friedman, J.H.: Greedy function approximation: a gradient boosting machine. Ann. Stat. **29**(5), 1189–1232 (2001)
9. John, G.H., Langley, P.: Estimating continuous distributions in Bayesian classifiers. In: Eleventh Conference on Uncertainty in Artificial Intelligence, pp. 338–345 (1995)
10. Kolari, P., Java, A., Finin, T., Oates, T., Joshi, A.: Detecting spam blogs: a machine learning approach. In: National Conference on Artificial Intelligence and the Eighteenth Innovative Applications of Artificial Intelligence Conference, Boston, 16–20 July 2006
11. Lahiri, S., Mitra, P., Lu, X.: Informality judgment at sentence level and experiments with formality score. In: Gelbukh, A. (ed.) CICLing 2011. LNCS, vol. 6609, pp. 446–457. Springer, Heidelberg (2011). doi:10.1007/978-3-642-19437-5_37
12. Ntoulas, A., Najork, M., Manasse, M., Fetterly, D.: Detecting spam web pages through content analysis. In: World Wide Web Conference, pp. 83–92 (2006)
13. Potthast, M., Köpsel, S., Stein, B., Hagen, M.: Clickbait detection (2016)

Recommendation Systems

A Novel Hybrid Friends Recommendation Framework for Twitter

Yan Zhao[1], Jia Zhu[2](✉), Mengdi Jia[1], Wenyan Yang[1], and Kai Zheng[1]

[1] School of Computer Science and Technology, Soochow University, Suzhou, China
[2] School of Computer Science, South China Normal University, Guangzhou, China
jzhu@m.scnu.edu.cn

Abstract. As one of the key features of social networks, friends recommendation is a kind of link prediction task with ranking that was extensively investigated recently in the area of social networks analysis as users would like to follow people who have similar interests to them. We use Twitter as a case study and propose a novel hybrid friends recommendation framework that is not only based on friends relationship but also users' location information, which are recorded by Twitter when they posted their tweets. Our framework can recommend friends to users who have similar interests based on location features by using collaborative filtering to effectively filter out those common places which are meaningless, e.g., bus station; and focuses on those places that have high probability that people are there more likely to become friends, e.g., dance studio. In addition, we propose a multiple classifiers combination method to leverage the information contained in friends and locations features in order to get better outcomes. We evaluate our framework on two real corpora from Twitter, and the favorable results indicate that our proposed approach is feasible.

Keywords: Social network · Recommendation systems

1 Introduction

With the fast growing of Web 2.0, social networking sites like Twitter are becoming increasingly popular. For example, people can use Twitter to find their friends and catch up their recent status. Friends recommendation is a kind of link prediction task with ranking that was extensively investigated recently in the area of social networks analysis [5,7,31]. It is also very important in information management. As a typical example like Twitter, which allows its users to send and read text-based posts of up to 140 characters, known as tweets. It is not sufficient for a system to provide recommendation based on friends relationship or the most popular users because users normally only want to follow people who have similar interests [10].

In addition, Twitter have millions of users that means they are often quite sparse with low density of links among users. As a result, the link prediction

© Springer International Publishing AG 2017
L. Chen et al. (Eds.): APWeb-WAIM 2017, Part II, LNCS 10367, pp. 83–97, 2017.
DOI: 10.1007/978-3-319-63564-4_7

space is huge and highly imbalanced. Existing approaches merely focus on finding friends in the 2-hop social neighborhood, i.e., friends-of-friends of a user [23]. It may likely result in an exponentially larger set of increasingly less likely candidates if we extend the range to 3 or more hops neighborhood. As a consequence, the friends recommendation problem appears heavily influenced by network distance between users.

To overcome this problem, we propose a friends recommendation framework to use location information that is the places visited by each user apart from friends relationship information. The increasing availability of location-acquisition technologies, e.g., GPS, nowadays enable people to log the location histories with spatial-temporal data. Such real-world location histories provide us with the correlations of users' interests and the places they have visited [1,7,19,25,30,31,33]. We use Twitter as a case study in this paper not only because Twitter is one of biggest social networks in the world but also the way of communication in Twitter is quite standard compared to other social networks.

However, Twitter users have the right to determine if they want to share their current location and the location information is not always available. Fortunately, we can retrieve some location information from Twitter as it added an explicit GPS tag that can be specified for each tweet in early 2010 and is continually improving the location-awareness of its service [23]. In addition, we can also extract location information from users' tweets even if users turn off the GPS tagging. For example, users who posted tweets with the keywords "step up dance studio" in the same period of time might be friends or may become friends as they all have been to "step up dance studio" or at least interested in this place. Therefore, we may recommend friends to these users based on the extracted information.

Our framework can recommend friends to users who have similar interests to them based on location features by using collaborative filtering [4,24,32] to effectively filter out those common places which are meaningless, e.g., bus station; and focuses on those places where have high possibility that people are there more likely to become friends, e.g., dance studio. We then pick top 10 places based on the score generated by our methods. Additionally, to achieve better recommendation outcomes, we propose a multiple classifiers combination method to combine the outputs from different classifiers. Our main contributions are summarized below:

1. We propose a hybrid friends recommendation framework, which uses collaborative filtering to effectively filter out those common places which are meaningless so that the recommendation performance can be improved.
2. We further propose a multiple classifiers combination (MCC) method that combines outputs of multiple classifiers, which leverages the information contained in the features of friends relationship and location information.
3. Extensive experiments have been performed on two real data sets we retrieved from Twitter. We show that our framework performs significantly better than baseline method in different scenarios.

The rest of this paper is organized as follows: In Sect. 2, we discuss related works in link prediction, particularly friends recommendation. In Sect. 3, we formulate our problem and describe the details of our framework. In Sect. 4, we present our experiments, evaluation metrics, and results. We also conclude this study and discuss future work in Sect. 5.

2   Related Work

Hoff et al. [11] introduced a class of latent class models from the perspective of social networks analysis which tries to project all the networked objects to a latent space, and the decision for link existence is based on their spatial positions. However, the authors have not discussed in detail about the choice of a prior distribution for latent positions so that the outcome is not really impressive. Taskar et al. [26] proposed a model by applying the Relational Markov Network (RMN) framework to define a joint probabilistic model over the entire link graph. The application of the RMN algorithm provided significant improvements in accuracy over flat classification. Though we have applied RMN and obtain success in our earlier work for link prediction [35], but we found that RMN is good at predict linking over relational data, e.g. friends relationship, but it is not suitable for uncertain relational data like location information as some relationships among objects are not useful.

Basilico and Hofmann [2] proposed to use the inner product of two nodes and their attributes as similarity measure for collaborative filtering. Their method showed how generalization occurs over pairs with a kernel function generated from either attribute information or user behaviors. They did not consider the relationship might be changed, which means the similarity measure might be different along with the time. Therefore, we do not use similarity measure for the collaborative filtering in our approach.

Huang et al. [12] introduced the time-series link prediction model problem and taking into consideration temporal evolutions of link occurrences to predict link occurrence probabilities at a particular time. We learn time-series link prediction from their model and apply additional location information into our approach.

Hannon et al. [10] focused on the creation of relationships between users and attempt to harness the real-time web as the basis for profiling and recommendation. They evaluated a range of different profiling and recommendation strategies based on a large dataset of Twitter users and their tweets. Their profiling algorithm is interesting but again they ignored those important location information which make their system recommend a list of popular persons rather than a list of persons who might have similar interest to users.

Scellato et al. [23] designed a Link prediction systems in a location-based social network called Gowalla with periodic snapshots to capture its temporal evolution. They defined new prediction features based on the properties of the places visited by users which are able to discriminate potential future links among them and found about 30% of new links are added among place-friends, i.e., among users who visit the same places. They proved the location features certainly can help people find friends though some of properties they described in the paper are not available in Twitter, which is one of the issues we try to solve in this paper.

Sadileka [21] presented a recommendation system called Flap which infers social ties by considering patterns in friendship information, the content of messages, and user location. Each component is a weak predictor of friendship alone, but combining them results in a strong model, accurately identifying the majority of friendships. They evaluated Flap on a large sample of highly active users from two distinct geographical areas and show high accuracy on the reconstruction graph even when no edges are given. However, they did not consider some places are useless, e.g. bus stop, which should not be used as location features.

Regards to extract useful location information, Zhou et al. [34] used a collaborative filtering recommenders based only on users' check-in data for location recommendation. Though their research is not for friends recommendation but their user-based model can be adapted for our approach to get common places.

Biancalana et al. [3] proposed a recommendation system to identify users' needs. Though their methodology is not for friends recommendation, but their information filtering process for points of interests are good reference for our work. Later, Gurini et al. [9] proposed a user recommendation method based on a novel weighting functions, which uses users' sentiments to build user profiles to employ in the recommendation process. Though users' sentiments are proved to be useful for friends recommendation by the authors, it can not resolve the geographical gap between users. Trattner et al. [27] recently evaluated the social proximity of users via supervised and unsupervised learning approaches and establish that location-based social networks have a great potential for the identification of a partner relationship.

3   Proposed Framework

As we mentioned earlier, our recommendation framework focuses on combining location information and friends relationship information to recommend friends to users because people who often visit the same place are normally have similar interest and likely to become friends [16].

The proposed friends recommendation framework is shown in Fig. 1. We first implement a PHP script to retrieve data from Twitter by using Twitter Stream API[1], and generate a list locations based on information from GPS tags and tweets by adopting collaborative filtering methods. We then use these locations information plus friends relationship information as features to a classifier and generate Top-K friends for users from test dataset as recommended friends according to the assignment by the classifier. There are several key components in this framework, and we will discuss each of them in the following sections.

---

[1] https://dev.twitter.com/docs/streaming-apis.

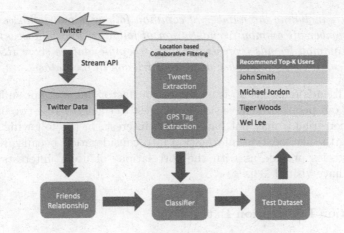

Fig. 1. Friends recommendation framework

3.1 Friends Relationship Features

Friends relationship also known as friends-of-friends, is always an important element to be considered in a social network friends recommendation system. Users are likely to add more friends through friends relationship as the recommended friends are users may know [17]. In the context of twitter, we define friends relationship as the common followees and followers of two users.

For example, if two users share a large amount of followers, then they are likely to have similar interests or occupation, e.g., two popular singers. It makes sense to recommend them as friends of each other to create more business opportunities. Therefore, we treat both users' followees and followers as friends of users, and we do not consider any further distance between them, e.g. 3-hop neighborhood, due to the possibility of resulting in an exponentially larger set of increasingly less likely candidates [23].

Assume a Twitter user U in the time snapshot t has a list of followees $FE(U_t)$ and followers $FL(U_t)$' union set $FR(U_t)$ as friends of U_t, we then have the recommendation score RS for user U_t' to be a friend of U_t as shown in Eq. (1):

$$RS(U_t, U_t') = \frac{|FR(U_t) \bigcap FR(U_t')|}{|FR(U_t) \bigcup FR(U_t')|} \qquad (1)$$

where

$$FR(U_t) = FE(U_t) \bigcup FL(U_t) \qquad (2)$$

$$FR(U_t') = FE(U_t') \bigcup FL(U_t') \qquad (3)$$

We then define the friends relationship features as below:

Definition 1. *Friends Relationship Features: We call $F(U_t, U_t')$ is a set of friends relationship features for a Twitter user U_t' at the time snapshot t to be* *a friend of U_t including the number of common followers and followees between users, the number of common friends(union of followers and followees), and the fraction of common friends represents as the recommendation score $RS(U_t, U'_t)$, which is calculated by the number of mutual friends in the dataset.*

Though friends relationship is an important factor but it is not sufficient for users to find out people who have similar interest to them because two users have many common friends may not have similar interest. In the following sections, we are going to take location information into consideration because real world location histories provide us with the correlations of users' interests and the places they have visited [33].

3.2 Location Information Extraction

There are two types of location information we are going to use from users' tweets in our framework. One is the GPS tags associated with each tweet, the other is the location information in each tweet. However, the GPS tags are not always available if users manually turn off the option or users use a laptop to post tweets. In this case, we try to explore the information in the tweets to collect possible location information. For instance, if an user posts a tweet "I am watching shows in Hilton.", then "Hilton" certainly is a meaningful location information in this context though it is possible that there are many locations with this name but we can use collaborative filtering to solve this issue. Details are given in the next section.

To extract the location related information from tweets, we used the Stanford Named Entity Recognizer[2] to perform the extraction. This recognizer provides a general implementation of Conditional Random Field (CRF) sequence models [15], which is a class of statistical modelling method to recognize the words in a text which are names, e.g. location names. CRF can take context into account compared to other ordinary classifiers which require labelling for a single sample without regard to "neighboring" samples. We choose the good 3 class (PERSON, ORGANIZATION, LOCATION) named entity recognizers for English and its performance has been described in [8].

3.3 Location Features Construction

Though we can retrieve a set of GPS tags and location names from tweets but not all of them are useful in our approach. For example, location like bus station is meaningless, people might be always in the same bus station every single working day but they might have never talked to each other in their whole life. Thus, the probability for them to become friends are not high from our perspective because this kind of location information (e.g., XXX bus station) does not show common interest from their tweets, particularly compared to places like "XXX dance studio". Therefore, we need to adopt collaborative filtering to

---

[2] http://nlp.stanford.edu/software/CRF-NER.shtml.

effectively filter out those common places which are meaningless and focus on places that may represent users' interest. We choose two popular collaborative filtering approaches, memory-based and model-based, to perform the filtering process.

We first give a definition for the term "Common Places":

Definition 2. *Common Places: We call $SP(U_t, U_t')$ is a list of common places that a Twitter user U_t shared with U_t' if the GPS tags in each tweet they posted at time snapshot t are in the range of 3 km or location names extracted from each tweet they posted at time snapshot t are similar.*

The reason we selected 3 km range as threshold is because 3 km is a standard industry setting for many popular location based service mobile apps, e.g. AroundMe[3]. The location names similarity is calculated based on the popular Jaccard coefficient approach for string similarity calculation. The definition and performance of Jaccard coefficient can be found at [22].

We then have a list of features for location information:

(1) *Number of Common Places* - The number of same place two users U_t and U_t' share at time snapshot t, represents as $|SP(U_t, U_t')| = |P(U_t) \bigcap P(U_t')|$ where $P(U_t)$ and $P(U_t')$ are the list of places extract for U_t and U_t'' tweets at time snapshot t [23].

(2) *Fraction of Common Places* - Similar to number of common place, we have fraction of common place proposed in [23] as shown in Eq. (4):

$$|SFP(U_t, U_t')| = \frac{|P(U_t) \bigcap P(U_t')|}{|P(U_t) \bigcup P(U_t')|}. \qquad (4)$$

(3) *Memory-Based Collaborative Filtering Common Places* - As we discussed earlier, some locations like bus stations are meaningless. Therefore we apply collaborative filtering to filter out these locations as collaborative filtering is a method of making automatic predictions about the interests of a user by collecting preferences from many users.

In our case, we first used memory-based collaborative filtering [24] to calculate a score $Score_f(p)$ for each common place p of U_t and U_t' as shown in Eq. (5) which indicates place frequency:

$$Score_f(p) = \frac{1}{N} \Sigma_{U \in FR} K_p \cdot \frac{1}{N'} \Sigma_{U' \in FR'} K_p', \qquad (5)$$

where N and N' are the number of friends, FR and FR' are the set of friends of U_t and U_t'. K_p and K_p' are the number of times they have been to the place p at time snapshot t.

However, the high $Score_f(p)$ does not indicate that the location has been visited by many friends because the place may be visited by a small proportion of friends with high frequency. To address this issue, we adopt the

---

[3] http://www.aroundmeapp.com/.

inverse place frequency from document processing [13] as shown in Eq. (6):

$$Score_{if}(p) = log\frac{N + N'}{N_p + N'_p},\tag{6}$$

where N and N' are the number of friends, and N_p and N'_p are the number of friends of U_t and U'_t who have visited the place p. We then have Eq. (7), which is the product of place frequency and inverse place frequency:

$$Score(p) = Score_f(p) \times Score_{if}(p).\tag{7}$$

We then pick the top 10 places with the highest score $Score_{(p)}$ as location features. By using this method, we can solve the issue we have addressed above as it is common sense that people will not post tweets like "I am in bus stop." many times, therefore useless places like "bus stop" will be filtered out.

(4) *Model-Based Collaborative Filtering Common Places* We also use model-based clustering collaborative filtering algorithm [29] as it has been shown to be useful for classification tasks. Similar to (3), $Score(p)$ is the score for each common place p of U_t and U'_t at time snapshot t, but in clustering collaborative filtering algorithm, $Score(p)$ is calculated according to Eq. (8):

$$Score(p) = \sqrt[2]{\Sigma_{i=1}^n |x_i - y_i|^2}\tag{8}$$

where x_i, y_i are the number of friends of U_t and U'_t have been to the place p i times at time snapshot t, $i \leq n$. We also pick the top 10 places with the highest score as location features.

3.4   Multiple Classifiers Combination Method

Once a set of features has been obtained, we then need to choose a categorization algorithm to build a classifier that can produce the probability of two users to become friends. Note that we do not consider feature selection in this paper as it is not the main focus. Most friends recommendation algorithms are based on machine learning techniques, Support Vector Machine and Bayesian Network are two popular machine learning techniques among being widely used.

In our approach, we have four classifiers based on friends relationship and location information, namely, SVM Friends, SVM Locations, BN Friends and BN Locations. SVM and BN Friends classifiers are based on friends relationship information while SVM and BN Location classifiers use location information in each tweet as features. We did not combine friends relationship and location information features into one classifier as these two features are based on different types of information and our preliminary analysis showed the combination of homogeneous classifiers can achieve better results which will be discussed in the coming sections.

For each classifier, We then adopt an algorithm based on the approach proposed in [23].

For every snapshot t, we compute features for each pair of users who are not friends at t, and assign a positive label to each pair if they become friends at $t + 1$, and a negative label otherwise. Thus, training and testing sets are built so that the features from a given time interval are mapped to class labels in a future time interval. Hence, given M snapshots, we can create $M - 1$ learning sets, each one with labels drawn from the next snapshot. Classifiers can then be trained to build models and recognize positive and negative items from their features.

Based on this algorithm, we further propose a multiple classifiers combination (MCC) method to combine all four classifiers together because the intuition is that the combination of homogeneous classifiers using heterogeneous features can improve the final result [18].

Assume each classifier produces a unique decision regarding the recommended friends of each user U in the test dataset, we then compare the results among all the four classifiers and the final output depends on the reliability of the decision confidences delivered by the participating classifiers. We apply the concept of Decision Template (DT) to avoid the case in which the classifier make independent errors [14] and calculate the confidence score.

Assume each classifier produces a output $E_i(U) = [d_{i1}(U), ..., d_{i|G|}(U)]$ where $d_{ij}(U)$ is the membership degree given by classifier E_i for the recommended friend j to a user U in the test dataset, $j \in G$, G is the set of friends recommended or not recommended by classifier E_i. The outputs of all classifiers can be represented by a decision matrix DP, which is defined as follows:

$$DP(U) = \begin{pmatrix} d_{11}(U) & ... & d_{1|G|}(U) \\ d_{21}(U) & ... & d_{2|G|}(U) \\ d_{31}(U) & ... & d_{3|G|}(U) \\ d_{N1}(U) & ... & d_{N|G|}(U) \end{pmatrix}$$

The membership degree $d_{ij}(U)$ is calculated using the data T_f in training set, T_f indicates a person, $f = 1, 2, ..., |G|$ in each time snapshot as follows:

$$d_{ij}(U) = \frac{\sum_{j=1}^{|G|} Ind(T_j, i)}{|G|} \tag{9}$$

where $Ind(T_j, i)$ is an indicator function with value 1 if T_j is a recommended friend and 0 otherwise. At this stage, we have the membership degree for a recommended friend j to each user U and store in a matrix $DP(U)$.

We then calculate the confidence score $Score_j(U)$ for each user U using various rules from the $DP(U)$ for each recommended friend j and pick the top most recommended friends with the highest confidence score. Assume N is the number of classifiers, we apply minimum, maximum and average rules for the matrix below to consider the diversity among multiple classifiers:

$$MinimumRule : Score_j(U) = Min_{i=1}^{N}(d_{ij}(U)) \tag{10}$$

$$MaximumRule : Score_j(U) = Max_{i=1}^{N}(d_{ij}(U)) \tag{11}$$

$$AverageRule : Score_j(U) = Mid_{i=1}^{N}(d_{ij}(U)) \tag{12}$$

4   Evaluations

4.1   Corpora and Data Preparation

In our experiment, we collected public tweets from two cities by using Twitter Stream API, New York and Sydney, from March 10, 2013 to May 10, 2013. We implemented a PHP script to extract tweets that have GPS tags enabled and store into database every 15 min. For evaluation purpose, we only recommend friends to the users who have posted more than 5 tweets as they are active users. We also remove the users who have more than 1000 friends as they are most likely "advertisement users". We randomly selected 10% of these active users and constructed a set of pairs of users according to the algorithm we described in Sect. 3. The statistics of our data are shown in Table 1.

4.2   Evaluation Metrics and Baseline Method

We evaluated our framework based on Receiver-Operating-Characteristic (ROC) curves [20], which is the indicator adopted by most of researchers including the work on friends recommendation we introduced in Sect. 2.

ROC curve is a graphical plot which illustrates the performance of a binary classifier system as its discrimination threshold is varied. It is created by plotting true positive rate (TPR is also known as sensitivity) and false positive rate (FPR is also known as specificity) at various threshold settings, $TPR = \frac{TP}{TP+FN}, FPR = \frac{FP}{FP+TN}$.

In our case, TP stands for True Positive which means the number of pairs of users correctly labeled as belonging to the positive class, FP stands for False Positive which means the number of pairs of users incorrectly labeled as belonging to the positive class, TN stands for True Negative which means the number of pairs correctly labeled as belonging to the negative class and FN stands for False Negative which means the number of pairs of users who were not labeled as belonging to the positive class but should have been.

Table 1. Data Corpora

	New York	Sydney
No. of unique users	87400	39567
No. of tweets with location information	295388	157605
No. of unique active users	52644	29328
Avg tweets by active users	5.61	5.37
Avg places of active users	7.73	7.29
No. of followees of active users	3211130	2133880
No. of followers of active users	4853070	2924550
No. of pairs of users being evaluated	126520	49334

We also use the area under the ROC curve (AUC) which is equal to the probability that a classifier will rank a randomly chosen positive instance higher than a randomly chosen negative one [6]. Assume our classifiers output K positive pair of users $x_i, i = 1...KP$ and KN negative pair of users $y_j, j = 1...KN$. An unbiased estimator of the AUC is $\frac{1}{KP*KN}\Sigma_i\Sigma_j|f(\frac{(x_i)}{(i,j)}) - f(\frac{(y_j)}{i,j})|$, where function f denotes a classifier trained without the i-th and j-th training example.

For baseline method, we used the friends of friends relationship that is Twitter and other micro blog services that are currently using to recommend friends to users, e.g., Weibo[4], which is the number of common friends between users. Then the recommendation score $RS(U_t, U_t')$ in baseline method is calculated as $RS(U_t, U_t') = |FR(U_t) \bigcap FR(U_t')|$ derived from Eq. (1). Rather use them as features, we simply pick the top K pair of users with the highest score.

Since our approach is for friends recommendation, we randomly selected K pairs of users for each active user in individual classifiers, and picked top K pairs of users with the highest score in the MCC method. We also evaluated various cases when $K = 5, 10, 20, 30$, details are given in the following sections.

4.3   Evaluation Results

This section is to discuss the comparisons of individual classifiers as mentioned in Sect. 3.4. All classifiers are implemented by Weka API [5], and we use RBFKernel [28] for SVM. We set $M = 5$, which means the dataset for classifiers is split to training and testing with the proportion 80% and 20%, respectively. All classifiers are trained by 5-fold cross validation.

Results of Individual Classifier. We first evaluated each individual classifier by apply ROC curve on both New York and Sydney corpora with K = 5, 10, 20, 30 as shown in Figs. 2 and 3 from top left to bottom right respectively.

As the results shown in both New York and Sydney corpora, SVM Friends and Locations classifiers generally outperform BN Friends and Locations classifiers and the baseline method. In addition, we also notice that with the increase of number of recommended friends, the performance of all methods are dropped.

From the AUC value in Fig. 4, we observe that location information alone is not sufficient to accurately provide friends recommendation though both location classifiers still perform better than baseline but not as good as friends classifiers.

Results of MCC Method. In this section, we evaluated the MCC method. The main purpose of this method is to see if location information can help to improve predication accuracy compared to only use friends relationship information. Figure 5 shows the ROC curve on New York corpora with K = 5, 10, 20, 30 based on various rules. Compared to the results of individual classifier, our MCC method achieve better results, particularly on average rule which proves

---

[4] http://weibo.com.
[5] http://www.cs.waikato.ac.nz/ml/weka/.

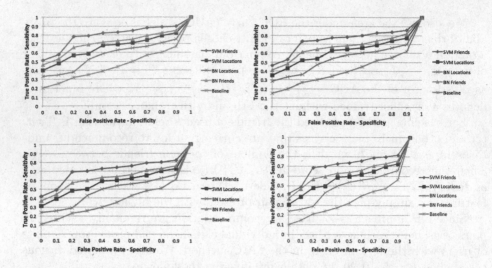

Fig. 2. ROC for individual classifier with K = 5, 10, 20, 30 on New York corpus

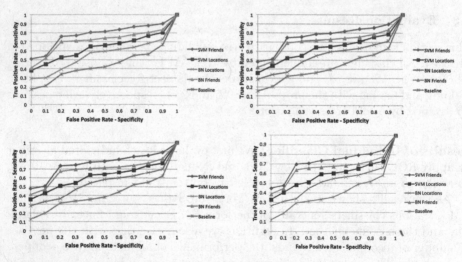

Fig. 3. ROC for individual classifier with K = 5, 10, 20, 30 on Sydney corpus

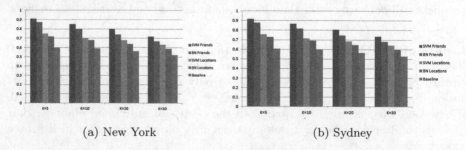

(a) New York   (b) Sydney

Fig. 4. Overall AUC for individual classifier

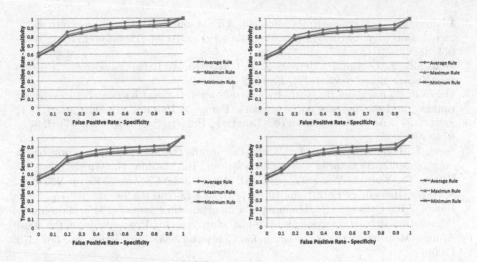

Fig. 5. ROC for MCC method with K = 5, 10, 20, 30 on New York corpus

that location features are useful for friends recommendation while keep the link prediction space as small as possible.

5 Conclusions

In this paper, we propose a hybrid friends recommendation framework for Twitter. Our framework not only takes existing friends relationship of Twitter users as consideration but also combines location features generated from collaborative filtering methods. In addition, we also contributed a method to extract location information from Tweets and a multiple classifiers combination method to leverage the information contained in our features, either friends relationship or locations. We evaluated our framework on two corpora from real world with comparisons between different classifiers and baseline method. The experiments indicated that our framework is feasible.

Acknowledgments. This work was supported by Natural Science Foundation of Guangdong Province, China (No. 2015A030310509), and the S&T Projects of Guangdong Province (No. 2016A030303055, No. 2016B030305004, 2016B010109008), Natural Science Foundation of China (No. 61532018 and No. 61502324).

References

1. Bao, J., Zheng, Y., Wilkie, D., Mokbel, M.: Recommendations in location-based social networks: a survey. GeoInformatica **19**(3), 525–565 (2015)
2. Basilico, J., Hofmann, T.: Unifying collaborative and content-based filtering. In: ICML, pp. 9–17 (2004)

3. Biancalana, C., Gasparetti, F., Micarelli, A., Sansonetti, G.: An approach to social recommendation for context-aware mobile services. ACM Trans. Intell. Syst. Technol. **4**(1), 1–31 (2013)
4. Bobadilla, J., Ortega, F., Hernando, A., Gutierrez, A.: Recommender systems survey. Knowl. Based Syst. **46**(1), 109–132 (2013)
5. DeScioli, P., Kurzban, R., Koch, E., Liben-Nowell, D.: Best friends alliances, friend ranking, and the myspace social network. Perspect. Psychol. Sci. **6**(1), 6–8 (2011)
6. Fawcett, T.: An introduction to ROC analysis. Pattern Recognit. Lett. **27**(8), 861–874 (2006)
7. Feng, S., Huang, D., Song, K., Wang, D.: Online friends recommendation based on geographic trajectories and social relations. In: Motoda, H., Wu, Z., Cao, L., Zaiane, O., Yao, M., Wang, W. (eds.) ADMA 2013. LNCS, vol. 8346, pp. 323–335. Springer, Heidelberg (2013). doi:10.1007/978-3-642-53914-5_28
8. Finkel, J.R., Grenager, T., Manning, C.: Incorporating non-local information into information extraction systems by gibbs sampling. In: Proceedings of the 43rd Annual Meeting of the Association for Computational Linguistics, pp. 363–370 (2005)
9. Gurini, D., Gasparetti, F., Micarelli, A., Sansonetti, G.: A sentiment-based approach to twitter user recommendation. In: Proceedings of the 5th ACM RecSys Workshop on Recommender Systems and the Social Web, pp. 1–4 (2013)
10. Hannon, J., Bennett, M., Smyth, B.: Recommending twitter users to follow using content and collaborative filtering approaches. In: Proceedings of the Fourth ACM Conference on Recommender Systems, pp. 199–206 (2010)
11. Hoff, P., Raftery, A., Handcock, M.S.: Latent space appraches to social network analysis. J. Am. Stat. Assoc. **97**(460), 1090–1098 (2002)
12. Huang, Z., Lin, D.K.J.: Time-series link prediction problem with applications in communication surveillance. INFORMS J. Comput. **21**(1), 286–303 (2009)
13. Jones, K.S.: A statistical interpretation of term specificity and its application in retrieval. J. Doc. **28**(1), 11–21 (1972)
14. Kuncheva, L.I., Bezdek, J.C., Duin, R.P.: Decision templates for multiple classifier fusion. Pattern Recognit. **34**(2), 299–314 (2001)
15. Lafferty, J., McCallum, A., Pereira, F.: Conditional random fields: probabilistic models for segmenting and labeling sequence data. In: Proceedings of the18th International Conference on Machine Learning, pp. 282–289 (2001)
16. Li, Q., Zheng, Y., Xie, X., Ma, W.: Mining user similarity based on location history. In: Proceedings of the ACM SIGSPATIAL International Conference on Advances in Geographical Information Systems, pp. 247–256 (2008)
17. Liben-Nowell, D., Kleinberg, J.: The link prediction problem for social networks. In: Proceedings of CIKM, pp. 556–559 (2003)
18. Orrite, C., Rodríguez, M., Martínez, F., Fairhurst, M.: Classifier ensemble generation for the majority vote rule. In: Ruiz-Shulcloper, J., Kropatsch, W.G. (eds.) CIARP 2008. LNCS, vol. 5197, pp. 340–347. Springer, Heidelberg (2008). doi:10.1007/978-3-540-85920-8_42
19. Ozsoy, M., Polat, F., Alhajj, R.: Multi-objective optimization based location and social network aware recommendation. In: International Conference on Collaborative Computing: Networking, Applications and Worksharing, pp. 233–242 (2014)
20. Provost, F.J., Fawcett, T., Kohavi, R.: The case against accuracy estimation for comparing induction algorithms. In: Proceedings of ICML, pp. 445–453 (1998)
21. Sadileka, A., Kautz, H., Bigham, J.P.: Finding your friends and following them to where you are. In: Proceedings of the Fifth ACM International Conference on Web Search and Data Mining, pp. 723–732 (2012)

22. Salton, G.: Introduction to Modern Information Retrieval. McGraw-Hill, New York (1983)
23. Scellato, S., Noulas, A., Mascolo, C.: Exploiting place features in link prediction on location-based social networks. In: Proceedings of the 17th ACM SIGKDD International Conference on Knowledge Discovery and Data Mining, pp. 1046–1054 (2011)
24. Su, X.Y., Khoshgoftaar, T.M.: A survey of collaborative filtering techniques. Adv. Artif. Intell. **2009**(4), 1–19 (2009)
25. Tang, J., Hu, X., Liu, H.: Social recommendation: a review. Soc. Netw. Anal. Min. **3**(4), 1113–1133 (2013)
26. Taskar, B., Wong, M.F., Abbeel, P., Koller, D.: Link prediction in relational data. In: NIPS, pp. 1–9 (2003)
27. Trattner, C., Steurer, M.: Detecting partnership in location-based and online social networks. Soc. Netw. Anal. Min. **5**(1), 1–15 (2015)
28. Vapnik, V.: The Nature of Statistical Learning. Springer, New York (1995)
29. Veloso, M., Jorge, A., Azevedo, P.J.: Model-based collaborative filtering for team building. In: Proceedings of ICEIS, pp. 241–248 (2004)
30. Xiao, X., Zheng, Y., Luo, Q., Xie, X.: Inferring social ties between users with human location history. J. Ambient Intell. Humaniz. Comput. **5**(1), 3–19 (2014)
31. Yu, X., Pan, A., Tang, L.A., Li, Z., Han, J.: Geo-friends recommendation in GPS-based cyber-physical social network. In: International Conference on Advances in Social Networks Analysis and Mining (ASONAM), pp. 361–368 (2011)
32. Zheng, V.W., Zheng, Y., Xie, X., Yang, Q.: Collaborative location and activity recommendations with GPS history data. In: World Wide Web Conference Series, pp. 1029–1038 (2010)
33. Zheng, Y., Zhang, L., Ma, Z., Xie, X., Ma, W.: Recommending friends and locations based on individual location history. ACM Trans. Web **5**(1), 1–44 (2011)
34. Zhou, D., Wang, B., Rahimi, S.M., Wang, X.: A study of recommending locations on location-based social network by collaborative filtering. In: Kosseim, L., Inkpen, D. (eds.) AI 2012. LNCS, vol. 7310, pp. 255–266. Springer, Heidelberg (2012). doi:10.1007/978-3-642-30353-1_22
35. Zhu, J., Xie, Q., Chin, E.J.: A hybrid time-series link prediction framework for large social network. In: Liddle, S.W., Schewe, K.-D., Tjoa, A.M., Zhou, X. (eds.) DEXA 2012. LNCS, vol. 7447, pp. 345–359. Springer, Heidelberg (2012). doi:10.1007/978-3-642-32597-7_30

A Time and Sentiment Unification Model for Personalized Recommendation

Qinyong Wang[1], Hongzhi Yin[2(✉)], and Hao Wang[1]

[1] Institute of Software, Chinese Academy of Sciences,
University of Chinese Academy of Sciences, Beijing, China
{qinyong2014,wanghao}@iscas.ac.cn
[2] School of Information Technology and Electrical Engineering,
The University of Queensland, Brisbane, Australia
h.yin1@uq.edu.au

Abstract. With the rapid development of social media, personalized recommendation has become an essential means to help people discover attractive and interesting items. Intuitively, users buying items online are influenced not only by their preferences and public attentions, but also by the crowd sentiment (i.e., the word of mouth) to the items. Specifically, users are likely to refuse an item whose most reviews are negative from the crowd. Therefore, a good personalized recommendation model also needs to take crowd sentiment into account, which most current methods do not. In light of this, we propose TSUM, a model that jointly integrates time and crowd sentiment, for personalized recommendation in this paper. TSUM simultaneously models user-oriented topics related to user preferences, time-oriented topics relevant to temporal context, and crowd sentiment towards items. TSUM combines the influences of user preferences, temporal context and crowd sentiment to model user behavior in a unified way. Extensive experimental results on two large real world datasets show that our recommender system significantly outperforms the state-of-the-arts by making more effective personalized recommendations.

Keywords: Temporal recommendation · User behavior modeling · Crowd sentiment

1 Introduction

As social media platforms gain prominence, user-generated contents (UGC) become valuable resources [14,15] to analyze user behavior and capture user preferences, which is an important foundation for personalized recommendation [19,21,24,25]. Lots of existing researches assume that users choose an item only due to their intrinsic preferences, and thus model user preferences to help them find interesting items. However, in real world case, user behavior is influenced not only by user preferences, but also by public attentions. In a recent research, [22] investigated user behavior on multiple social media datasets and indicated

© Springer International Publishing AG 2017
L. Chen et al. (Eds.): APWeb-WAIM 2017, Part II, LNCS 10367, pp. 98–113, 2017.
DOI: 10.1007/978-3-319-63564-4_8

that user behavior is significantly influenced by two factors: user preferences and public attentions. Thus it is more desirable to model user behavior by incorporating both factors [22,23].

Another observation is that user behavior is also influenced by the word-of-mouth or crowd sentiments. For instance, if an item whose overall history reviews by previous buyers is negative is recommended to a user, it is less likely that the user is going to accept it. In this case, although the item might match the user's preferences or draw the public attentions, it is not supposed to be an ideal choice to recommend. Therefore, in order to model user behavior in a more accurate way, it is necessary to couple user preferences, public attentions as well as crowd sentiments all together. However, it is very challenging to model them in a unified framework because of the following problems: Firstly, in a social media platform, most users often don't leave any reviews for items. How to capture users' implicit sentiments according to their other explicit behaviors such as clicks, votes and purchase records? Secondly, user preferences, public attentions and crowd sentiments usually mix together in the same context How to find a way to distinguish user preferences from public attentions with crowd sentiments?

Figure 1 shows the acceptance rate of items recommended to users from user preferences or public attentions under positive sentiment or negative sentiment, which is from a large real-world dataset. By observing the crowd sentiment, we see that the items with positive crowd sentiment are more likely to be accepted.

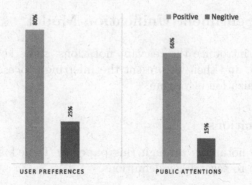

Fig. 1. The acceptance rate

To this end, we propose a **T**ime and **S**entiment **U**nification **M**odel (TSUM) to mimic user behavior in a process of decision making. In TSUM, we represent user preferences and public attentions by user-oriented topics and time-oriented topics respectively. User-oriented topics refer to users' preferences, which are reflected in their routine posts. They are relatively independent of temporal context and often evolve slowly. On the other hand, time-oriented topics are hot issues happening in real time, and they usually trigger heat discussion and wide propagation in social media at a period of time. However, it is not enough to only consider these to precisely model user behavior, it needs to model the implicit sentiments of the crowd over items. To the best of our knowledge, we are the first to model user behavior by integrating user preferences, temporal context and crowd sentiment in a unified way.

The major contributions of our research are summarized as follows.

- We propose a novel probabilistic generative model to account for user behavior by simultaneously considering the influences of user preference, public attention and crowd sentiment in a unified way.
- We design an effective and practical personalized recommender system based on TSUM. With the added information of temporal context and sentiments, we can alleviate the data sparsity problem.
- We conduct extensive experiments to evaluate the performance of the TSUM-based recommender system on two large-scale real-world datasets from different social media. The experimental results demonstrate that our recommender system outperforms the state-of-the-arts in the task of personalized recommendation.

The rest of this paper is organized as follows: Sect. 2 formulates the problem and describes the model, and then presents the inference algorithm. Section 3 discusses the experimental performance of the TSUM-based recommender system. Section 4 reviews the related work. Section 5 concludes this paper and presents our future work.

2   Time and Sentiment Unification Model

In this section, we introduce the relevant notations, some key definitions and problems of TSUM, and then we present the inferring process for TSUM with Expectation-Maximization algorithm.

2.1   Model Definitions

We summarize the notations through this paper in Table 1 and then make a formal introduction to several key definitions.

Definition 1 *(Feedback). A feedback is denoted by a triple (u, t, v), which represents user u's behaviors on item v at time slice t, such as purchasing, clicking, tagging and rating.*

In *Definition* 1, a time slice t is actually a timestamp obtained by dividing the original raw timestamps (e.g., "974765950" and "2015-07-29T20:28:42.000+08:00") according to a predefined granularity (e.g., daily or weekly granularity).

Definition 2 *(Intensity). An intensity associated with a feedback (u, t, v) indicates how strong interests user u shows to item v at time slice t.*

Table 1. Notations used through this paper

Symbol	Description
u, t, v, s	User, time slice, item, crowd sentiment
z, x	User-oriented topic, time-oriented topic
N, T, V, S	# of users, time slices, items and sentiments
K_1	# of user-oriented topics
θ_u	u's preferences
θ_{uz}	Probability of z chosen by u
Ω_z	Distribution of s under z
Ω_{zs}	Probability of s chosen under z
ϕ_{zs}	Distribution of v under z and s
ϕ_{zsv}	Probability of v generated by z and s
K_2	# of time-oriented topics
θ'_t	The temporal context at t
θ'_{tx}	Probability of x chosen at t
Ω'_x	Distribution of s under x
Ω'_{xs}	Probability of s chosen under x
ϕ'_{xs}	Distribution of v under x and s
ϕ'_{xsv}	Probability of v generated by x and s

Definition 3 (Intensity Cube). *An intensity cube IC with size of $N \times T \times V$ is a data structure where each cell stores the intensity for the feedback (u, t, v).*

For datasets that are generated by asking users to rate scores for items, we take the scores (assumed to be integers in our settings) to be the intensity. However, for datasets that only keep track of user behaviors (e.g. purchasing, clicking and tagging), we treat the frequency of interactions between user u and item v at time slice t as the intensity. For instance, we can use the total number of times a user visits a product on an online store during a particular time period as the intensity.

Definition 4 (User-Oriented Sentimental Topic). *A user-oriented sentimental topic is defined as a multinomial distribution over items set I, namely, $\phi_{zs} = \{\phi_{zsv}\}_{v=1}^{V}$, where $I = \{I_v\}_{v=1}^{V}$. Furthermore, a user-oriented topic z is defined by the summation over the sentiments of user-oriented sentimental topic.*

Definition 5 (Time-Oriented Sentimental Topic). *A time-oriented sentimental topic is defined as a multinomial distribution over items set I, namely, $\phi'_{xs} = \{\phi'_{xsv}\}_{v=1}^{V}$, where $I = \{I_v\}_{v=1}^{V}$. Furthermore, a time-oriented topic z is defined by the summation over the sentiments of time-oriented sentimental topic.*

With the definitions, we present the principal problem we try to solve.

Problem 1. Given an *intensity cube IC* that contains a large number of *feedbacks* and their *intensities*, we are asked to build a model to find out the user-oriented sentimental topics, time-oriented sentimental topics, user preferences and temporal context.

2.2 Model Description

In TSUM, the procedure of generating an item v by user u at time slice t given an intensity cube IC is demonstrated as follows.

First, TSUM flips a coin l following the Bernoulli distribution parameterized by λ_u to determine whether the item is generated from u's preferences or the temporal context at t. If it is from u's preferences (i.e., $l = 1$), then u chooses a user-oriented topic z from the multinomial distribution θ_u. Following that, z picks a crowd sentiment s from multinomial distribution Ω_z. Finally, item v is generated from the user-oriented sentimental topic with multinomial distribution ϕ_{zs}. On the other hand, if the item is from the temporal context θ'_t (i.e., $l = 0$), then at time slice t, the time-oriented topic x is sampled from the temporal context with multinomial distribution θ'_t. After that, x picks a crowd sentiment s from the multinomial distribution Ω'_x. Similarly, item v is picked from the time-oriented sentimental topic with multinomial distribution ϕ_{xs}.

The generative process of TSUM is summarized in Algorithm 1 and the formal definition of the generative process which corresponds to the Bayesian network is shown in Fig. 2.

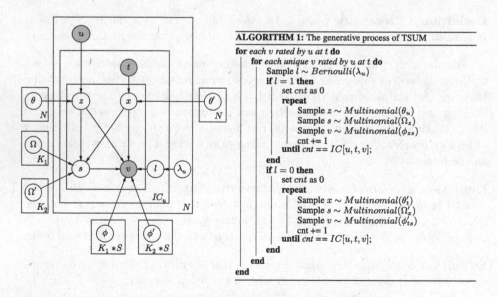

ALGORITHM 1: The generative process of TSUM

for *each v rated by u at t* **do**
    for *each unique v rated by u at t* **do**
        Sample $l \sim Bernoulli(\lambda_u)$
        if $l = 1$ **then**
            set *cnt* as 0
            repeat
                Sample $z \sim Multinomial(\theta_u)$
                Sample $s \sim Multinomial(\Omega_z)$
                Sample $v \sim Multinomial(\phi_{zs})$
                cnt += 1
            until *cnt* $== IC[u, t, v]$;
    end
    if $l = 0$ **then**
        set *cnt* as 0
        repeat
            Sample $x \sim Multinomial(\theta'_t)$
            Sample $s \sim Multinomial(\Omega'_x)$
            Sample $v \sim Multinomial(\phi'_{ts})$
            cnt += 1
        until *cnt* $== IC[u, t, v]$;
    end
    end
end

Fig. 2. The graphical representation of TSUM

2.3   Model Inference

Given an intensity cube IC, the learning procedure of TSUM is to estimate the unknown model parameter set $\Psi = \{\theta, \Omega, \phi, \theta', \Omega', \phi', \lambda\}$. The log-likelihood of the model is as follows:

$$L(\Psi|IC) = \sum_{u=1}^{N} \sum_{t=1}^{T} \sum_{v=1}^{V} IC[u,t,v] \log P(v|u,t,\Psi) \tag{1}$$

where $P(v|u,t,\Psi) = \lambda_u P(v|\theta_u) + (1-\lambda_u)P(v|\theta'_t)$

$P(v|\theta_u)$ is the probability that item v is generated from user u's preferences and $P(v|\theta'_t)$ is the probability that item v is generated from the temporal context at time slice t. $P(v|\theta_u)$ and $P(v|\theta'_t)$ are formulated as:

$$P(v|\theta_u) = \sum_{z=1}^{K_1} \sum_{s=1}^{S} P(v|\phi_{zs})P(z|\theta_u); P(v|\theta'_t) = \sum_{x=1}^{K_2} \sum_{s=1}^{S} P(v|\phi'_{xs})P(x|\theta'_t) \tag{2}$$

Our goal is to maximize the log-likelihood in Eq. (1). However, it cannot be solved directly by applying Maximum Likelihood Estimation (MLE), we employ the EM algorithm to obtain the approximate estimation instead.

In the **Expectation-step** of the EM algorithm, our target is to obtain the two posterior probabilities of latent random variable z and s. With the help of Jensen inequality, we obtain:

$$P(z,s|l=1,u,t,v) = \frac{P(z,s,l=1|u,v,t)}{\sum_{z'=1}^{K_1} \sum_{s'=1}^{S} P(z',s',l=1|u,t,v)} \tag{3}$$

where $P(z,s,l=1|u,v,t) = P(v|\phi_{zs})P(s|\Omega_z)P(z|\theta_u)$.

$$P(x,s|l=0,u,t,v) = \frac{p(x,s,l=0|u,t,v)}{\sum_{x'=1}^{K_2} \sum_{s'=0}^{S} P(x',s',l=0|u,t,v)} \tag{4}$$

where $P(x,s,l=0|u,t,v) = P(v|\phi'_{xs})P(s|\Omega'_x)P(x|\theta'_t)$.

In the **Maximization-step**, we maximize the complete data log-likelihood or Q function $Q(\Psi)$ for TSUM with respect to the parameters, where:

$$\begin{aligned}
Q(\Psi) = \sum_{u=1}^{N} \sum_{t=1}^{T} \sum_{v=1}^{V} IC[u,t,v] &\times \{P(l=1|u,t,v;\hat{\Psi}) \sum_{z=1}^{K_1} \sum_{s=1}^{S} P(z,s|l=1,u,t,v;\hat{\Psi}) \\
&\times \log[\lambda_u P(v|\phi_{zs})P(s|\Omega_z)P(z|\theta_u)] + P(l=0|u,t,v;\hat{\Psi}) \sum_{x=1}^{K_2} P(x,s|l=0,u,t,v;\hat{\Psi}) \\
&\times \log[(1-\lambda_u)P(v|\phi'_{xs})P(s|\Omega'_x)P(x|\theta'_t)]\}
\end{aligned} \tag{5}$$

where

$$P(l|u,t,v) = \frac{l\lambda_u P(v|\theta_u) + (1-l)(1-\lambda_u)P(v|\theta'_t)}{\lambda_u P(v|\theta_u) + (1-\lambda_u)P(v|\theta'_t)} \tag{6}$$

In addition, $Q(\Psi)$ is constraint by $\sum_{v=1}^{V} P(v|\phi_{zs}) = 1$, $\sum_{v=1}^{V} P(v|\phi'_{xs}) = 1$, $\sum_{s=1}^{S} P(s|\Omega_z) = 1$, $\sum_{s=1}^{S} P(s|\Omega'_x) = 1$, $\sum_{z=1}^{K_1} P(z|\theta_u) = 1$ and $\sum_{x=1}^{K_2} P(x|\theta'_t) = 1$.

With some simple derivations, we update the parameter set with the following formulas:

$$P(z|\theta_u) = \frac{\sum_{v=1}^{V} \sum_{t=1}^{T} P(z)}{\sum_{z'=1}^{K_1} \sum_{v=1}^{V} \sum_{t=1}^{T} P(z')}; P(s|\Omega_z) = \frac{\sum_{v=1}^{V} \sum_{t=1}^{T} \sum_{u=1}^{N} P(s)}{\sum_{s'=1}^{S} \sum_{v=1}^{V} \sum_{t=1}^{T} \sum_{u=1}^{N} P(s')} \quad (7)$$

where $P(z) = IC[u,t,v]P(z, l = 1|u,t,v;\hat{\Psi})$, $P(s) = IC[u,t,v]P(z,s,l = 1|u,t,v;\hat{\Psi})$

$$P(v|\phi_{zs}) = \frac{\sum_{u=1}^{N} \sum_{t=1}^{T} P(v)}{\sum_{v'=1}^{V} \sum_{u=1}^{N} \sum_{t=1}^{T} P(v')}; P(x|\theta'_t) = \frac{\sum_{v=1}^{V} \sum_{u=1}^{N} P(x)}{\sum_{x'=1}^{K_2} \sum_{v=1}^{V} \sum_{t=1}^{T} P(x')} \quad (8)$$

where $P(v) = IC[u,t,v]P(z,s,l = 1|u,t,v;\hat{\Psi})$ and $P(x) = IC[u,t,v]P(x,l = 0|u,t,v;\hat{\Psi})$

$$P(s|\Omega'_x) = \frac{\sum_{v=1}^{V} \sum_{t=1}^{T} \sum_{u=1}^{N} P(s)}{\sum_{s'=1}^{S} \sum_{v=1}^{V} \sum_{t=1}^{T} \sum_{u=1}^{N} P(s')}; P(v|\phi'_{xs}) = \frac{\sum_{u=1}^{N} \sum_{t=1}^{T} P(v)}{\sum_{v'=1}^{V} \sum_{u=1}^{N} \sum_{t=1}^{T} P(v')} \quad (9)$$

where $P(s) = IC[u,t,v]P(x,s,l = 0|u,t,v;\hat{\Psi})$ and $P(v) = IC[u,t,v]P(x,s,l = 0|u,t,v;\hat{\Psi})$

$$\lambda_u = \frac{\sum_{t=1}^{T} \sum_{v=1}^{V} M[u,t,v]P(l = 1|u,t,v;\hat{\Psi})}{\sum_{t=1}^{T} \sum_{v=1}^{V} \sum_{l_0}^{1} M[u,t,v]P(l|u,t,v;\hat{\Psi})} \quad (10)$$

The introduction of λ_u aims to automatically adapt the model parameter estimation to various users to enable personalized treatment.

2.4 TSUM-Based Recommender System

In this section, we present a temporal-sentiment-aware recommender system which includes two major components: *modeling component* and *recommendation component*. The offline modeling component employs the TSUM to mimic user behavior in the process of decision making and then the online recommendation component generates personalized *top-k* recommendation results for every query (u,t), where u and t stand for the querying user u and the querying time slice t respectively. Given a query (u,t), the online recommender component responds by combining u's preferences and the temporal context at t, which are learned by the TSUM. Then the recommender system returns the *top-k* items with highest ranking scores. The architecture of the proposed recommender system is shown in Fig. 3.

More specifically, we present the methods to compute the ranking scores in details. When the recommender system receives a query $q = (u,t)$, a new multinomial distribution θ_q is constructed, which combines both θ_u and θ'_t. Since θ_u and θ'_t have different dimensions (i.e., user preference distribution θ_u with

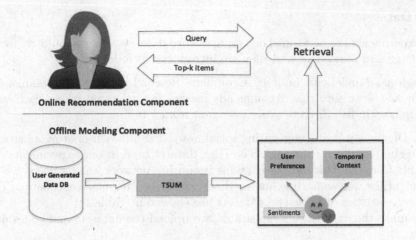

Fig. 3. The architecture of the proposed recommender system

dimension K_1 and temporal context distribution θ'_t with dimension K_2) respectively, we need to expand the dimension of the query multinomial distribution θ_q to K, where $K = K_1 + K_2$. To utilize the mixing coefficient λ_u for topic fusion, we define the joint multinomial distribution θ_q as follows:

$$\theta_q = <\lambda_u\theta_{u1}, \lambda_u\theta_{u2}, \cdots, \lambda_u\theta_{uK1}, (1-\lambda_u)\theta'_{tK1+1}, (1-\lambda_u)\theta'_{tK1+2}, \cdots, (1-\lambda_u)\theta'_{tK2}> \quad (11)$$

To represent the item v in the expanded space as the query q, we introduce a new notation for the multinomial distribution of item v, $\phi_{\hat{z}s_0v}$ and $\phi_{\hat{z}s_1v}$, which assumes $\|s\| = 2$, that is, we adopt a binary sentiment setting: positive and negative. In this way, the query q and the item v are in the same dimension space. It is defined as follows:

$$\phi_{\hat{z}s_1v} = \begin{cases} \phi_{zs_1v}, & \hat{z} \leq K_1 \\ \phi_{xs_1v}, & \hat{z} > K_1 \end{cases} \quad (12)$$

Given a query q, the ranking score $S(u,t,v,s)$ for item v is computed as the inner product of the two vectors, of which we are only interested in the items with positive crowd sentiment (i.e., $s = 1$).

$$S(u,t,v,s=1) = \sum_{\hat{z}=1}^{K_1+K_2} \theta_{q\hat{z}}\phi_{\hat{z}s_1v} \quad (13)$$

3   Experimental Setup

In this section, we evaluate the effectiveness of the proposed TSUM on two real-world large-scale datasets.

3.1 Datasets

Our experiments are performed on two real-word datasets that are publicly available: Digg and Movielens, and they are introduced as follows:

Movielens. MovieLens, built by Grouplens Research Lab, is a recommender system and a website that recommends movies for its users, where each user votes movies in five discrete ratings ranging from 1 to 5.

Digg. Digg is a website aggregating social news that allows users to vote stories up (digging) or down (burying). The Digg dataset used in our experiments are available from [7]. This dataset contains friendship network among users, which we can utilize for some baseline methods that need explicit sentiments. The overview statistics of those two datasets are showed in Table 2.

To make the experiments repeatable, we upload the datasets and codes at[1].

Table 2. Basic statistics of the datasets

Dataset	Number of users	Number of items	Number of feedbacks	Time span (year)
Digg	139,409	3,553	3,018,197	2009–2010
Movielens	71,567	10,681	10,000,054	1998–2009

3.2 Comparative Approaches

We compare TSUM with the following three competitive methods that represent the state-of-the-art recommendation techniques, and they are TCAM, eTOT, BPR and TimeSVD++.

TCAM: Temporal context-aware mixture model (TCAM) proposed by [22] simultaneously models both user preferences and temporal context and then combines the two factors to model user behavior in a unified way. This model outperforms most of the existing recommendation models that only consider user preferences or temporal information. In TCAM, an item v is generated by the following formula:

$$P(v|u,t;\Psi) = \lambda_u P(v|\theta_u) + (1 - \lambda_u)P(v|\theta'_t) \tag{14}$$

The notations in Eq. 14 are identical to those we use in TSUM, that is, $P(v|\theta_u)$ and $P(v|\theta'_t)$ represent that the item v is generated by user-oriented topic or temporal context topic, λ_u is the probability of choosing from them.

eTOT: eTOT is a topic model that extends TOT [20] model, and eTOT assumes the topics as a Beta distribution over time and a multinomial distribution over emotions. eTOT captures both the temporal information and sentiments but it ignores user preferences. For eTOT, the sentiments are exploited as an explicit

---

[1] https://goo.gl/Z4A8Lh.

variable. However, most datasets like Digg and MovieLens do not contain explicit sentiment information. Hence, in order to perform recommendation with eTOT on Movielens and Digg we need to make some assumptions and make use of other information (e.g., friendship network in Digg) to make sentiment explicitly available. Taking Movielens as an example, we assume the ratings above a user's average score to be positive and negative otherwise.

BPR: BPR is a state-of-the-art matrix factorization model for item prediction (ranking). BPR model outperforms most existing recommender systems that perform the task of *top-k* item recommendation. We utilize the BPR algorithm provided by MyMediaLite [5], a free recommender system library for the Common Language Runtime.

TimeSVD++: TimeSVD++ [10] is a model that extends SVD++, which shows temporal dynamics through a specific parameterization with factors drifting from a central time, but it can only deal with a limited amount of time slices. SVDFeature [3] toolkit provides an easy access for us to run experiments with TimeSVD++.

3.3 Evaluation Methods and Metrics

In this part, we introduce the evaluation methods for generating online recommendations and evaluation metrics for the suggested items from recommender systems.

To make an overall evaluation of the recommendation effectiveness, we follow the methodological description framework proposed in [2], and then we present the following principles: (1) We split each dataset into a training set and a testing set independently on each user's ratings, ensuring that all users have ratings in both sets. (2) We first sort each user's ratings by timestamp, and then use the $80th$ percentile as the cutoff point, such that the part before this point is used for train set and the rest is used for testing set. Specifically, $S(u)$ is divided into the train set $S^{train}(u)$ and the test set $S^{test}(u)$. (3) In the evaluation, given a target user u, the recommender system should find the *top-k* items from all available items except those in the train set $S^{train}(u)$. (4) The recommender system should find the items relevant to the querying user u, where the set of relevant items for u is formed by the items in u's test set. $S^{test}(u)$.

According to those principles, we perform the following steps:

1. Given a query $q = (u, t)$, for user u we compute the scores of items that occur at time slice t in train set S^{train}, denoted as $scores(q)$ with the output of TSUM.
2. A ranked list is formed by ordering $scores(q)$ reversely, then we pick the k top ranked items from the list.
3. We measure the count of hits within those k items, where the hits denote relevant items in test set S^{test}. We formally define the measurement metrics to be $Presicion@k$ and $nDCG@k$, which are presented as:

$$Precision@k = \frac{\#relevances}{k} \tag{15}$$

where $\#relevances$ is the number of relevant items in the *top-k* recommended items.

$$nDCG@k = \frac{1}{IDCG} \times \sum_{i=1}^{k} \frac{2^{r_i} - 1}{\log(i+1)} \tag{16}$$

where r_i is 1 if the item at position i is a relevant item and 0 otherwise. $IDCG$ is chosen for the purpose of normalization so that the perfect ranking has an $nDCG$ value of 1.

3.4   Recommendation Effectiveness

In this part, we present the optimal performance with well tuned parameters. The comparison of *precision@k* and $nDCG@k$ of recommendations performed on Movielens and Digg datasets by TSUM and its competitors are showed in Fig. 4, where we only show the performance when k is 1, 5 and 10 since a larger k is usually beyond necessity in reality.

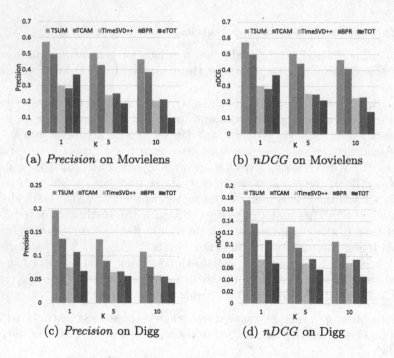

(a) *Precision* on Movielens   (b) *nDCG* on Movielens

(c) *Precision* on Digg   (d) *nDCG* on Digg

Fig. 4. Recommendations on Movielens and Digg

Clearly, from the reported results, we can see that TSUM outperforms other competitors (i.e., TCAM, eTOT and BPR) significantly on both datasets w.r.t. *precision* and $nDCG$. After closer observations, we make the following conclusions:

1. By simultaneously consider users' preferences, temporal context and crowd sentiments, the accuracy of recommendation can be greatly improved. TSUM outperforms TCAM on both datasets, which proves the significance of implicit crowd sentiments when doing recommendation.
2. Comparing to BPR, TimeSVD++ exploits the temporal context information when doing recommendation. TimeSVD++ has better performance than BPR on Digg dataset and performs equally well as BPR on Movielens, this result verifies the fact that Digg dataset is more time sensitive than Movielens since the time information in Movielens dataset plays a relatively less significant role.
3. TSUM outperforms TimeSVD++, this is partly because TimeSVD++ relies heavily on explicit feedback data (e.g., user's rating scores), which are not always available. On the contrary, TSUM is able to deal with both explicit and implicit feedback data.
4. eTOT shows the poorest performance. Maybe it results from the inappropriate explicit sentiment assumptions in Movielens and Digg datasets, which suggests that it is better for us to treat the crowd sentiment as implicit and latent factor as TSUM dose.

3.5 Impact of Different Factors

There are three hyper-parameters in TSUM needed to be tuned, namely, the length of time slice, the number of user-oriented topics ($K1$) and time-oriented topics ($K2$), one thing to mention is that the length of crowd sentiments is fixed to be 2, namely, we assume there exist 2 types of crowd sentiments for an item: positive and negative. Tuning these parameters is critical to the model performance. We study those impacts in details in the next parts.

Impacts of the Length of Time Slice. Intuitively, a larger length of time slice indicates that the recommendation results are less time context aware. This experiment explores the impacts of the length of time slice since it controls the time granularity of temporal recommendation. We report *Precision* and *nDCG* of k at 1, 5 and 10 on the Digg dataset in Table 3. We can see from the tables that as the length of time slice increases, both the *Precision* and the *nDCG* first increase and then decrease. Finally, 3-day is the best time slice for Digg to do recommendation. We explain the figure as follows: at first, as the length gets larger, the data at each time slice become denser. After the length is over some point, the temporal information is tremendously reduced at each time slice.

Impacts of the Number of Topics. This experiment studies the impacts of the number of user-oriented topics ($K1$) and the number of time-oriented topics ($K2$). Figure 5 reports the *Precision*@5 and *nDCG*@5 as we vary K_2 (i.e., the number of time-oriented topic) at the range of $[30, 70]$ while fixing K_1 (i.e., the number of user-oriented topic) as 30, 50, 70, which are denoted as TSUM-30, TSUM-50 and TSUM-70, respectively. We have the observation that for both

Table 3. Effect of length of time slice to precision and nDCG on Digg

Length	Precision						nDCG					
	1 day	2 day	3 day	4 day	5 day	6 day	1 day	2 day	3 day	4 day	5 day	6 day
k = 1	0.16	0.157	**0.196**	0.159	0.153	0.155	0.168	0.166	**0.175**	0.155	0.158	0.151
k = 5	0.117	0.117	**0.135**	0.117	0.111	0.112	0.126	0.121	**0.130**	0.114	0.117	0.114
k = 10	0.098	0.097	**0.109**	0.099	0.091	0.093	0.102	0.099	**0.105**	0.095	0.096	0.095

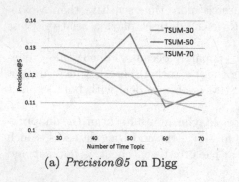

(a) *Precision@5* on Digg

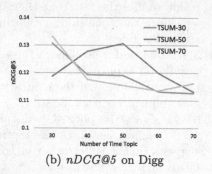

(b) *nDCG@5* on Digg

Fig. 5. Impacts of varying topic numbers on Digg

Precision@5 and *nDCG*@5, TSUM performs best when K_1 and K_2 are both set as 50.

4   Related Work

Topic models like latent Dirichlet allocation [1] and probabilistic latent semantic indexing [6] are usually used to detect topics from textual datasets. Typically in topic models, documents are modeled as a distribution over a fixed number of topics and those topics are modeled as a distribution over words. Many researchers utilize topic models to analyze user behaviors.

Early work focuses on the inference of the user behaviors given sufficient information. Author-topic model [13] was developed to simultaneously model the content of documents and the interests of authors. In that model, each author is a multinomial distribution over topics and each topic is a multinomial distribution over words. A document with multiple authors allows the mixture weights for different topics to be determined by those users. [8] proposed Topic Tracking (TTM) Model, which is a probabilistic consumer purchase behavior model based on latent Dirichlet allocation adaptively tracking changes in interests and trends based on current purchase logs and previously estimated interests and trends.

Temporal information is incorporated into the models. Diao et al. [4] designed TimeUserLDA to find bursty topics from microblogs. It assumes that the user posting behaviors are influenced by both user preferences and global topical trends. However, Yin et al. pointed out that the existing work dose not distinguish between underlying user-oriented topics and time-oriented topics, and leads to some noises and confusions [23]. Therefore, Yin et al. [22] proposed TCAM that simultaneously models the topics related to users' preferences and the topics related to temporal context and then combines the influences from the two factors to model user behavior in a unified way.

More recently, sentiments are introduced to model user behavior [9,11,12,16–18,26]. Multi-Aspect Sentiment (MAS) model [9] aims at modeling topics to the predefined aspects explicitly rated by users in reviews, where the sentiment is modeled on the aspect level by the sentiment distribution from the mixture weight of extracted topics and words. User-Sentiment Topic Mode (USTM) [26] is an unsupervised generative model that captures user's sentiment in a topic level by considering topic and sentiment simultaneously. This model can not only discover whether users like or dislike the topics but also reveal why they show these opinions by refining topics with sentiment information. However, Li et al. [11] found that different users may use different sentiment expressions for different items, thus they proposed a Supervised User-Item based Topic (SUIT) model to simultaneously utilize the textual topic and user item factors for sentiment analysis.

Different from those existing work, TSUM jointly considers user-oriented and time-oriented topics while incorporating the crowd sentiments to model user behaviors in social media in a unified way. TSUM distinguishes from other models as follows: (1) TSUM has the ability to detect sentimental user-temporal topics and sentimental time-oriented topics simultaneously. (2) TSUM is fully unsupervised. (3) TSUM can handle both explicit and implicit feedbacks.

5   Conclusion and Future Work

In this paper, we proposed a joint temporal and user preference generative model, TSUM, for temporal item recommendation, which effectively addresses the challenges arising from ambiguous feedbacks, temporal-sentiment awareness. specifically, we proposed a model that introduces two kinds of latent topics to model user preference and temporal context respectively. Moreover, we utilize the implicit crowd sentiment to indicate the items with positive reviews. A series of experimental results show the superiority of TSUM over existing methods in the task of *top-k* temporal item recommendation, which verifies our motivation.

The relatively simplicity of TSUM allows for several extensions in order to better conduct the temporal recommendation in the era of Web 2.0. Our model studies the rating pattern in social media, however, contents such as reviews and tags contain richer sentimental information, and thus the mixture of contents for sentiment analysis is practical. Complex social network structures reveal much valuable information (e.g., common community and friends relationship), so it can be integrated for a more precise user behaviors model.

Acknowledgement. The work described in this paper is supported by National Natural Science Foundation of China (61602453, 61672501, 61603373). It was also partially supported by ARC Discovery Early Career Researcher Award (DE160100308), ARC Discovery Project (DP170103954).

References

1. Blei, D.M., Ng, A.Y., Jordan, M.I.: Latent Dirichlet allocation. JMLR **3**, 993–1022 (2003)
2. Campos, P.G., Díez, F., Cantador, I.: Time-aware recommender systems: a comprehensive survey and analysis of existing evaluation protocols. User Model. User Adapt. Interact. **24**(1–2), 67–119 (2014)
3. Chen, T., Zhang, W., Lu, Q., Chen, K., Zheng, Z., Yu, Y.: SVDFeature: a toolkit for feature-based collaborative filtering. JMLR **13**(1), 3619–3622 (2012)
4. Diao, Q., Jiang, J., Zhu, F., Lim, E.P.: Finding bursty topics from microblogs. In: ACL, pp. 536–544. Association for Computational Linguistics (2012)
5. Gantner, Z., Rendle, S., Freudenthaler, C., Schmidt-Thieme, L.: Mymedialite: a free recommender system library. In: RecSys, pp. 305–308. ACM (2011)
6. Hofmann, T.: Probabilistic latent semantic indexing. In: SIGIR, pp. 50–57. ACM (1999)
7. Hogg, T., Lerman, K.: Social dynamics of Digg. EPJ Data Sci. **1**(1), 1–26 (2012)
8. Iwata, T., Watanabe, S., Yamada, T., Ueda, N.: Topic tracking model for analyzing consumer purchase behavior. IJCAI **9**, 1427–1432 (2009)
9. Jo, Y., Oh, A.H.: Aspect and sentiment unification model for online review analysis. In: WSDM, pp. 815–824. ACM (2011)
10. Koren, Y.: Collaborative filtering with temporal dynamics. Commun. ACM **53**(4), 89–97 (2010)
11. Li, F., Wang, S., Liu, S., Zhang, M.: Suit: a supervised user-item based topic model for sentiment analysis. In: AAAI (2014)
12. Lin, C., He, Y.: Joint sentiment/topic model for sentiment analysis. In: CIKM, pp. 375–384. ACM (2009)
13. Rosen-Zvi, M., Griffiths, T., Steyvers, M., Smyth, P.: The author-topic model for authors and documents. In: UAI, pp. 487–494. AUAI Press (2004)
14. Su, B., Ding, X.: Linear sequence discriminant analysis: a model-based dimensionality reduction method for vector sequences. In: ICCV, pp. 889–896 (2013)
15. Su, B., Ding, X., Liu, C., Wu, Y.: Heteroscedastic max-min distance analysis. In: CVPR, pp. 4539–4547 (2015)
16. Wang, H., Xu, F., Hu, X., Ohsawa, Y.: Ideagraph: a graph-based algorithm of mining latent information for human cognition. In: SMC, pp. 952–957. IEEE (2013)
17. Wang, H., Zhang, C., Wang, W., Hu, X., Xu, F.: Human-centric computational knowledge environment for complex or ill-structured problem solving. In: SMC, pp. 2940–2945. IEEE (2014)
18. Wang, H., Zhang, C., Yin, H., Wang, W., Zhang, J., Xu, F.: A unified framework for fine-grained opinion mining from online reviews. In: HICSS, pp. 1134–1143. IEEE (2016)
19. Wang, W., Yin, H., Sadiq, S., Chen, L., Xie, M., Zhou, X.: Spore: a sequential personalized spatial item recommender system. In: ICDE, pp. 954–965. IEEE (2016)
20. Wang, X., McCallum, A.: Topics over time: a non-Markov continuous-time model of topical trends. In: SIGKDD, pp. 424–433. ACM (2006)
21. Xie, M., Yin, H., Wang, H., Xu, F., Chen, W., Wang, S.: Learning graph-based POI embedding for location-based recommendation. In: CIKM, pp. 15–24. ACM (2016)
22. Yin, H., Cui, B., Chen, L., Hu, Z., Zhou, X.: Dynamic user modeling in social media systems. TOIS **33**(3), 10 (2015)

23. Yin, H., Cui, B., Lu, H., Huang, Y., Yao, J.: A unified model for stable and temporal topic detection from social media data. In: ICDE, pp. 661–672. IEEE (2013)
24. Yin, H., Cui, B., Zhou, X., Wang, W., Huang, Z., Sadiq, S.: Joint modeling of user check-in behaviors for real-time point-of-interest recommendation. TOIS **35**(2), 11 (2016)
25. Yin, H., Zhou, X., Cui, B., Wang, H., Zheng, K., Nguyen, Q.V.H.: Adapting to user interest drift for POI recommendation. ICDE **28**(10), 2566–2581 (2016)
26. Zhao, T., Li, C., Ding, Q., Li, L.: User-sentiment topic model: refining user's topics with sentiment information. In: SIGKDD Workshop on Mining Data Semantics, p. 10. ACM (2012)

Personalized POI Groups Recommendation in Location-Based Social Networks

Fei Yu[✉], Zhijun Li, Shouxu Jiang[✉], and Xiaofei Yang

School of Compute Science and Technology, Harbin Institute of Technology,
Harbin 150001, China
{yf,lizhijun_os,jsx}@hit.edu.cn, xiaofei.hitsz@gmail.com

Abstract. With development of urban modernization, there are a large number of hop spots covering the entire city, defined as Pionts-of-Interest (POIs) Group consist of POIs. POI Groups have a significant impact on people's lives and urban planning. Every person has her/his own personalized POI Groups (PPGs) based on preferences and friendship in location-based social networks (LBSNs). However, there are almost no researches on this aspect in recommendation systems. This paper proposes a novel PPGs Recommendation algorithm, and models the PPGs by expanding the model of DBSCAN. Our model considers the degree to each PPG covering the target users' POI preferences. The system recommends the target user with the PPGs which have the top-N largest scores, and it is one NP-hard problem. This paper proposes the greedy algorithm to solve it. Extensive experiments on the two LBSN datasets illustrate the effectiveness of our proposed algorithm.

Keywords: POI group recommendation · Personalization · Geo-social distance · Density-based clustering

1 Introduction

With the development of mobile community, persons' demands on location-based services become ever more. Wherein Point-of-Interest (POI) recommendation is one typical application for location-based social networks (LBSNs). Users check in many different POIs in LBSNs, and these checking-in historical trajectories can represent users' checking-in behaviour features and POI preferences. Meanwhile, with the rapid development of the modern city, a large number of hop spots cover the total city space. Each city has its own characteristics of urban culture, and generally they are represented by frequently checked-in points-of-interest (POIs). The hop spots recommendation is that the system recommends the tourists with these frequently checked-in points-of-interest (POIs), however, the tourists' POI preferences are different from that of the city. So, the existing urban hop spot recommendation cannot recommend users with locations which are in accordance with users' POI category preferences. Currently, POI recommendation methods mainly focused on the recommendation quality, that are the

L. Chen et al. (Eds.): APWeb-WAIM 2017, Part II, LNCS 10367, pp. 114–123, 2017.
DOI: 10.1007/978-3-319-63564-4_9

accuracy and recall ratio of the recommended POI list. Since the recommended list cannot cover the total city, and these POIs don't own the whole urban characters. So, existing POI recommendation methods [4–6] cannot provide the tourists with the city's hop spots.

In this paper, we mainly research the hop spots with the target user's POI category preferences. Since these hop spots consist of many POIs covering the city, our goal is to recommend users with POI groups sufficiently covering their POI preferences. Then we call these POI groups as the personalized POI groups. There are four steps in our researches: First, we propose the definition of POI groups and formula POI groups. Second, this paper studies the intra-cluster correlation in each POI group, and models personalized POI groups (PPGs). Third, we propose the measure of the degree to each POI group covering users' POI category preferences. Finally, this system recommends the target user with the top-K PPGs ranking of the degree of POI category preference coverage.

This paper proposes one novel recommendation problem: Personalized POI group recommendation problem. Our research can solve the following question: which hot spots in one city we may interest based on my own POI preferences? Which regions of a city my friends check in, and these regions are places I may be interested in

2   Problem Definition

In this section, we will give a formal definition of Personalized POI Group recommendation problem (**PPG-Rec**) in LBSN, and this paper formulates this problem.

Definition 1 (LBSN). An LBSN $<G, C>$ consists of a social network $G = <U, E>$, where U is the users set, E is the set of edges, and check-in records $C = \{(u, l, t)\}$, (u, l, t) represents one check-in record where user u checks in the location l at time t. A location l denotes: $l = (lon, lat, a)$, wherein lon is longitude, lat is latitude, a is one POI category.

Definition 2 (PPG-Rec). Given an LBSN $<G, C>$, a target user u_T and his/her friend set U_F, given check-in records (POI set) $POI = \{I_1, I_2, ...\}$, each item I_i is a triple in the form of $<lon, lat, time, I_i.a>$, wherein lon, lat and $time$ respectively denotes longitude, latitude and check-in time, and $I_i.a$ represents the POI I_i's category. Personalized POI group recommendation problem is to select the set of clustering $C^* = \{C_1^*, C_2^*, ...\}$, wherein C_i^* is a set of POIs which satisfies u_T's preference demand and represents the semantics typical semantics and geography features corresponding to the clustering C_i^*.

3   PPGs Recommendation

In this section, we firstly model POI groups (**PG**). This paper extends the model of the DBSCAN [2] for modeling PG. For each POI I_i in the given LBSN,

PG finds the geo-social distance ϵ-neighborhood $N_\epsilon(I_i)$ of I_i, which includes all POIs I_j such that $p(I_i, I_j) \leq \epsilon$. If ϵ-neighborhood of I_i contains at least $MinPts$ POIs, then I_i is a *core* POI; in this case, I_i and all POIs in ϵ-neighborhood should belong to a POI group $C(I_i)$. If another core I_j belongs to $C(I_i)$, then $C(I_i) = C(I_j)$, in one words, the two POI group are merged. When the all core POIs are identified and merged to the corresponding POI groups (PG), PG ends up with a set of POI groups and a set of outliers. These outliers are the POIs who cannot belong to these POI groups, as Fig. 1 shown.

3.1   Modeling POI Groups

In this paper, we describe POI group based geo-social distance between locations. Wherein the geo-social distance is denoted as $D_{GS}(I_i, I_j)$, and it merges the geography distance $D_G(I_i, I_j)$ and $D_S(I_i, I_j)$. $D_{GS}(I_i, I_j)$ denotes as the following equation:

$$D_{GS}(I_i, I_j) = \lambda \cdot D_G(I_i, I_j) + (1 - \lambda) \cdot D_S(I_i, I_j) \tag{1}$$

The parameter $\lambda \in (0, 1)$ is the tradeoff of the geography distance and social distance, and it depends on the user's personal interests.

In this paper, the geography and social distance between POIs I_i, I_j respectively denotes as the followings:

Definition 3 (Geography Distance). Given LBSN $<G, C>$, two POIs I_i, I_j in given city, the geography distance between I_i and I_j is defined as the normalized Euclidean distance:

$$D_G(I_i, I_j) = \frac{E(I_i, I_j)}{maxD} \tag{2}$$

where $E(I_i, I_j)$ is Euclidean distance, $D_G(I_i, I_j) \in [0, 1]$, in this paper, the researched users set consists of the target user and his/her friends $\{u_t \cup U_F(u_T)\}$. The social distance $D_S(I_i, I_j)$ between POIs I_i, I_j naturally depends on the social network relationships between the set $U(I_i)$ and $U(I_j)$ of users who checked in I_i, I_j, respectively. Our social distance $D_S(I_i, I_j)$ is based on the set of contributing users $CU(I_i, I_j)$ between POIs I_i and I_j.

Definition 4 (Contributing Users). Given two POIs I_i, I_j with checking in users set (concluding the target user u_T and $U_F(u_T)$), the set of contributing users $CU(I_i, I_j)$ for the POI pair (I_i, I_j) is defined as the following:

$$CU(I_i, I_j) = \{u_a \in U(I_i)|u_a \in U(I_j) \text{ or } u_b \in U(I_j), u_a,$$
$$u_b \in \{u_T \cup U_F(u_T)\}\} \cup \{u_a \in U(I_j)|u_a \in U(I_i) \tag{3}$$
$$\text{or } u_b \in U(I_i), u_a, u_b \in \{u_T \cup U_F(u_T)\}\}$$

Definition 5 (Social Distance). Given LBSN $<G, C>$, two POIs I_i, I_j with visiting users U_{I_i}, U_{I_j}, the social distance between I_i and I_j is defined as:

$$D_S(I_i, I_j) = 1 - \frac{|CU(I_i, I_j)|}{|U(I_i) \cup U(I_j)|} \tag{4}$$

3.2   Intra-PG Correlation Analysis

In LBSN, each user has oneself own POI preference distributions. Generally, the distribution is the power-law distribution [1]. This paper focuses on the personalized POI groups (**PPGs**) recommendation, and the personal features are the target user u_T and his/her friends $U_F(u_T)$'s checking-in behaviours preferences. So, we synthesize u_T and $U_F(u_T)$'s POI preferences' features to formula the personalized POI groups (**PPGs**). In this paper, the system recommends the target user u_T with the PPGs, which cover the target user u_T's POI preferences as far as possible.

Preference Coverage. Preference coverage considers how the POI group $C_i \in \mathcal{C}$ in Sect. 3.1, concludes different POI categories. In fact, each POI category may label many POIs, and one POI may be labelled with multiple POI categories. In this paper, we analyse the degree to the every POI group concluding the target user's POI preferences, then select the subset of POIs C_k^* in the POI group C_k instead of all POIs in the POI group. The larger degree to which C_k^* covering u_T's POI preferences $A_{u_T} = \{a_1, a_2, ..., a_{M_{u_T}}\}$, the more information C_k^* can provide for u_T. In this paper, we consider the intra-PG correlation based on the preference coverage.

Let $A_{u_T} = \{a_1, a_2, ..., a_{M_{u_T}}\}$ is the target user u_T's POI categories preferences, and this paper computes the preference coverage of POI group in Eq. 5.

$$PreC(C) = \frac{1}{|C|} \cdot \sum_{a_k \in A_{u_T}} cov_{a_k}(C) \tag{5}$$

where $cov_{a_k}(C)$ measures the degree to a_k is covered by at least one POI in C. This section formulas $cov_{a_k}(C)$ with the following equation:

$$cov_{a_k}(C) = 1 - \prod_{I_i \in C} [1 - cov_{a_k}(I_i)] \tag{6}$$

where $cov_{a_k}(I_i)$ represents the degree to POI I_i covers a_k. The popular degree to the POI I_i labelled by the POI category a_k is described by the number of checking-in I_i with the label a_k. The popular degree $cov_{a_k}(I_i)$ denotes: $cov_{a_k}(I_i) = \frac{Num(I_i, a_k)}{\sum_{I_j \in C} Num(I_j, a_k)}$, $Num(I_i, a_k)$ represents the number of users checking-in the POI I_i with the label a_k, $\sum_{I_j \in C} Num(I_j, a_k)$ is the number of users checking-in the POIs with the label a_k in the POI group C.

Modeling Personalized POI Group. In this section, we select a subset of POIs C^* in each POI group C, the POIs in the subset of POIs C^* cover the target user u_T as far as possible, as Fig. 4. This paper expects to optimize the subset of POIs. Given the target user and his/her friends $\{u_T \cup U_F(u_T)\}$, the POI groups $\mathcal{C} = \{C_1, C_2, \cdots\}$ from Sect. 4.1, the problem is to select a subset of POIs in each POI group $C_i \in \mathcal{C}$ that maximizing the preference coverage function $PreC(C^*)$ in the check-in data. We regard the subset of POIs C^* as the personalized POI group (PPG), and this problem can be called as PPG-Rec. Meanwhile, we model this problem as one multi-objective optimization problem:

$$Given : \mathcal{C}, K, \{u_T \cup U_F(u_T)\}$$

$$Objective\ Function : \max_{C^* \subset C} PreC(C^*)$$

$$s.t.\quad |C^*| = K.$$

Algorithm 1. Select_PPGs algorithm

Input: POI Groups $\mathcal{C} = \{C_1, C_2, \cdots\}$, the target user u_T and $U_F(u_T)$, A_{u_T}, K.
Output: A subset of POIs $C^* \subseteq C$, $|C^*| = K$, the value $PreC(C^*)$.
  1: Initialize $C^* \Leftarrow \phi$
  2: **for** $j = 1$ **to** K **do**
  3:      $I_j \leftarrow \arg\max_{I_j \in C}[PreC(C^* \cup I_j) - PreC(C^*)]$;
  4:      $C^* \leftarrow C^* \cup I_j$
  5: **return** C^*, $PreC(C^*)$.

Greedy Algorithm: Due to the objective function $PreC(C^*)$'s monotone and submodular property [1], we give a greedy algorithm to compute the problem. The algorithm is called as Select_PPGs algorithm, and its detail description is as the following Algorithm 1.

Top-N PPGs Recommendation Algorithm. This paper is in order to find some personalized POI groups $\{C_1^*, C_2^*, \cdots, C_N^*\}$ to recommend the target user u_T in LBSN. Comparing with the value $PreC(C_i^*), C_i^* \subseteq C_i$ in POI groups $\mathcal{C}$, we select the top-N personalized POI groups $C_{(i)}^*, (i = 1, 2, ..., N)$ as the recommended PPGs.

4   Experimental Evaluation

Dataset Description. This paper utilizes the two real LBSN datasets, such as Foursquare, Gowalla datasets. They respectively consist of 36,907 users, 4,163 users; 26,907 locations, 121,142 locations; 1048,575 check-in times, 483.813 check-in times; the time span: 4/14/2010–1/17/2011, 1/18/2010–8/11/2011; friendship pairs: 23,148 pairs, 32,512 pairs; POI categories: 6,636 categories, 7,835 categories.

Comparative Approaches. To illustrate the effectiveness of our method, we compare GSD-PPG against anther several methods. GD-based POI group (**GD-PGs**) recommendation method [2] utilizes the DBSCAN to model the POI group. This POI groups represent the space clustering only, and it is without considering POI category preference and social relationship information. GSD-based POI group (**GSD-PGs**) recommendation method [3] takes advantage of the DCPGS, and denotes the social and geography information. Due to the size of the PG is big, this situation leads to the smaller value of the personalization in each PG. Region-based POI group **(Region-PGs)** recommendation method [4] is in order to recommendation the target user with the POIs in a given query region. This recommendation and our approach are not from the same perspective. So, we do not compare this method and our method in this paper's experiment part.

Evaluation Metrics. This paper utilizes the objective function is effective and reasonable, and the recommended Personalized POI Groups (PPGs) based on the function $PreC(\cdot)$ can describe the personalization of the PPGs. We utilize the recall ratio: Recall@K [7,8] to evaluate the quality of POI groups recommendation, and it is important to find out how many recommended POI categories actually belong to the target user POI category preferences, $Recall@K = \frac{|RC_{re}(\bigcup_{i=1}^{K} C^*_{(i)}) \cap A_{u_T}|}{|A_{u_T}|}$. Wherein A_{u_T} are user u_T's POI category preferences set. These metrics for the entire POI groups recommendation system are computed by averaging the above two metrics value for 2000 users (as the target users u_T) respectively.

Experimental Results. Visualization-based Analysis Fig. 1(a) shows the POI groups are spacial clusterings, but these POI groups disregards the social network behind POIs. These POI groups can not answer the above mentioned application issues. Figure 1(b) displays the POI groups can represent how close POIs are in the aspect of the spacial and social distance. By tuning the two parameters of spacial clustering method, this paper cannot find these POI groups found by GSD-based clustering method. As Fig. 1(a), (b) shown, the POIs in the region **A** belong to one POI group. And its corresponding area is the region **A*** in Fig. 1(b), there are two POI groups in this region. Which reason is that these POIs in this region **A*** are partitioned based on the social relationship. From Fig. 1(c), (d), we can see the comparison results between two clustering methods is similar to that presented in Fig. 1(a), (b).

Personalization Analysis for PG. Each POI group has its own POI features for maximizing covering the u_T POI preferences. In this paper, we regard these POIs with the maximization score $PreC(C^*)$ in each POI group as the POI features in the POI group. Then the Personalized POI group denotes these POIs, $C^* = \{I_{(1)}, I_{(2)}, \cdots, I_{(k)}\}$, $C^* \subseteq C$. From Fig. 2(b), we observe that there is black dots distribution in each POI group. These black dots are the features in POI groups. Some POI groups have many black dots, and some POI groups have less black dots.

As Fig. 2(b) shown, there are some POIs in each POI group, and these POIs (as the black dots in the figure) belong to the target user's POI preferences, called as the personalized POI (**PP**) in this paper. From Fig. 2(b), we see that some POI groups with many POIs have less personalized POIs. Hence, the size of POI

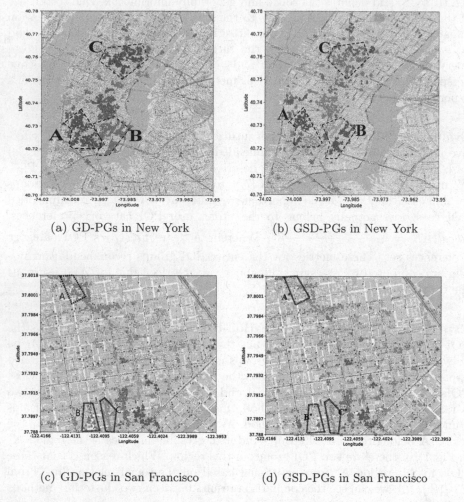

(a) GD-PGs in New York   (b) GSD-PGs in New York

(c) GD-PGs in San Francisco   (d) GSD-PGs in San Francisco

Fig. 1. POI groups of check-in datasets found in two cities

groups do not determine the number of PPs in each POI group. In this section, we utilize the function $PreC(C^*) = \frac{1}{|C^*|} \sum_{a_k \in A_{u_T}} cov_{a_k}(C^*)$ to measure the degree to the personalization in each PPGs. Our recommend method is to recommend the target user with the top-N PPGs according to the personalization degree of PPGs. This method outperforms other competitor methods significantly as Fig. 3 shown.

Effectiveness of Methods. This paper uses the Recall@K to evaluate the effectiveness between our method and other methods. Figure 4 reports the performance of our personalized POI groups recommendation method on Foursquare

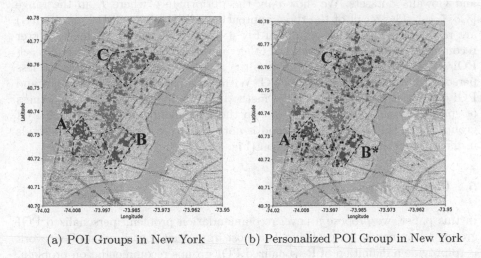

(a) POI Groups in New York   (b) Personalized POI Group in New York

Fig. 2. POI groups and personalized POI groups. In (b), the black dots are POI categories which belong to the target user's POI preferences

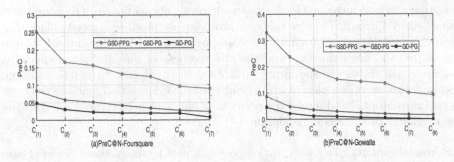

Fig. 3. Comparison of PreC of three methods

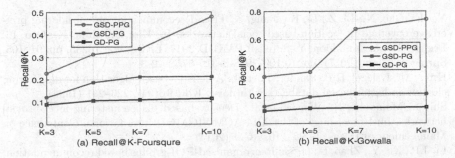

Fig. 4. Recall@K of different methods

and Gowalla datasets. We show only the performance where K in the range $[3, 5, 7, 10]$. The recall of the three approaches (GSD-PPG, GSD-PG, GD-PG) are reported in Fig. 4. Observe from Fig. 4, GSD-PPG method achieves better recommendation accuracy than the other method. Due to the big size of each POI group, GSD-PG method provides the recommended POI groups with less personalization score than GSD-PPG. Without considering social distance and POI categories preferences, GD-PG method mainly focuses on the compactness of space between locations. And due to the big amount of POIs in each POI group, the recall ratio of this method is not as good as the above two methods in terms of recommendation precision (Fig. 4).

5   Conclusions

In this paper, we propose a novel recommendation problem: personalized POI groups recommendation problem. The most important contribution of our work is to provide a definition of Personalized POI groups recommendation problem, and to formula the POI group with the geo-social clustering method Then this paper extracts the personalized features in each POI group, these features (PPG) consist of POIs in each POI group and replace the group's personalization. We design a metric to measure the degree to personalization of each POI group/ personalized POI group. To solve this new problem, we propose a greedy algorithm with $(1 - \frac{1}{e})$ theoretical bound. The enormous scale of LBSNs dataset verifies the degree to personalization and the effectiveness of our method. Especially, our approach shows the significant advantage in the degree to personalization. As the future work, we plan to utilize multi-source information (such as the temporal information), and model the periodic of check-in behaviours for improving the efficiency of the recommendation.

Acknowledgments. This work was supported in part by the National Science Foundation grants NSF-61672196, NSF-61370214, NSF-61300210.

References

1. Yu, F., Che, N., Li, Z., Li, K., Jiang, S.: Friend recommendation considering preference coverage in location-based social networks. In: Kim, J., Shim, K., Cao, L., Lee, J.-G., Lin, X., Moon, Y.-S. (eds.) PAKDD 2017. LNCS, vol. 10235, pp. 91–105. Springer, Cham (2017). doi:10.1007/978-3-319-57529-2_8
2. Ester, M., Kriegel, H.P., Sander, J., et al.: A density-based algorithm for discovering clusters in large spatial databases with noise. Kdd **96**(34), 226–231 (1996)
3. Shi, J., Mamoulis, N., Wu, D., et al.: Density-based place clustering in geo-social networks. In: Proceedings of the 2014 ACM SIGMOD International Conference on Management of Data, pp. 99–110. ACM (2014)
4. Li, J.P., Xu, Y., Zhao, L.: OPGs-Rec: organized-POI-groups based recommendation. In: Li, F., Shim, K., Zheng, K., Liu, G. (eds.) APWeb 2016. LNCS, vol. 9932, pp. 521–524. Springer, Cham (2016). doi:10.1007/978-3-319-45817-5_56

5. Li, Y., Wu, D., Xu, J., et al.: Spatial-aware interest group queries in location-based social networks. In: ACM International Conference on Information and Knowledge Management, pp. 2643–2646. ACM (2012)
6. Wang, X., Donaldson, R., Nell, C., et al.: Recommending groups to users using user-group engagement and time-dependent matrix factorization. In: Thirtieth AAAI Conference on Artificial Intelligence, pp. 1331–1337. AAAI Press (2016)
7. Ye, M., Yin, P., Lee, W.C., et al.: Exploiting geographical influence for collaborative point-of-interest recommendation. In: Proceeding of the International ACM SIGIR Conference on Research and Development in Information Retrieval, SIGIR 2011, Beijing, July 2011, pp. 325–334 (2011)
8. Cheng, C., Yang, H., King, I., et al.: Fused matrix factorization with geographical and social influence in location-based social networks. Aaai **12**, 17–23 (2012)

Learning Intermediary Category Labels for Personal Recommendation

Wenli Yu, Li Li[✉], Jingyuan Wang, Dengbao Wang, Yong Wang,
Zhanbo Yang, and Min Huang

Faculty of Computer and Information Science, Southwest University,
Chongqing 400715, China
m13101332539@163.com, {lily,hmin}@swu.edu.cn,
wangdengbao0620@gmail.com,
{wjykim,wy8654,perphyoung}@email.swu.edu.cn

Abstract. In many recommender systems, category information has been used as additional features for recommender for quite some time, whose application has tended to be understand relationships between products in order to surface recommendations that are relevant to a given context. Nevertheless, the categories as intermediary are labels for not only attributes of products but also preference characteristics of people, is ignored. Here we propose a framework to learn the intermediary role of categories acting as a bridge between users and items. The framework includes two parts. Firstly, we collect the intermediary factors that category labels affect attributes of items and user preferences respectively. Secondly, we integrate the category medium of assemble item attributes and user preferences to online recommender systems to help users discover similar or complementary products. We evaluate our framework on the Amazon product catalog and demonstrate hierarchy categories can capture characteristics of users and items simultaneously.

Keywords: Collaborative Filtering · Category labels · Latent factors

1 Introduction

Recommender Systems (RSs) help users to navigate a huge selection of items with unprecedented opportunities to meet a variety of special needs and user tastes. RSs based Matrix Factorization (MF) suffer from cold-start issues due to the lack of observations to estimate the latent factors of new users and items. Making use of side-signals on top of MF approaches can provide auxiliary information in cold-start settings while still maintaining MFs strengths. Such signals include the content of the items themselves, ranging from the timbre and rhythm for music [1], textual reviews that encode dimensions of opinions [2–4], or social relations [5,6]. Moreover, knowing which items are 'similar', substitutable or complementary, is key to building systems that can understand user's context, recommend alternative items from the same style [7], or generate bundles of items that are compatible [8–10].

© Springer International Publishing AG 2017
L. Chen et al. (Eds.): APWeb-WAIM 2017, Part II, LNCS 10367, pp. 124–132, 2017.
DOI: 10.1007/978-3-319-63564-4_10

There has been some effort to investigate taxonomy-aware recommendation models, including earlier works extending neighborhood-based methods [11] and more recent endeavors to extend MF using either explicit [12] or implicit [13–15] taxonomies. McAuley et al. [16] models substitutable and complementary product graphs with topics associated with category tree and leaf categories. He et al. [17] improves the sparse hierarchical embeddings, where the items from different parts of a category hierarchy may vary considerably.

Nevertheless, the fact that not only items have category information but also users show complementary preferences for categories in hierarchical structure, is ignored. The category labels, which users are interested in, are the latent information offering more personalized recommendations for users in RSs. For items, diverse categories label the different aspects of items contents, which are important latent factors to acquire multiple comprehensive relatedness among items to generate diverse set of recommendations. Users will choose their favorite combination among diversity categories firstly. Despite the important intermediary role of category acting between item and user, when we explore intermediary information of category labels, there are several interesting questions: How could we establish the role of category labels as the medium associating properties of products with preference characteristics of users? How would we utility the intermediary information to capture the variance across diverse categories in order to offer users functionally complementary or visually compatible products.

In the paper, a framework named Bayesian Probabilistic Matrix Factorization with Category (Category-BPMF) is proposed to explore the intermediary role of categories jointing users and items. The framework consists of two parts. Firstly, we introduce MF to factor item-category information and category-user information matrix to extract the intermediary features of category labels unifying the hierarchical structures of users' preferences and classifying principles of items correspondingly, which provide priors for user behaviors and item characteristics simultaneously. Secondly, we incorporate user-category and item-category factors as the priors of user and items inherently features matrices correspondingly, in order to produce recommendations that not only meet our needing, but also collect multiple category factors complementary for user preferences.

The paper is organized as follows: Sect. 2 is the problem formulation, model learning and inferring is detailed. The experiments are presented in Sect. 3. Section 4 is the conclusion.

2   Category BPMF

In this paper, we have three observed matrices: the rating matrix $R \in R^{M \times N}$ the latent information of categories and items $C \in R^{P \times M}$ matrix and users' tastes for categories $RC \in R^{N \times P}$ matrix. And some other notation used in the paper described in Table 1.

Traditional methods such as Matrix Factorization [18] are usually based on a low-rank assumption. They project users and items to a low-rank latent space (D-dimensional) such that the coordinates of each user within the space capture

Table 1. Notations.

Symbol	Description
N, M, P	N users, M items, P categories
I_{ij}, D	Whether user i rated item j, the dimension of the latent factors space
R_{ij}, C_{pj}	Rating of item j by user i, item j labeled by category p
U_i, V_j	User i feature vector, item j feature vector
RC_{ip}	Probability of user i preferences of category p
VP_j, UP_i	Item-category feature for item j(1 × D), user-category feature for user i(1 × D)
CC_p	Category feature for category p(1 × D)
Θ_U, Θ_V	The hyperparameters for user features, the hyperparameters for item features

the preferences towards these D latent dimensions. The affinity $\hat{R}_{ui}$ between user u and item i is then estimated by the inner product of the vector representations of u and i:

$$\hat{R}_{ui} = \langle U_u, V_i \rangle \tag{1}$$

Categories bind together the hierarchical structure of items and hierarchical structures of users' preferences. In order to recommend alternative items that are relevant to a given context from the same style, or generate bundles of items that are compatible, our objective in this section is to detect user needs across-categories and offer substitutable and complementary products fitting the user mostly. Although theoretically latent factors in Eq. 1 are able to uncover any relevant dimensions, one major problem it suffers from is the existence of cold items in the system, about which there are too few associated observations to estimate their latent dimensions. Using category features extracted from category structures can alleviate this problem by providing an auxiliary signal in such situations and generate bundle recommendations of items that are compatible.

2.1 Weight Matrix of Category

Our goal is to learn the category labels furnishing features for items and users so as to recommend substitutable or complementary products for users. However, categories are organized as a hierarchical structure, seen in Fig. 2, and the category of each product is a node in the tree. Here, there is a question: How would we build matrix C and RC in the framework?

We build matrix C using the following scheme: First, each product is represented by a path, or more simply a set of nodes in the category tree. For products belonging to multiple categories, we will take the union of those paths. Second, each node of the hierarchy structure has a number. The largest category, as the root node in the tree, is the number.1 in the serial number and nodes on the same layer are assigned a number one by one from high to low. The $c_{pj} = 1$ in matrix C indicate that the path of item j has the p-th node in category tree, otherwise $c_{pj} = 0$. Third, in order to give prominence to their own sub-category factors, we make the higher layer has lower weight and the leaf nodes of the

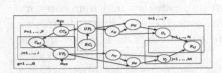

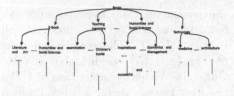

Fig. 1. The graphical model for Category-BPMF

Fig. 2. Part of the product hierarchy for Amazon books products

path have maximum weight. For example, if item j has a category label p at the fourth layer, $c_{pj} = 1 \times 4$ and 4 is the weight for fourth layer.

Matrix RC manifests the users' attentions for multiple categories. RC is structured with the program: due to the hierarchical structure of categories and purchase history of users, users' preferences for category labels can be protruding by items' paths mentioning above in category tree. In spite of there would have high level categories in comment, the lower the levels have greater weight as mentioned above. This method can succeed in finding most appropriate relationships between users and categories.

2.2 Extracting the Category Information

In the section, we will present the details of extracting category factors, which are side-signals of MF approaches providing auxiliary information in cold-start settings. As we can see on the Fig. 1 left, category weights matrix C is factored to explore category feature CC_p and item-category feature VP_j. And our formulation assumes the following model to predict the importance of a category p toward an item i and user preferences for category structure:

$$\hat{C}_{pj} = \langle CC_p, VP_j \rangle, \; and \; , UP = RC^T \cdot CC \qquad (2)$$

where user-category factor UP is the variety of the set of category features CC_p and the matrix RC, which represents the user preferences for categories. The item-category factor VP represents item features connecting directly with categories ($e_{pj} = C_{pj} - \hat{C}_{pj}$). The minimum optimization of CC_p and VP_j can be approached using gradient descent method as below:

$$\min_{VP,CC} \sum \left(C_{pj} - \hat{C}_{pj} \right)^2 + \lambda_1 \|VP\|_F^2 + \lambda_2 \|CC\|_F^2 \qquad (3)$$
$$VP_j = VP_j + \eta(e_{pj} - \lambda_1 VP_j), CC_p = CC_p + \eta(e_{pj} - \lambda_1 CC_p)$$

2.3 BPMF with Category Factors

In the section, we will present the details of the second part of the framework. For the purpose of provide the abundant alternative fitting the heterogeneous needs of users, seen on the Fig. 1 right, the core in this paper is that we derive the parameters and hyperparameters ($\Theta_U = \{\mu_U, \Lambda_U\}$, $\Theta_V = \{\mu_V, \Lambda_V\}$), of features matrices (U, V), automatically from user-category and item-category factors UP, VP. The conditional distribution over the user hyperparameters $\Theta_U = \{\mu_U, \Lambda_U\}$ conditioned on the user feature matrix U is given by the Gaussian-Wishart distribution of UP and U. The conditional distribution over the user feature vector U_i, conditioned on the item features V_j, observed user rating matrix R, and the values of $\Theta_U = \{\mu_U, \Lambda_U\}$ is:

$$p(U_i|R, V, \Theta_U, \alpha) = \prod_{j=1}^{M} \left[R_{ij}|\hat{R}_{ui}, \alpha^{-1} \right]^{I_{ij}} p(U_i|\mu_U, \Lambda_U) \qquad (4)$$

$$p(\mu_U, \Lambda_U|U, UP, \Theta_0) = N\left(\mu_U|\mu_0^*, (\beta_0^* \Lambda_U)^{-1} \right) W(\Lambda_U|W_0^*, \nu_0^*) \qquad (5)$$

where $\beta_0^* = \dfrac{\beta_0 \mu_0 + N\bar{U}}{\beta_0 + N}, \beta_0^* = \beta_0 + N, \nu_0^* = \nu_0 + N, e_U = \left(1 - \gamma^{1-t}\right), e_{UP} = \gamma^{1-t}$

$$[W_0^*]^{-1} = W_0^{-1} + N\bar{S} + \frac{\beta_0 N}{\beta_0 + N}\left(\mu_0 - \bar{U}\right)\left(\mu_0 - \bar{U}\right)^T, \gamma > 1$$

$$\bar{U} = \frac{1}{N}\left(e_U \sum_{n=1}^{N} U_i^{t-1} + e_{UP} \sum_{i=1}^{N} UP_i \right), \bar{S} = \frac{1}{N}\left(e_U \sum_{n=1}^{N}(U_i^{t-1})^2 + e_{UP} \sum_{i=1}^{N} UP_i^2 \right)$$

t equals to t, if this is the No. t iteration sampling in the Gibbs sampling algorithm. e_U and e_{UP} are the coefficients U_i and UP_i. e_U is a increasing function and e_{UP} is a decreasing function with raising of t. The coefficients for U_i and UP_i originate from two reasons:

- First, when $t = 1$ at the first iteration, $e_U = 0$, $e_{UP} = 1$ ensure the conditional distribution over $\Theta_U = \{\mu_U, \Lambda_U\}$ is given only by the Gaussian-Wishart distribution of UP as: $\bar{U} = \frac{1}{N}\sum_{i=1}^{N} UP_i$ and $\bar{S} = \frac{1}{N}\sum_{i=1}^{N} UP_i^2$.
- Second, when t $>= 2$, $0 < e_U$, $e_{UP} < 1$ ensure the conditional distribution over $\Theta_U = \{\mu_U, \Lambda_U\}$ given by the Gaussian-Wishart distribution of comprise of UP and U^{t-1}. The component of U^{t-1} will become big with raising of t. Using U^{t-1} as prior for U^t during iterations is necessary for the convergence of user feature matrix U to deal with overfitting for regularization parameters.

The conditional distributions over the item feature vectors and $\Theta_V = \{\mu_V, \Lambda_V\}$ have exactly the same form. The samples $\left\{U_i^{(t)}, V_j^{(t)}\right\}$ are generated by running Gibbs sampling whose stationary distribution is the posterior distribution over the model parameters and hyperparameters $\{VP, UP, \Theta_V, \Theta_U\}$. The Gibbs sampling algorithm then takes the following form:

1. Initialize model parameters $\left\{U^0 V^0\right\}$
2. For t = 1,...,T
   - Sample the hyperparameters (Eq. 5)
     $\Theta_U^t \sim p\left(\Theta_U|\left(e_U U_i + e_{UP} UP_i\right), \Theta_0\right), \Theta_V^t \sim p\left(\Theta_V|\left(e_V V_j + e_{VP} VP_j\right), \Theta_0\right)$

- For each i = 1, ..., N sample user features and for each j = 1, ..., M sample item features in parallel (Eq. 4):
$U_i^{t+1} \sim p(U_i|R, (e_V V_j + e_{VP} VP_j), \Theta_U^t), V_j^{t+1} \sim p(V_i|R, (e_U U_i + e_{UP} UP_i), \Theta_V^t)$

3   Experimental Results

3.1   Dataset and Evaluation Metric

To fully evaluate the ability of Category-BPMF to handle real-world tasks, we want to experiment on the largest dataset available. In this section, we adopt the dataset from Amazon[1] recently, which includes review texts and time stamps spanning from May 1996 to July 2014 and each top-level category of products on Amazon.com has been constructed as an independent dataset. Statistics are shown in Table 2. Across the entire dataset, such relationships are noisy, sparse, and not always meaningful. To address issues of noise and sparsity to some extent, its sensible to focus on the relationships within the scope of a particular high-level category. Two popular metrics, the Root Mean Square Error (RMSE) and the Mean Absolute Error (MAE), are chosen to evaluate the prediction performance.

Table 2. Dataset statistics for a selection of categories on Amazon.

Dataset	Subcategories	Items	Users	Rating	Viewed items
Movies	44	208 K	2.11 M	6.17 M	1.42 M
Electronics	56	2498 K	4.25 M	11.4 M	7.3 M
Books	187	2.73 M	8.2 M	25.9 M	12.46 M
Women's clothing	116	838 K	1.82 M	14.5 M	6.35 M

3.2   Baselines

- MF [18]: It is a low-rank approximation based on minimizing the sum-of-squared-errors and does not employ other information for users and items.
- Category Tree (CT): This method computes a matrix of co-occurrences between subcategories from the training data. Then a pair (x, y) is predicted to be positive if the subcategory of y is one of the top 50% most commonly connected subcategories to the subcategory of x.
- Bayesian Personalized Ranking with Category Tree (BPR-C): Introduced by [19], BPR is the state-of-the-art method for personalized ranking. BPR-C makes use of category tree to extend BPR by associating a bias term to each finegrained category on the hierarchy.

---

[1] https://www.amazon.com/.

- Item-to-Item Collaborative Filtering (CF) [8]: This baseline identifying items that had been browsed or purchased by similar sets of users, follows the same procedure, actual browsing or purchasing data we consider sets of users who have reviewed each item.

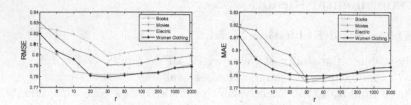

Fig. 3. Impact of parameter γ

3.3   Experimental Results

Impact of Parameter Analysis the Base of Exponential Function γ: γ is base of exponential function γ^{1-t} regulating proportion of UP_i and U_i, VP_j and V_j. Figure 3 shows the impacts of γ on MAE/RMSE with D = 20 for four datasets. As γ increases, the two error metrics decrease, but when γ surpasses a value 30, all the metrics increase. We observe that error metrics have the minimum for about $\gamma = 30$ simultaneously. So $\gamma = 30$ can control the strength of Category-BPMF well.

Analysis of Recommendation Performance: All the experiments are conducted using 20 latent factor dimensionality, $\gamma = 30$ and 50 iterations. We can make a few observations to explain and understand our finding as follows (Table 3):

Table 3. RMSE of the predictions on all items or cold start on four datasets.

Dataset	Setting	CT	MF	CF	BPR-C	Category BPMF	% impr. C-BPMF vs CF	% impr. C-BPMF vs BPR-C
Movies	All items	0.917	0.842	0.854	0.801	**0.771**	8.54	3.74
	Cold start	0.945	0.856	0.868	0.812	**0.774**	8.52	4.67
Electronics	All items	0.912	0.839	0.833	0.798	**0.768**	7.8	3.71
	Cold start	0.935	0.855	0.857	0.805	**0.764**	10.82	4.67
Women's clothes	All items	0.908	0.835	0.821	0.792	**0.776**	5.7	2.02
	Cold start	0.931	0.840	0.842	0.810	**0.778**	3.8	3.95
Books	All items	0.906	0.832	0.844	0.795	**0.775**	8.17	2.51
	Cold start	0.941	0.848	0.838	0.812	**0.774**	7.6	4.67

(1) CT is particularly inaccurate for our task. We also observed relatively high training errors with this method for most experiments. This confirms our conjecture that raw similarity is inappropriate for our task, and that in order to learn the relationships across categories, some sort of expressive transforms are needed for manipulating the raw features. (2) MF performs considerably worse than other methods. This reveals that the predictive information used by the other models goes beyond the features of the products, i.e., that the category based models are learning relationships between finer grained attributes. (3) BPR-C does significantly outperform CT, presumably because the category signals are already encoded by those features. This suggests that improving BPR requires more creative ways to leverage such signals. (4) Category-PMF, BPR-C compared with MF and CF indicate that category information is very important information and can not be ignored. (5) Category-BPMF consistently outperforms baseline methods. These results suggest that hierarchical categories of users and items contain complementary information and capturing them simultaneously can further improve the recommendation performance.

4   Conclusion

In this paper, we exploit the hierarchical category structures of items and users for recommendation when they are not explicitly available and propose a novel recommendation framework Category-BPMF, which captures the hierarchical structures of items and users into a coherent model. Experimental results on real-world dataset demonstrates the importance of the implicit hierarchical categories of both items and those of users in the recommendation systems.

Acknowledgments. This work is supported by NSFC (No. 61170192) and National Science and Technology Support Program (No. 2015BAK41B01).

References

1. Wang, X., Wang, Y., Hsu, D., Wang, Y.: Exploration in interactive personalized music recommendation: a reinforcement learning approach. TMCCA **11**, 7 (2014)
2. Mcauley, J., Leskovec, J.: Hidden factors and hidden topics: understanding rating dimensions with review text. In: RecSys, pp. 165–172 (2013)
3. Zhang, Y.F., Lai, G., Zhang, M.: Explicit factor models for explainable recommendation based on phrase-level sentiment analysis. ACM (2014)
4. Ling, G., Lyu, R., King, I.: Ratings meet reviews, a combined approach to recommend, pp. 105–112 (2014)
5. Zhao, T., Mcauley, J., King, I.: Leveraging social connections to improve personalized ranking for collaborative filtering, pp. 261–270 (2014)
6. Pan, W., Chen, L.: GBPR: group preference based Bayesian personalized ranking for one-class collaborative filtering. In: IJCAI, pp. 2691–2697 (2013)
7. Hu, D., Hall, R., Attenberg, J.: Style in the long tail: discovering unique interests with latent variable models in large scale social e-commerce. In: SIGKDD, pp. 1640–1649. ACM (2014)

8. Linden, G., Smith, B., York, J.: Amazon.com recommendations: item-to-item collaborative filtering. IEEE Internet Comput. **7**(1), 76–80 (2003)
9. McAuley, J., Targett, C., Shi, Q.: Image-based recommendations on styles and substitutes. In: SIGIR, pp. 43–52. ACM (2015)
10. Zheng, J., Wu, X., Niu, J.: Substitutes or complements: another step forward in recommendations. In: ACM, pp. 139–146. ACM (2009)
11. Ziegler, C., Lausen, G., Schmidt-Thieme, L.: Taxonomy-driven computation of product recommendations. In: ACM, pp. 406–415. ACM (2004)
12. Mnih, A.: Taxonomy-informed latent factor models for implicit feedback. In: KDD Cup, pp. 169–181 (2012)
13. Zhang, Y., Ahmed, A., Josifovski, V.: Taxonomy discovery for personalized recommendation. In: ACM, pp. 243–252. ACM (2014)
14. Mnih, A., Teh, Y.: Learning label trees for probabilistic modelling of implicit feedback. In: ANPIS, pp. 2816–2824 (2012)
15. Wang, S., Tang, J., Wang, Y., Liu, H.: Exploring implicit hierarchical structures for recommender systems. In: IJCAI, pp. 1813–1819 (2015)
16. McAuley, J.J., Pandey, R., Leskovec, J.: Inferring networks of substitutable and complementary products. In: KDD (2015)
17. He, R., Packer, C., McAuley, J.: Learning compatibility across categories for heterogeneous item recommendation. In: ICDM (2016)
18. Salakhutdinov, R., Mnih, A.: Probabilistic matrix factorization. In: Nips, vol. 1, pp. 1–2 (2007)
19. Rendle, S., Freudenthaler, C., Gantner, Z.: BPR: Bayesian personalized ranking from implicit feedback. In: CUAI, pp. 452–461 (2009)

Skyline-Based Recommendation Considering User Preferences

Shuhei Kishida, Seiji Ueda, Atsushi Keyaki(✉), and Jun Miyazaki

Tokyo Institute of Technology, Tokyo, Japan
{kishida,ueda,keyaki}@lsc.cs.titech.ac.jp, miyazaki@cs.titech.ac.jp

Abstract. In this paper, we propose a skyline-based recommendation and ranking function. We suppose that some recommender systems, such as hotel recommender systems, are based not only on user preferences but also cost performance. For these kinds of applications, We first extract items with good cost performance and then identify items that users prefer, which reduce the computational cost of the online process. Based on the results of our preliminary experiments, we propose user feedback-based scoring and density-aware scoring methods where items that are highly similar to a user's latent requirements are recommended and attribute values in a dense area are quantized into a single value. The result of the experiments suggest that the density-aware scoring provides equal to or greater accuracy than the basic scoring.

Keywords: Skyline operator · Recommender system · User feedback · Density

1 Introduction

A vast amount of data has been generated and is now stored on the Web. It is expected that much more data will be generated in the future. In this situation, a recommender system is required because finding information a user needs from such big data is laborious. Hence, many studies on recommender systems have been conducted. In particular, content filtering [8] and collaborative filtering [10,12] has emerged as one of the most beneficial techniques. As the number of items becomes large, the cost of calculating the similarity between items becomes very expensive.

We suppose that some kinds of applications of recommender systems are based not only on user preference but also cost performance. For example, suppose a user would like to be recommended some useful hotels for his/her business trip. Some users may want to give priority to the price of the hotels (group A in Fig. 1), while others may prioritize the distance from the hotels to the nearest station (group B in Fig. 1). Others may prefer hotels with a balance between price and distance (group C in Fig. 1). This implies that no user prefers to stay in hotels that are both expensive and far from the nearest station (group D in Fig. 1). As a result, preferable hotels, namely, the items of groups A, B, and C, tend to be located in

© Springer International Publishing AG 2017
L. Chen et al. (Eds.): APWeb-WAIM 2017, Part II, LNCS 10367, pp. 133–141, 2017.
DOI: 10.1007/978-3-319-63564-4_11

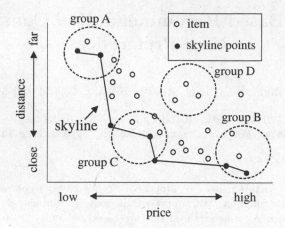

Fig. 1. Skyline operator and skyline points

the lower left of the figure. We regard these items as hotels with good cost performance. Considering the current user's behavior, we first extract only items with good cost performance, and then identify items that the user prefers. An advantage of this approach is that the computational cost of the online process can be reduced.

The hotels at the lowest and furthest left in Fig. 1 (indicated by black dots) are hotels with the best cost performance. The *skyline operator* is an algorithm that efficiently extracts these hotels [1]. These items are called *skyline points*, and formally defined as hotels that are not dominated by any other hotel in all dimensions. The line connecting the skyline points is called the *skyline*. Features of an item such as price and distance are called *attributes*.

The skyline operator helps to efficiently extract the best cost performance items. Thus, a skyline-based recommender system is expected to be useful in some specific scenarios. Problems on achieving a useful skyline-based recommender system are as follows:

- The original skyline operator itself does not support a ranking function of extracted items, although a ranking function is a highly desirable feature of a recommender system.
- Some researches expanded the skyline operator and embedded the scoring function [2,6,7,9,11,14]. However, some of them suppose a situation where user's attribute weights are given and the others does not take account of a user's preference in scoring function. Moreover, to our knowledge, no research conducted a user study to evaluate effectiveness of the scoring function.

Our preliminary experiments for confirming the potential of a skyline-based recommender system revealed that the priority that users give to their preferences does not always agree with the actual importance of the preferences in decision making. To resolve this issue, we propose the user feedback-based scoring and density-aware scoring methods to the skyline-based recommender system.

2 Related Work

Researches on a ranking/scoring function of skyline operation tend to suppose *fuzzy query* [3–5]. Fuzzy query is a query where a user expresses his/her latent requirement in not concrete value but fuzzy expression, e.g., "inexpensive" for price and "close" for distance. Thus, a user's preference in this study is related to fuzzy query, although our approach suppose that a user just selects attributes and not implicitly express largeness of an attribute value.

An initial research [9] of embedding a scoring function to the skyline operator requires that the scoring function is predefined. Latter research [11] exploits context information to choose an appropriate scoring function. However, an effective scoring function for skyline-based recommender system is not obvious. Some researches [7,14] proposed a skyline operation with a scoring function. These researches rank items according to the number of dominating items. Other research [2] ranks items based on occurrence probability of an event. These ranking function is too naive to satisfy a user's requirements because these reflect only distribution of items and does not reflect a user's preference. Then, flexible fuzzy skyline [6] argues the necessity of a ranking function according to a user's preference, however, a concrete ranking function is not proposed.

3 Basic Scoring Function for Skyline Points

In this study we focus on ranking of only skyline points, although items close to skyline are expected to be useful. This is because the first priority of this research is to confirm the potential of effectiveness of a skyline-based recommender system. Items, i.e., hotel data in this paper, are normalized in advance. $base(h, a)$, a base score of a hotel h in terms of an attribute a, is calculated as follows:

$$base(h, a) = \frac{original(h, a) - worst_a}{best_a - worst_a} \qquad (1)$$

where $original(h, a)$ is a original value of h in terms of a, $worst_a$ is the worst value of a of all items, and $best_a$ is the best value of a in all items. In this study we do not take account of categorical data because categorical data is used as a filter rather than used in scoring items.

We introduce weighted linear sum into scoring skyline points because a base score of each attribute and importance of the attribute in terms of a user's preference in this study can be regarded as each score and a weight of the score in information retrieval, respectively. The basic score $S_{basic}(h)$ of hotel h is calculated as follows:

$$S_{basic}(h) = \sum_{a \in A} weight(a) \cdot base(h, a) \qquad (2)$$

where A is a set of attributes used for the skyline operator, a is an attribute in A, $weight(a)$ is a weight of a, and $base(h, a)$ is a base score of h in terms of a.

4 Improvement of the Skyline-Based Recommendation

4.1 Preliminary Experiment

We conducted preliminary experiments which confirmed (1) weighted linear sum scoring function is effective, and (2) a skyline-based recommendation does not decrease accuracy by extracting only hotels with good performance. We should note that accuracy improves when the attribute weights are set equally with the exception of price. Specifically, the attribute weight of price should be set large for accurate recommendation. One very interesting result is that half of the users gave a different order of importance to the attributes than they did at the beginning. It is not always effective for a user to set the attribute weights by him/herself. It also indicated that stated user preferences do not always agree with actual importance in decision making.

4.2 User Feedback-Based Scoring

The preliminary experiments suggest that users cannot always assign proper weights to attributes. Thus, we propose a method that leverages user feedback to interpret a user's latent requirements. Items that are highly similar to a user's latent requirements are recommended. This approach is a kind of content filtering because we calculate similarities between items.

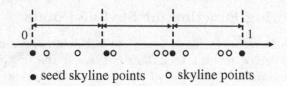

● seed skyline points   ○ skyline points

Fig. 2. Seed hotels for user feedback-based scoring

Figure 2 draws arranged skyline points according to the base score of a certain attribute. First, we split the space into k subspaces (k is 3 in this case). We then extract $k+1$ skyline points located at the nearest to each border of the subspaces. We define these skyline points as seed skyline points (indicated by black dots in Fig. 2). Note that seed skyline points are extracted for every attribute used by the skyline operator.

Next, we ask a user to choose the most preferable skyline points among the seed skyline points. We assume that the chosen seed skyline points represent the user's latent requirement. Hence, similar skyline points to the chosen seed skyline points are reasonable to be recommended. Therefore, a feedback score $S_{Feedback}(h)$ of a hotel h is calculated by similarities between the chosen seed skyline points and other skyline points as follows:

$$S_{Feedback}(h) = \sum_{c \in C} sim(h, c) \qquad (3)$$

where C is a set of chosen seed skyline points, c is an chosen seed skyline point in C, $sim(h, c)$ is a similarity between h and c.

We use cosine similarity as a similarity measure in this paper; however, any similarity measure can be applied. There are also some options for calculating similarity because each skyline point has multiple attributes. We vary the number of attributes to be used when calculating a similarity. Concretely, we use three variations in the experiment: only the first attribute, three attributes used by the skyline operator, and all eight available attributes for each item, as will be introduced in Sect. 5.1. Thus, a similarity measure is calculated as follows:

$$sim(h, c) = \frac{\sum_{a' \in A'} v_{h,a'} v_{c,a'}}{\sqrt{\sum_{a' \in A'} v_{h,a'}^2} \sqrt{\sum_{a' \in A'} v_{c,a'}^2}} \tag{4}$$

where A' is a set of attributes to be chosen for calculating similarity, a' is an attribute in A', $v_{h,a'}$ is a base score of h in terms of a', and $v_{c,a'}$ is a base score of c in terms of a'. In this study, we set k to 3 because we assume that a three-way split is enough to capture a user's latent requirements[1]. We also decided to use only one item for feedback to avoid a user's labor.

4.3 Density-Aware Scoring

Judging from the interview results of the preliminary experiments, it seems that dense areas occur in some attributes. When evaluating skyline points in these areas, it is expected that users will give more importance to other attributes because there is little difference that can be used to determine candidates along an attribute with many skyline points with a similar score. In other words, there is a gap between the scoring method of base score and decision making because even trivial differences of the base scores among the skyline points are considered. To address this issue, we quantize the base scores of skyline points in dense areas.

We investigated the distributions of the skyline points along each attribute of the skyline operators for many combinations of attributes using kernel density estimation [13]. As a result, the density curves of the distributions have a largely similar shape, that is, a gentle curve. Figure 3 illustrates the kernel density distribution for location. In contrast, price is an exception to this trend. Its shape is sharp, as shown in Fig. 4.

It seems that the features of the distributions are appropriately captured using kernel density estimation, as Figs. 3 and 4 show. Therefore, we use the kernel density distribution to identify dense areas. An example of quantization for price is depicted in Fig. 5. A dense area is defined as a range where the density exceeds a threshold value. The base scores of skyline points located within the dense area are quantized to a single value, which is the average value of these

---

[1] As k increases, the system may be able to express the user's latent requirements more precisely. However, a user's labor increases when the number of seed skyline points increases.

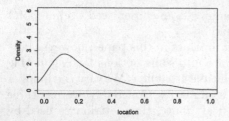
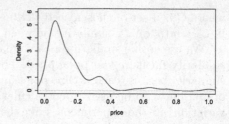

Fig. 3. Location kernel density distribution   **Fig. 4.** Price kernel density distribution

base scores. Consequently, a density score $S_{Density}(h)$ of a hotel h is calculated as follows:

$$S_{Density}(h) = \sum_{a \in A} weight(a) \cdot density(h, a) \tag{5}$$

$$density(h, a) = \begin{cases} base(h, a) & (base(h, a) < \tau) \\ quant(h, a) & (otherwise) \end{cases} \tag{6}$$

where A is a set of attributes, a is an attribute in A, $weight(a)$ is a weight of a, τ is a threshold value, and $quant(h, a)$ is a quantized value of h in terms of a.

The difference between the density-aware scoring and the basic scoring is that the density-aware scoring quantizes the base scores in the dense areas. Accordingly, the subsequent processes for density-aware scores are the same as those for the basic ones. Note that the best threshold value will be examined in the next section.

5   Experimental Evaluation

5.1   Data Set

In the experiments, we used the Rakuten Travel Data Set, which contains Japanese hotel data. We used 692 hotels located in Tokyo. Each hotel has eights

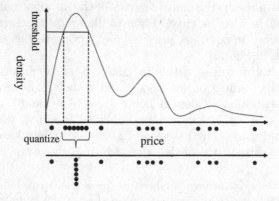

Fig. 5. Quantization of base scores in dense areas

attributes, that is, price, distance, and six user review ratings (service, location, facility, room, bath/spa, and food). The users review ratings are between 1 to 5; a higher value indicates better rating.

5.2 Experimental Setting

The number of the participants in these experiments was 30, and they are graduate students. We operated two experiments: one did *not* include the price attribute for the skyline operator and the other included the price attribute. This is because the tendency between them was quite different, as the results of the preliminary experiments showed. Evaluation measures are nDCG at 30. The users choose three attributes for the skyline operator and the weight combinations of attribute are shown in Table 1.

We examined five methods. Two of them were proposed to further improvement of accuracy in Sect. 4, i.e., a user feedback-based scoring (*Feedback*) and a density-aware scoring (*Density*). The other methods are *Basic*, *Linear*, and *Sort* where hotels are sorted by the base score of the first axis as a naive method.

Table 1. Attribute weights

	First attribute	Second attribute	Third attribute
p1	0.33	0.33	0.33
p2	0.50	0.50	0.0
p3	0.90	0.05	0.05

Table 2. Results without price

	$nDCG_{30}$
p1-Basic	0.990
p1-Linear	0.990
p1-Density-3	0.988
p2-Density-3	0.962
p2-Basic	0.960
p2-Linear	0.956
Feedback-8	0.945
p3-Density-2	0.940
p3-Basic	0.932
p3-Linear	0.926
Feedback-3	0.912
Feedback-2	0.892
Sort	0.851

Table 3. Results including price

	$nDCG_{30}$
p3-Density-5	0.918
Feedback-8	0.916
Feedback-3	0.909
p1-Density-5	0.901
p2-Density-5	0.895
p2-Basic	0.888
p2-Linear	0.878
p3-Basic	0.877
p3-Linear	0.876
p1-Basic	0.866
p1-Linear	0.863
Feedback-1	0.848
Sort	0.368

5.3 Experimental Results

Experiment Without Price. Table 2 shows the result of the experiment without price. The number of attributes used in the similarity calculation for *Feedback* and the threshold value for *Density* are added to the end of the method labels. Moreover, we only include the most accurate version of *Density* per weight combination. For example, p1-Density-1 is omitted because it is less accurate than p1-Density-2. It is effective to count every attribute in the scoring function when the attributes for the skyline operator do not include price.

In contrast, *Feedback* is less effective than all other methods except *Sort*. The accuracy of *Density* is slightly inferior to that of *Linear* and *Basic*. As a result, *Feedback* and *Density* do not improve accuracy in the experiment without price.

Experiment with Price. *Density* is the most accurate method in the experiment that includes price, as illustrated in Table 3. We assume that this is achieved because the attribute of price has a wider range of dense areas and Density works well in this situation. Further, *Feedback* improved accuracy compared with *Basic* and *Linear* when eight or three attributes were used.

Discussion. According to the above results, *Density* attains accurate and stable recommendations. In this method, weights for attribute do not have to be tuned and can be just assigned an equal value when there is no attribute that overwhelms the other attributes in decision making. In contrast, it is effective to assign a high weight to an attribute that overwhelms the others.

6 Conclusion

In this paper, we discussed a skyline-based recommendation with a ranking function taking account of a user's preference and conducted a user study. We proposed the *Feedback* and *Density* methods for improving existing ones. Experimental evaluations showed that *Density* recommends items accurately with stability. With *Density*, automatically tuning the threshold value is a part of our future work. Applying more advanced score fusion methods instead of weighted linear sum is also our future work.

Acknowledgements. This work was partly supported by JSPS KAKENHI Grant Numbers 15H02701, 15K20990, 16H02908, 26280115, and 25240014. The Rakuten Travel data set was provided according to the contract between NII and Rakuten, Inc.

References

1. Börzsönyi, S., Kossmann, D., Stocker, K.: The skyline operator. In: Proceeding of the 17th ICDE, pp. 421–430 (2001)
2. Bosc, P., Hadjali, A., Pivert, O.: On possibilistic skyline queries. In: Christiansen, H., Tré, G., Yazici, A., Zadrozny, S., Andreasen, T., Larsen, H.L. (eds.) FQAS 2011. LNCS (LNAI), vol. 7022, pp. 412–423. Springer, Heidelberg (2011). doi:10.1007/978-3-642-24764-4_36
3. Bosc, P., Pivert, O.: Fuzzy queries and relational databases. In: Proceeding of SAC, pp. 170–174 (1994)
4. Bosc, P., Pivert, O.: SQLf: a relational database language for fuzzy querying. IEEE Trans. Fuzzy Syst. **3**, 1–17 (1995)
5. Bosc, P., Pivert, O., Mokhtari, A.: Top-k queries with contextual fuzzy preferences. In: Bhowmick, S.S., Küng, J., Wagner, R. (eds.) DEXA 2009. LNCS, vol. 5690, pp. 847–854. Springer, Heidelberg (2009). doi:10.1007/978-3-642-03573-9_72
6. Hadjali, A., Pivert, O., Prade, H.: On different types of fuzzy skylines. In: Kryszkiewicz, M., Rybinski, H., Skowron, A., Raś, Z.W. (eds.) ISMIS 2011. LNCS (LNAI), vol. 6804, pp. 581–591. Springer, Heidelberg (2011). doi:10.1007/978-3-642-21916-0_62
7. Lee, H.-H., Teng, W.-G.: Incorporating multi-criteria ratings in recommendation systems. In: Proceeding of the IEEE IRI, pp. 273–278 (2007)
8. Lops, P., de Gemmis, M., Semeraro, G.: Systems, content-based recommender: state of the art and trends. In: Ricci, F., Rokach, L., Shapira, B., Kantor, P.B. (eds.) Recommender Systems Handbook, pp. 73–105. Springer, New York (2011). doi:10.1007/978-0-387-85820-3_3
9. Papadias, D., Tao, Y., Fu, G., Seeger, B.: An optimal and progressive algorithm for skyline queries. In: Proceeding of SIGMOD, pp. 467–478 (2003)
10. Resnick, P., Iacovou, N., Suchak, M., Bergstrom, P., Riedl, J.: GroupLens: an open architecture for collaborative filtering of netnews. In: Proceeding of CSCW, pp. 175–184 (1994)
11. Sacharidis, D., Arvanitis, A., Sellis, T.: Probabilistic contextual skylines. In: Proceeding of the 26th ICDE, pp. 273–284 (2010)
12. Sarwar, B., Karypis, G., Konstan, J., Riedl, J.: Item-based collaborative filtering recommendation algorithms. In: Proceeding of the 10th WWW, pp. 285–295 (2001)
13. Silverman, W.B.: Density Estimation for Statistics and Data Analysis. CRC Press, London (1986)
14. Yiu, M.L., Mamoulis, N.: Efficient processing of top-k dominating queries on multi-dimensional data. In: Proceeding of the 33rd VLDB, pp. 483–494 (2007)

Improving Topic Diversity in Recommendation Lists: Marginally or Proportionally?

Xiaolu Xing[✉], Chaofeng Sha, and Junyu Niu

Shanghai Key Laboratory of Intelligent Information Processing,
School of Computer Science, Fudan University, Shanghai, China
{14212010021,cfsha,jyniu}@fudan.edu.cn

Abstract. Diversifying the recommendation lists in recommendation systems could potentially satisfy user's needs. Most diversification techniques are designed to recommend the top-k relevant and diverse items, which take the coverage of the user preferences into account. The relevance scores are usually estimated by methods such as latent matrix factorization. While in this paper, we model the users' interests with the topic distributions on the rated items. And then we investigate how to improve the topic diversification within the recommendation lists. We first estimate the topic distributions of users and items through training Latent Dirichlet Allocation (LDA) on the rating set. After that we propose two topic diversification methods based on submodular function maximization and proportionality respectively. Experimental results on MovieLens and FilmTrust datasets demonstrate that our approach outperforms state-of-the-art techniques in terms of distributional diversity.

Keywords: Recommender system · Diversity · LDA

1 Introduction

Most traditional recommendation systems usually recommend items with high predicted scores to users by using some standard recommendation algorithms. However, many recent studies have shown theoretically and empirically that it is more beneficial to take the diversity within the recommendation lists into account as well [1,7,16], particularly when different users with diverse and ambiguous interests. Recommending a diverse set of the most relevant items is more likely to satisfy all potential needs of a given user. So finding a top-k itemset for a given user is of crucial importance to recommendation systems.

In order to resolve the user interest ambiguity and avoid the redundancy in the recommendation lists, a number of diversification frameworks [1,7,9,12,16] have been proposed, which optimize the top-k items collectively in terms of both relevance and diversity. Most of these former works, e.g., Entropy Regularized [16], define some objective functions to balance between maximizing relevance and maximizing diversity. These methods estimate the relevance score through methods such as Latent Matrix Factorization [13]. However, we should be aware

© Springer International Publishing AG 2017
L. Chen et al. (Eds.): APWeb-WAIM 2017, Part II, LNCS 10367, pp. 142–150, 2017.
DOI: 10.1007/978-3-319-63564-4_12

that most of the fundamental methods like LMF are proposed to achieve predictive ratings, which are not so suitable for getting the top-k lists that meet all of the potential needs of users.

Inspired by search result diversification methods from information retrieval field [3,14], we also propose a two-stage diversification framework. In the first stage, we estimate the topic distributions of users and items through training LDA on rating set as we mentioned above. And in the second stage, we employ two methods to "cover" the topic distributions of users and items. In the second method, we use the election-based approach proposed in [14] to diversify recommendation lists proportionally, which has advantages on exploiting the two topic distributions.

The contributions of the paper can be summarized as follows: 1. We model each user's rating set as a document and train LDA on the "document" set to get the topic distributions, which are incorporated into the proposed objective functions. 2. We propose two diversification methods based on submodular function maximization and proportionality respectively. 3. We conduct extensive experiments on MovieLens dataset and FilmTrust dataset, which include performance comparisons with state-of-the-art methods. The experimental results demonstrate the effectiveness of our proposed solutions.

2   Related Work

Our work is related to the search result diversification in the information retrieval field and item diversification in the recommendation systems, and also related to the submodular function maximization. We give a brief review of those works in this section.

In the diversification based on item similarity, There is a trade-off between the utility and diversity of items. In the most popular framework MMR [4], the item i is greedily or marginally added into the result set S through maximize $(1 - \lambda)w_i - \lambda \max_{j \in S} sim(i, j)$. The conditional differential entropy $h(r_S | r_\Omega, V)$ is employed in [16] as a diversity regularizer to the utility function $\sum_{i \in S} p_u^T q_i$. The adaptive attribute-based diversification model proposed in [10] follows the MMR framework. When computing similarity, they consider the diversification attribute of each user with respect to different item attributes. The propensity towards diversity is measured by entropy for each user. In [15], the authors propose a general framework which tradeoffs the utility of the items, the coverage of the user preferences and the diversity between items.

In [8], the diversity of recommendation lists is improved by maximizing the explicit topic attribute differences among items. In our first method, we follow the framework of IA-Select [3] which maximizes a probabilistic coverage function defined through the distribution of explicit topics. A probabilistic framework xQuAD (eXplicit Query Aspect Diversification) is proposed in [18] for Web search result diversification, which explicitly accounts for the various aspects of an under-specified query. The ranking algorithm is designed to satisfy the uncovered aspect by the selected documents. The explicit relevance model is also studied in [17]. In the diversity-weighted utility maximization (DUM) algorithm proposed in [9], the item is greedily added into to the recommendation list which try to maximize the objective function $\sum_{i=1}^{k}[f(A_i) - f(A_{i-1})]w_i$.

Our second diversification method follows the election-based approach proposed in [14] which diversifies the search result by proportionality. For each position in the recommendation list, We iteratively determine the topic that best maintains the overall proportionality, and then select the best item on this topic for this position. Therefore, proportional topics in the recommendation list could satisfy the preferences of the user.

3   Proposed Methods

In this section, we introduce a two-stage diversification framework. In the first stage, we estimate the topic distributions through training LDA on the transformed rating set. Then we propose two diversification methods based on the inferred distributions.

3.1   Problem Statement

Before delving into our methods, we introduce notations used in the paper. The set of users and items are denoted as U and V. The rating set is denoted as R. The set of topics we infer from LDA is denoted as T. The notation u, i, and t denote an user, an item, and a topic respectively. The diversification problem in this paper is to recommend an itemset S with size constraint $|S| = k$ to user u with diverse topics.

3.2   Topic Distribution Estimation

To improve the topic diversity within the recommendation lists, we should estimate the topic distributions of users and items. In our framework, we retort to Latent Dirichlet Allocation. We treat each user as a document and the items rated by users as words in documents. In our method, we transform the rating values of items to the number of occurrences of the word, i.e., the tuple $t_{u,i} = (u, i, r_{u,i})$ will be transformed into $\hat{t}_{u,i} = (u, i, i, i...)$ where the number of occurrence of i is $r_{u,i}$.

After the transformation, we use Latent Dirichlet Allocation [6] to estimate two distributions: user-topic distributions and item-topic distributions, which are denoted as θ and φ. Let θ_u denotes the topical distribution of the user u and $\theta_{u,t}$ denotes the probability of that the user u chooses the topic t. Similarly, let φ_t denotes the distribution of topic t over the set of items and $\varphi_{t,i}$ denotes the probability of that the item i appears in the topic t. We employ symmetric Dirichlet priors $Dir(\alpha)$ and $Dir(\beta)$ with the hyper-parameters of α and β on θ and φ respectively. Then the user preferences can be computed by using θ and φ as follows:

$$P_{u,i} = \sum_{t \in T} \theta_{u,t} \varphi_{t,i}, \tag{1}$$

Algorithm 1. `LDA-Greed`

Input: The itemset V and an integer k
Output: Subset of items $S \subseteq V$ with $|S| = k$

1: $S \leftarrow \{\arg\max_{i \in V} P_{u,i} + \lambda \sum_{t \in T} \theta_{u,t} \phi_{i,t}\}$;
2: **while** $|S| < k$ **do**
3:      $i^* \leftarrow \arg\max_{i \in V-S} P_{u,i} + \lambda \sum_{t \in T} \theta_{u,t} \phi_{i,t} \prod_{i' \in S}(1 - \phi_{i',t})$
4:      $S \leftarrow S \cup \{i^*\}$
5:      $V \leftarrow V - \{i^*\}$
6: **end while**
7: **return** S

We treat $P_{u,i}$ as the propensity of u selecting i. In other words, it is treated as user u's preferences.

Then we can recommend the top-k unrated items for u by reordering u's preferences $[P_{u,i}$ for $i \in V]$. Actually this is one of baseline method in our experiment. For ease of use, we also introduce the item-topic distribution ϕ as a reverse of φ, then $\phi_{i,t}$ means the probability that item i covers topic t.

3.3 Diversify Marginally

Inspired by IA-Select [3], we recommend an itemset $S(|S| = k)$ to user u which maximizes the following objective function:

$$f(S) = R(S) + \lambda \sum_{t \in T} \theta_{u,t} \left(1 - \prod_{i \in S}(1 - \phi_{i,t})\right), \tag{2}$$

where $R(S) = \sum_{i \in S} P_{u,i}$.

It is easy to verify that this problem is NP-hard (reduction from the Set Cover problem). Fortunately, thanks to the well-known result of [2], we can resort to the submodularity of the probabilistic coverage function. The submodularity of the objective function provides a theoretical approximation guarantee of factor $1 - \frac{1}{e}$ for the greedy search algorithm. Formally,

Theorem 1. Let f be a submodular, monotone set function and $f(\emptyset) = 0$. Then, the greedy search algorithm finds a set S $(|S| = k)$ such that $f(S) \geq \left(1 - \frac{1}{e}\right) \max_{S':|S'| \leq k} f(S')$.

Theorem 2. The function defined in (2) is submodular and monotone.

Now we present the greedy algorithm outline in Algorithm 1, where we diversify the recommendation lists marginally.

3.4 Diversify Proportionally

In this section, we present a method to diversify items proportionally, inspired by the election-based approach which aims to diversify search result [14]. The main

---

Algorithm 2. LDA-PE

Input: The itemset V and an integer k
Output: Subset of items $S \subseteq V$ with $|S| = k$

1: $s_t \leftarrow 0, \forall t \in T$
2: **while** $|S| < k$ **do**
3:     **for** $t \in T$ **do**
4:         $quotient[t] = \frac{\theta_{u,t}}{2s_t+1}$,
5:     **end for**
6:     $t^* \leftarrow argmax_{t \in T} \ quotient[t]$.
7:     $i^* \leftarrow argmax_{i \in V-S} \quad \alpha \cdot quotient[t^*] \cdot \phi_{i,t^*} + (1-\alpha) \cdot \sum_{t \neq t^*} quotient[t] \cdot \phi_{i,t}$.
8:     $S \leftarrow S \cup \{i^*\}$
9:     $V \leftarrow V - \{i^*\}$
10:     **for** $t \in T$ **do**
11:         $s_t \leftarrow s_t + \frac{\phi_{i^*,t}}{\sum_{t' \in T} \phi_{i^*,t'}}$
12:     **end for**
13: **end while**
14: **return** S

---

idea behind the method is to find a representative itemset S from the unrating items, which has the same topic proportionality to the topic distribution inferred for user u, i.e., $\{s_t\}_{t \in T} \propto \theta_u$, where $\{s_t\}_{t \in T}$ is the topic distribution of items contained in S, and s_t is calculated as follows: $s_t = \sum_{i \in S} \frac{\phi_{i,t}}{\sum_{t' \in T} \phi_{i,t'}}$.

To achieve this proportionality, when we add an item into S, we first choose a topic which has the most lack of proportionality in S. We use $quotient[t]$ to denote the proportional lack of topic t in S which is calculated as $quotient[t] = \frac{\theta_{u,t}}{2s_t+1}$.

The topic t^* is chosen as follows: $t^* \leftarrow argmax_{t \in T} \ quotient[t]$. Then we choose the optimal item with respect to topic t^*. The direct way is to select the item i with the highest ϕ_{i,t^*}. However, adding items into S not only affects the concerned topic t^*, but also affects several other topics covered by these items. So we introduce the parameter α to balance these effects: $i^* \leftarrow argmax_{i \in V-S} \ \alpha \cdot quotient[t^*] \cdot \phi_{i,t^*} + (1-\alpha) \cdot \sum_{t \neq t^*} quotient[t] \cdot \phi_{i,t}$. The adopted diversification by proportionality algorithm is presented in Algorithm 2.

4   Experiments

In this section, we verify the performances of the proposed diversification frameworks on two data sets.

4.1   Datasets and Comparison Methods

The experiments are evaluated on the publicly accessible datasets: Movie-Lens(1M) and FilmTrust. The details of two data sets are shown in Table 1.

Table 1. Statistics of Data Sets

Dataset	Users	Items	Ratings	Density
MovieLens	6040	3952	1000209	4.47%
FilmTrust	1508	2071	35497	1.14%

We evaluate the LDA-PE and LDA-Greed methods proposed above in the experiments, and compare them with the following methods: SVD, LDA, MMR, DUM [9], and PMF++(PMF + α + β) [15]. As LDA cannot directly generate a top-k recommendation itemset, we rank the items by calculating $P_{u,i} = \sum_{t \in T} \theta_{u,t} \varphi_{t,i}$. Similarly we get the ranked lists by sorting with the ratings predicted by SVD.

4.2   Evaluation Metrics

Diversity. The diversity is measured based on the item similarities in the recommendation lists. A recommendation list is more diverse when it consists of dissimilar items. The diversity is defined as $Diversity = 1 - \dfrac{\sum\limits_{i_1,i_2 \in S, i_1 \neq i_2} sim(i_1,i_2)}{k(k-1)/2}$, where $sim(i_1, i_2)$ is the cosine similarity between i_1 and i_2, and is generated from the item-topic distribution ϕ in LDA.

Precision and Recall. We follow the evaluation protocol proposed in [19].

4.3   Experiment Design

Since DUM can only recommend a small number of items(less than 10), we first conduct other methods in MovieLens and FilmTrust datasets with $k \in \{10, 20, 30, 40, 50\}$, and then conduct DUM with LDA-PE and LDA-Greed in just MovieLens dataset with $k \in \{2, 4, 6, 8, 10\}$. All the methods take use of the "hit" protocol mentioned above. All the parameters are tuned manually. As LDA-PE, LDA-Greed are all based on LDA method, we run the LDA method first and make the further study on the results of LDA method. In addition, we also test the topic dimensionality d_T in LDA by studying the performances of LDA-Greed and LDA-PE on Movielens with $d_T \in \{25, 50, 75, 100\}$.

4.4   Results

For the parameter setting, we use $\alpha = 50.0/d_T$, $\beta = 0.01$ in LDA, and $\lambda = 0.02$, $\gamma = 0.005$ in LMF. These two methods are the basis of other methods. In addition, we set $\lambda = 0.5$ in LDA-PE, and $\lambda = 0.8$ in LDA-Greed to get the best performance. Topic dimensionality in LDA we set for MovieLens is $d_T = 50$, which we will explain in next subsection, and similarly we set $d_T = 25$ for FilmTrust.

Figure 1 shows the performances of our methods and the baseline methods in terms of diversity, precision and recall metrics with $k \in \{10, 20, 30, 40, 50\}$ on MovieLens dataset. As we can see, the diversities of all the methods except MMR increase as k grows, which means that for these methods, the larger recommended lists are more likely to have high diversity to satisfy all potential needs of users. Compared to other methods, LDA-PE has the best diversity. And LDA-Greed also has good performance, although its diversity is lower than MMR when $k <= 30$, but after that is higher than MMR. In particular, all other methods show an upward trend, except that LDA-PE's diversity curve remains stable declining, indicating that LDA-PE is more suitable for low-capacity lists than others. LDA has the best precision and recall, because it is designed not to predict the ratings of users but to capture the preferences of users, and our choice of precision and recall metrics just meets this feature. The precision and recall of LDA-PE and LDA-Greed are below LDA and MMR for the penalize of diversity promotion.

Figure 2 shows the performances of methods on FilmTrust dataset. Due to the sparsity of FilmTrust dataset and the atypical strategy of calculating precision and recall in our approach, SVD and PMF++ do not have good precision and recall even though they have good Root Mean Square Error(RMSE)(below 1), so

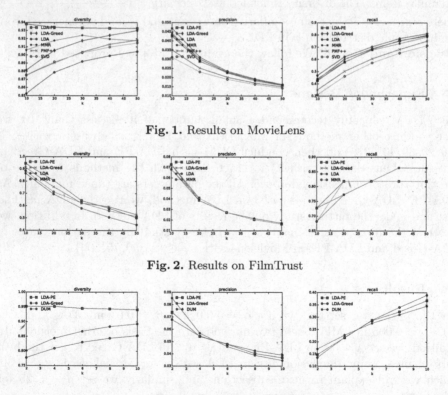

Fig. 1. Results on MovieLens

Fig. 2. Results on FilmTrust

Fig. 3. Compare to DUM

we just present the results of other methods here. As we can see, the performance of the diversity of all methods is rapidly deteriorating when k grows. We find that it's because the average number of users' ratings in FilmTrust is too small (about 20), so when k is too high, it's hard to get good results. Similar to Fig. 1, LDA-PE also has the best performance in terms of diversity, and LDA-Greed is slightly lower than LDA-PE. The precision and recall of these methods are broadly similar.

Figure 3 shows the comparison between DUM and our two proposed methods on the MovieLens dataset with $k \in \{2, 4, 6, 8, 10\}$. We can see that the diversity of LDA-PE shows a big advantage at this small range of k, while LDA-Greed has the worst performance, which indicates that LDA-Greed is not so suitable for a small k. As for precision and recall, DUM is the best when k is less than 6, but LDA-PE and LDA-Greed perform better later.

5 Conclusion

In this paper, we take a further adoption of LDA method to the top-k recommendation, which can better reflect user preferences. We propose two diversification methods based on margin and proportion respectively, and make a comparison between the two methods. Experimental results on the MovieLens and FilmTrust datasets demonstrate that our methods is superior to the most advanced methods in terms of distributional diversity, while the proportion-based method has better performance.

References

1. Zhang, M., Hurley, N.: Avoiding monotony: improving the diversity of recommendation lists, Lausanne, Switzerland, 23–25 October 2008 (2008)
2. Nemhauser, G.L., Wolsey, L.A., Fisher, M.L.: An analysis of approximations for maximizing submodular set functions - I. Math. Program. **14**, 265–294 (1978)
3. Agrawal, R., Gollapudi, S., Halverson, A., Ieong, S.: Diversifying search results. In: WSDM 2009, pp. 5–14 (2009)
4. Carbonell, J.G., Goldstein, J.: The use of MMR, diversity-based reranking for reordering documents and producing summaries. In: SIGIR 1998, Melbourne, Australia (1998)
5. Kim, Y., Shim, K.: TWILITE: a recommendation system for Twitter using a probabilistic model based on latent Dirichlet allocation. Inf. Syst. **42**, 59–77 (2014)
6. Blei, D.M., Ng, A.Y., Jordan, M.I.: Latent Dirichlet allocation. J. Mach. Learn. Res. **3**, 993–1022 (2003)
7. Zhou, T., Kuscsik, Z., Liu, J.-G., Medo, M., Wakeling, J.R., Zhang, Y.-C.: Solving the apparent diversity-accuracy dilemma of recommender systems. In: PNAS, vol. 107 (2010)
8. Ziegler, C., McNee, S., Konstan, J., Lausen, G.: Improving recommendation lists through topic diversification. In: WWW, pp. 22–32 (2005)
9. Ashkan, A., Kveton, B., Berkovsky, S., Wen, Z.: Optimal greedy diversity for recommendation. In: IJCAI 2015, pp. 1742–1748 (2015)

10. Di Noia, T., Ostuni, V.C., Rosati, J., Tomeo, P., Di Sciascio, E.: An analysis of users' propensity toward diversity in recommendations. In: RecSys 2014 (2014)
11. Wu, L., Liu, Q., Chen, E., Yuan, N.J., Guo, G., Xie, X.: Relevance meets coverage: a unified framework to generate diversified recommendations. ACM TIST **7**, 39 (2016)
12. Ashkan, A., Kveton, B., Berkovsky, S., Wen, Z.: Diversified utility maximization for recommendations. In: RecSys Poster Proceedings (2014)
13. Koren, Y., Bell, R.: Advances in collaborative filtering. In: Recommender Systems Handbook, pp. 145–186 (2011)
14. Van Dang, W., Croft, B.: Diversity by proportionality: an election-based approach to search result diversification. In: SIGIR 2012, Portland, OR, USA, 12–16 August 2012 (2012)
15. Sha, C., Wu, X., Niu, J.: A framework for recommending relevant and diverse items. In: IJCAI 2016, New York, NY, USA, 9–15 July 2016 (2016)
16. Qin, L., Zhu, X.: Promoting diversity in recommendation by entropy regularizer. In: IJCAI, pp. 2698–2704 (2013)
17. Vargas, S., Castells, P., Vallet, D.: Explicit relevance models in intent-oriented information retrieval diversification. In: SIGIR 2012, Portland, OR, USA, 12–16 August 2012 (2012)
18. Santos, R.L., Macdonald, C., Ounis, I.: Exploiting query reformulations for web search result diversification. In: WWW 2010 (2010)
19. Barbieri, N., Manco, G.: An Analysis of Probabilistic Methods for Top-N Recommendation in Collaborative Filtering. ECML PKDD 2011

Distributed Data Processing and Applications

Integrating Feedback-Based Semantic Evidence to Enhance Retrieval Effectiveness for Clinical Decision Support

Chenhao Yang, Ben He, and Jungang Xu[✉]

University of Chinese Academy of Sciences, Beijing, China
yangchenhao14@mails.ucas.ac.cn, {benhe,xujg}@ucas.ac.cn

Abstract. The goal of Clinical Decision Support (CDS) is to help physicians find useful information from a collection of medical articles with respect to the given patient records, in order to take the best care of their patients. Most of the existing CDS methods do not sufficiently consider the semantic evidence, hence the potential in improving the performance in biomedical articles retrieval. This paper proposes a novel feedback-based approach which considers the semantic association between a retrieved biomedical article and a pseudo feedback set. Evaluation results show that our method outperforms the strong baselines, and is able to improve over the best runs in the CDS tasks of TREC 2014 & 2015.

Keywords: Clinical Decision Support · Semantic association · Relevance feedback

1 Introduction

The goal of Clinical Decision Support (CDS) is to efficiently and effectively link relevant biomedical articles to meet physicians' needs for taking better care of their patients. In CDS, the patient records are considered as queries and the biomedical articles are retrieved in response to the queries. With the development of medical research, the volume of the published biomedical articles is growing rapidly, resulting in the difficulty in seeking out the most relevant and timely information for a particular clinical case.

Most of the existing CDS methods retrieve biomedical articles using the frequency-based statistical models [1,3,7,10]. Those methods extract concepts from queries and biomedical articles, and further utilize concepts to apply query expansion or document ranking. Then, the relevance score of a given article is assigned based on the frequencies of query terms or concepts. Despite the fact that the frequency-based CDS methods have been shown to be effective and efficient in the CDS task [22], they ignore the semantic evidence of relevance. We argue that the retrieval effectiveness of the CDS systems can be further improved by integrating the semantic information. For instance, suppose two short medical-related texts as follows:

© Springer International Publishing AG 2017
L. Chen et al. (Eds.): APWeb-WAIM 2017, Part II, LNCS 10367, pp. 153–168, 2017.
DOI: 10.1007/978-3-319-63564-4_13

- The child has symptoms of strawberry red tongue and swollen red hands.
- This kid is suffering from *Kawasaki disease*.

Though the two short texts have no terms in common, they convey the same meaning and are considered to be related to each other. However, the two sentences above are considered completely unrelated by the existing frequency-based CDS methods. In this paper, we aim to further enhance the retrieval performance of the CDS systems by taking the semantic evidence into consideration. Benefiting from recent advances in natural language processing (NLP), words and documents can be represented with semantically distributed real-valued vectors, i.e. *embeddings*, which are generated by neural network models [4,14,18,19]. The embeddings have been shown to be effective and efficient in many NLP tasks due to the ability in preserving semantic relationships in vector operations such as summation and subtraction [18]. In this study, we utilize the Word2Vec technique proposed by Mikolov et al. [14,18] to generate embeddings of words and biomedical articles, which is widely considered as an effective embedding method in NLP applications [9,17,26]. As a state-of-the-art topic model, latent Dirichlet allocation (LDA) [6] is also used for comparison with Word2Vec in generating distributed representations of biomedical articles in this study.

There have been efforts in utilizing the embeddings to improve IR effectiveness. For example, Vulić and Moens estimate a semantic relevance score by the cosine similarity between the embeddings of the query-document pair to improve the performance of monolingual and cross-lingual retrieval [27]. Similar idea is presented in [28], where the semantic similarity between the embeddings of the patient record and biomedical article is utilized to improve the CDS system. Note that the patient records are used as queries in CDS as described above. We argue that query is a weak indicator of relevance in that query is usually much shorter than the relevant documents, such that the use of semantic associations of the query-document pairs may only lead to limited improvement in retrieval performance. To this end, this paper proposes a feedback-based CDS method which integrates semantic evidence to further enhance retrieval effectiveness. Experimental results show that our proposed CDS method can have significant improvements over strong baselines. In particular, a simple linear combination of the classical BM25 weighting function with the semantic relevance score generated by our method leads to effective retrieval results that are better than the best TREC runs in both 2014 & 2015 CDS tasks.

2   Related Work

2.1   BM25 and PRF

As our CDS method is to integrate the semantic relevance score into the classical BM25 model with applying pseudo relevance feedback (PRF), we introduce BM25 model and PRF in this section. The ranking function of BM25 given a query Q and a document d is as follows [23]:

$$score(d, Q) = \sum_{t \in Q} w_t \frac{(k_1 + 1)tf}{K + tf} \frac{(k_3 + 1)qtf}{k_3 + qtf} \qquad (1)$$

where t is one of the query terms, and qtf is the frequency of t in query Q. tf is the term frequency of query term t in document d. K is given by $k_1((1 - b) + b \cdot \frac{l}{avg\_l})$, in which l and avg_l denote the length of document d and the average length of documents in the whole collection, respectively. k_1, k_3 and b are free parameters whose default setting is $k_1 = 1.2$, $k_3 = 1000$ and $b = 0.75$, respectively [23]. w_t is the weight of query term t, which is given by:

$$w_t = \log_2 \frac{N - df_t + 0.5}{df_t + 0.5} \qquad (2)$$

where N is the number of documents in the collection, and df_t is the document frequency of query term t, which denotes the number of documents that t occurs. PRF provides a feedback-based automatic method for improving the retrieval performance by expanding the original user input query [15].

2.2 State-of-the-Art CDS Methods

Due to the specificity of medical healthcare field, most of the existing CDS methods retrieve biomedical articles based on concepts, including unigrams, bigrams and multi-word concepts. These concepts are extracted from different resources, such as queries, biomedical articles, external medical databases, etc. These content-based CDS methods usually utilize concepts to apply query expansion or document ranking based on the frequencies of the concepts. Palotti and Hanbury proposed a concept-based query expansion method, increasing the weights of relevant concepts and expanding the original query with concepts extracted by MetaMap [20]. MetaMap is a highly configurable tool for recognizing the Unified Medical Language System (UMLS) concepts in text, which is usually utilized in the existing CDS methods. Song et al. proposed a customized learning-to-rank algorithm and a query term position based re-ranking model to improve the retrieval performance [25]. As biomedical articles are usually full-text scientific articles which are much longer than Web documents, Cummins et al. applied the recently proposed SPUD language model [11] to CDS for retrieving longer documents more fairly [10]. Abacha and Khelifi investigated several query reformulation methods utilizing Mesh and DBpedia. In addition, they applied rank fusion to combine different ranked document lists into a single list to improve the retrieval performance [1].

2.3 The Best Methods in the TREC 2014 and 2015 CDS Tasks

Choi and Choi proposed a three-step biomedical article retrieval method, which obtains the best run in the TREC 2014 CDS task [7]. Firstly, the method utilizes external knowledge resource to apply query expansion, and uses the query likelihood (QL) language model [21] to rank articles. Secondly, a text classification based method is used for the topic-specific ranking. Note that the topics used in the TREC CDS task are classified into three categories, i.e. *diagnosis*, *test* and *treatment*. Finally, the method combines the relevance ranking score and the topic-specific ranking score with Borda-fuse method [7].

The CDS methods proposed by Balaneshin-kordan et al. [3] obtained both the best automatic and manual runs in the TREC 2015 CDS task. Their method extracts unigrams, bigrams and multi-word UMLS concepts from queries, the pseudo relevance feedback documents or external knowledge resources, and then uses the Markov Random Field (MRF) model [16] for document ranking. The relevance score of a document d given a query Q is computed as follows [3]:

$$score(d, Q) = \sum_{c \in \mathbb{C}} \mathbb{1}_c score(c, d) \tag{3}$$

$$= \sum_{c \in \mathbb{C}} \mathbb{1}_c \sum_{T \in \mathbb{T}} \lambda_T f_T(c, d)$$

where $score(c, d)$ is the contribution of concept c to the relevance score of document d. $\mathbb{1}_c$ is an indicator function which determines whether the concept c is considered in the relevance weighting. $\mathbb{C}$ is the set of concepts. $\mathbb{T}$ is the set of all concept types, to which concept c belongs. Note that a concept can belong to multiple concept types at the same time. λ_T is the importance weight of concept type T, and $f_T(c, d)$ is a real-valued feature function.

The existing CDS methods retrieve biomedical articles based on the frequencies of concepts. As discussed in Sect. 1, the lack of semantic evidence of relevance may lead to limited retrieval performance. A recent work [28] integrates semantic similarity between the embeddings of the patient record and biomedical article to improve the CDS system, which is given by:

$$Sim(d, Q) = 0.5 \cdot \frac{\vec{d} \cdot \vec{Q}}{\| \vec{d} \| \times \| \vec{Q} \|} + 0.5 \tag{4}$$

where $\vec{d}$ and $\vec{Q}$ are the embeddings of biomedical article d and patient record Q, respectively. $Sim(d, Q)$ is the semantic similarity which is integrated into BM25 model [23] by a linear interpolation. As the patient records are usually much shorter than the full-text biomedical articles, they do not necessarily contain sufficient amount of semantic evidence of relevance. Therefore, the approach in [28] leads to limited improvement on the CDS task. To deal with this problem, in the next section, we propose a feedback-based approach that considers the semantic similarity between a retrieved article and a set of feedback articles, which is a better indicator of relevance than patient record.

3   Feedback-Based Semantic Evidence

The methods for generating the embeddings of biomedical articles are introduced in Sect. 3.1. The generated embeddings are utilized for enhancing the retrieval performance of CDS in Sect. 3.2.

3.1 Generating Embeddings of Biomedical Articles

The Word2Vec technique proposed by Mikolov et al. [14,18] is a state-of-the-art neural embedding framework, which has been shown to be effective and efficient in many NLP tasks. In this study, Word2Vec is also utilized to generate embeddings of words and biomedical articles. A unique advantage of Word2Vec is that the semantic relationships can be preserved in vector operations, such as addition and subtraction [18]. Therefore, the embeddings of biomedical articles can be generated through vector operations of word embeddings such that they are applicable to the CDS task. Considering the fact that informative words are usually infrequent in biomedical articles, we utilize the Skip-gram architecture of Word2Vec, which shows better performance for infrequent words than the CBOW architecture of Word2Vec in generating embeddings [18]. Besides, the negative sampling algorithm is used to train embeddings [18].

The Skip-gram architecture is composed of three layers, i.e. an input layer, a projection layer and an output layer. The basic idea of Skip-gram is to predict the context of a given word w. Considering the conditional probability $p(c(w)|w)$ given a word w and the corresponding context $c(w)$, the goal of Skip-gram model is to maximize the likelihood function as follows [12]:

$$\arg\max_{\theta} \prod_{(w,c(w))\in D} p(c(w)|w;\theta) \tag{5}$$

where w and $c(w)$ denote a word and the corresponding context, respectively. $(w, c(w))$ is a training sample, and D is the set of all training samples. θ is the parameter set that needs to be optimized. In addition, the conditional probability $p(c(w)|w)$ is modeled as Softmax regression, which is given as follows:

$$p(c(w)|w;\theta) = \frac{e^{v_w \cdot v_{c(w)}}}{\sum_{c(w)' \in C} e^{v_w \cdot v_{c(w)'}}} \tag{6}$$

where v_w and $v_{c(w)}$ are the embeddings of word w and the corresponding context $c(w)$, respectively. Substituting Eq. (6) back into Eq. (5), the final objective function of Skip-gram is given by:

$$\arg\max_{\theta} \prod_{(w,c(w))\in D} \log p(c(w)|w)$$

$$= \sum_{(w,c(w))\in D} \left(\log e^{v_w \cdot v_{c(w)}} - \log \sum_{c'} e^{v_w \cdot v_{c(w)'}} \right) \tag{7}$$

where the parameters in Eq. (7) are trained by stochastic gradient ascent.

A major challenge of the application of the word embeddings to CDS is how to generate effective embeddings for biomedical articles. In this paper, we adopt two ways of generating embeddings for biomedical articles, namely *Term Summation* and *Paragraph Embeddings*, abbreviated as *Sum* and *Para* respectively. As the semantic relationships are preserved in the embedding operations, one way of generating embeddings of biomedical articles is to sum up the word embeddings of the top-k most informative words in a given article, i.e. *Term Summation*, which is given by:

$$\vec{d} = \sum_{w \in W_k^d} tf\text{-}idf(w) \cdot \vec{w} \tag{8}$$

where $\vec{w}$ and $\vec{d}$ are the embeddings of word w and biomedical article d, respectively. W_k^d is the set of the top-k terms with the highest *tf-idf* weights in d. *tf-idf*(w) is used to measure the amount of information carried by word w, which is given by:

$$tf\text{-}idf(w) = tf \cdot \log_2 \frac{N - df_w + 0.5}{df_w + 0.5} \tag{9}$$

where tf is the term frequency of w in d. N is the total number of biomedical articles in the whole collection, and df_w is the document frequency of word w.

In addition to *Term Summation*, we adopt the *Paragraph Embeddings* technique [14] to generate embeddings of biomedical articles. *Paragraph Embeddings* is an improved version of Word2Vec, in which each document is marked with a special word called *Paragraph id*. The *Paragraph id* participates in the training of each word as part of each context, acting as a memory that remembers what is missing from the current context. The training procedure of *Paragraph Embeddings* is the same as *Word2Vec*. Finally, the embedding of the special word *Paragraph id* is used to represent the corresponding biomedical article. We denote embeddings of biomedical articles generated by *Term Summation* and *Paragraph Embeddings* as $\overrightarrow{d_{Sum}}$ and $\overrightarrow{d_{Para}}$, respectively.

3.2   Using Embeddings for CDS

In this section, we introduce our proposed feedback-based CDS method, which considers the semantic similarity between a biomedical article to be scored and a pseudo feedback set. As Mikolov et al. demonstrated that words can have multiple degrees of similarity [18], integrating semantic associations by directly measuring the similarity between the embeddings of patient records and biomedical articles may only lead to limited improvement in retrieval performance (as used in [28]). Instead, we estimate the semantic relevance of a biomedical article by measuring the semantic similarity between the article and a pseudo relevance feedback set. Once we obtain the preliminary retrieval results returned by BM25, the semantic relevance score of biomedical articles can be utilized to improve the retrieval performance, which is given as follows:

$$score(d, Q) = \lambda \cdot BM25(d, Q) + (1 - \lambda) \cdot SEM(d, D_{PRF}^k(Q)) \tag{10}$$

where $BM25(d, Q)$ is the ranking score of document d given by a baseline retrieval model, e.g. the classical BM25 model with PRF. $D_{PRF}^k(Q)$ is the pseudo relevance feedback set of biomedical articles, which is composed of the top ranked

k articles returned by the baseline model. It is usually assumed by the PRF technique that most of the documents in $D_{PRF}^{k}(Q)$ are relevant to query Q, thus $D_{PRF}^{k}(Q)$ can be considered as a better indicator of relevance than patient records. $SEM(d, D_{PRF}^{k}(Q))$ measures the semantic similarity between document d and the pseudo relevance feedback set $D_{PRF}^{k}(Q)$, which is given as follows:

$$SEM(d, D_{PRF}^{k}(Q)) = \sum_{d' \in D_{PRF}^{k}(Q)} w_{d'} \cdot Sim(d', d) \tag{11}$$

where d' is one of the biomedical articles in $D_{PRF}^{k}(Q)$. $w_{d'}$ is the importance weight of d', which is given as follows:

$$w_{d'} = BM25(d', Q) + \max_{d'' \in D_{PRF}^{k}(Q)} BM25(d'', Q) \tag{12}$$

$Sim(d', d)$ denotes the semantic similarity between d' and d, which is given by Eq. (4). In Eq. (12), the maximum relevance score is added to normalize the gap between the relevance scores of different articles. Note that both $BM25(d, Q)$ and $SEM(d, D_{PRF}^{k}(Q))$ in Eq. (10) are normalized by Min-Max normalization, such that the two scoring features are on the same scale.

4   Experimental Settings

In this section, we introduce the datasets used in the experiments and the experimental design.

4.1   Datasets

All our experiments are conducted on the standard datasets used in the TREC CDS tasks of 2014 and 2015. The target document collection used in the two years is an open access subset[1] of PubMed Central[2] (PMC), containing 733,138 full-text biomedical articles. We extract the *title*, *abstract*, *keywords* and *body* fields from each article as the source of the index. We use the open source Terrier toolkit version 4.1 [17] to index the collection with the recommended settings of the toolkit. Standard English stopwords are removed and the collection is stemmed using Porter's English stemmer. Using Porter's stemmer, inflected or derived words are reduced to their word stem, base or root forms.

There are 30 topics in each year, and each topic is a medical record narrative that serves as an idealized representation of actual patient record. These topics are classified into three categories, i.e. *diagnosis*, *test* and *treatment*, with 10 topics in each category. According to [24], there is little difference observed in retrieval performance when the three topic types are taken into account. Thus the topic types are not considered in our study. There are two versions of the

---

[1] http://www.ncbi.nlm.nih.gov/pmc/tools/openftlist/.
[2] http://www.ncbi.nlm.nih.gov/pmc.

medical record narratives, i.e. *Summary* and *Description* fields. Table 1 presents an example of the *Summary* and *Description* fields. The *Description* field is much longer than the *Summary* field, and has more detailed information about a patient. However, the *Description* field may contain more irrelevant information than the *Summary* field. In the experiments, both the *Summary* and *Description* fields are used as queries.

Table 1. Example of *Summary* and *Description* fields.

Topic type - diagnosis
Summary: A 62-year-old immunosuppressed male with fever, cough and intranuclear inclusion bodies in bronchoalveolar lavage.
Description: A 62 yo male presents with four days of non-productive cough and one day of fever. He is on immunosuppressive medications, including prednisone. He is admitted to the hospital, and his work-up includes bronchoscopy with bronchoalveolar lavage (BAL). BAL fluid examination reveals owl's eye inclusion bodies in the nuclei of infection cells

As described in Sect. 3.1, the Skip-gram model of Word2Vec[3] toolkit is utilized to generate embeddings of words and biomedical articles, which are trained using the negative sampling algorithm [18]. Note that the *title, abstract, keywords* and *body* fields of each biomedical article are extracted as the training set of Word2Vec, and the stopword removal and stemming are applied. As recommended in [18], the window size is set to 10 for Skip-gram model. As documents in the target collection are full-text long biomedical articles, the number of dimensions of the embeddings are set to 300, a value that is larger than the recommended 100 in [2].

4.2 Experimental Design

In our study, we evaluate our CDS method against two baselines. As described in Sect. 3.2, we use the BM25 model [23] with applying PRF as one of the baselines. In addition, we use the CDS method proposed in [28] as another baseline.

The parameters k_1 and k_3 of BM25 (See Eq. (1)) are set to default values and b is set to the optimal value on training data by grid search algorithm [5]. As described in Sect. 3.1, we adopt two methods for generating embeddings of biomedical articles, which are denoted as $\overrightarrow{d_{Sum}}$ and $\overrightarrow{d_{Para}}$ respectively. For convenience, we denote our proposed CDS method applying *Term Summation* and *Paragraph Embeddings* as $BM25 + SEM_{d_{Sum}-D^k_{PRF}}$ and $BM25 + SEM_{d_{Para}-D^k_{PRF}}$, respectively. Besides, the previously proposed CDS method [28] is denoted as $BM25 + Sim_{d_{Para}-Q}$, which only uses *Paragraph Embeddings* for generating embeddings of biomedical

---

[3] The learned embeddings of words and biomedical articles can be downloaded from http://gucasir.org/CDS.tgz.

articles. Note that our method has the following tunable parameters, i.e. hyperparameter λ (see Eq. (10)), top k terms to generate embeddings of biomedical articles when applying *Term Summation* (*# Terms*) and top k articles in $D_{PDF}^k(Q)$ (*# PRF Documents*). All the parameters are tuned on training data by grid search algorithm [5].

The evaluation results are obtained by a two-fold cross-validation, where the topics are split into two equal-size subsets by parity in odd or even topic numbers. In each fold, we use one subset of the topics for training, and the remaining subset for testing. There is no overlap between the training and testing topics. Then the overall retrieval performance is obtained by averaging over the two test subsets of topics. Apart from the official TREC measure inferred NDCG (infNDCG) [24], we also report on other popular evaluation metrics in the CDS task, including Mean Average Precision (MAP) [8], R-Precision (R-Prec) [8] and inferred Average Precision (infAP) [24]. All statistical tests are based on the t-test at the 0.05 significance level.

5   Evaluation Results

In this section, we present the evaluation results of our proposed CDS method. Tables 2 and 3 present the evaluation results of the TREC 2014 CDS task using the *Summary* and *Description* fields respectively, and Tables 4 and 5 present the evaluation results of the TREC 2015 CDS task A. Note that all the evaluation results are obtained by a two-fold cross-validation based on the parity of the topic numbers. As described in Sect. 4.2, $BM25 + SEM_{d_{Para}\text{-}D_{PRF}^k}$ and $BM25 + SEM_{d_{Sum}\text{-}D_{PRF}^k}$ denote two different applications of our proposed CDS method, in which the embeddings of biomedical articles are generated by *Paragraph Embeddings* and *Term Summation*, respectively. Tables 6 and 7 present the comparison between our CDS method and the $BM25 + Sim_{d_{Para}\text{-}Q}$ method proposed in [28]. BM25 is the baseline retrieval model used for verifying the effectiveness of our proposed feedback-based semantic relevance score. In addition, the comparisons between our approach and the best methods in the TREC

Table 2. Evaluation results on the *Summary* field on the TREC 2014 CDS task. The difference in percentage is measured against the baseline retrieval model *BM25*. A statistically significant difference is marked with a *. The best result of each evaluation metric is in **bold**.

Model	infNDCG	infAP	MAP	R-Prec
BM25	0.2524	0.0805	0.1537	0.2004
$BM25 + SEM_{d_{Para}\text{-}D_{PRF}^k}$	0.2698 + 6.89%*	0.0935 16.15%*	0.1628 + 5.92%*	0.2067 + 3.14%
$BM25 + SEM_{d_{Sum}\text{-}D_{PRF}^k}$	**0.2748 + 8.87%***	**0.0953 + 18.39%***	**0.1645 + 7.03%***	**0.2083 + 3.94%**

Table 3. Evaluation results on the *Description* field on the TREC 2014 CDS task. The difference in percentage is measured against the baseline retrieval model *BM25*. A statistically significant difference is marked with a *. The best result of each evaluation metric is in **bold**.

Model	infNDCG	infAP	MAP	R-Prec
BM25	0.2460	0.0700	0.1440	0.2065
$BM25 +$ $SEM_{d_{Para}-D^k_{PRF}}$	0.2751 +11.83%*	**0.0918 +** **31.14%***	0.1623 +12.71%*	0.2196 +6.34%*
$BM25 +$ $SEM_{d_{Sum}-D^k_{PRF}}$	**0.2830 +** **15.04%***	0.0911 + 30.14%*	**0.1661 +** **15.35%***	**0.2206 +** **6.83%***

Table 4. Evaluation results on the *Summary* field on the TREC 2015 CDS task. The difference in percentage is measured against the baseline retrieval model *BM25*. A statistically significant difference is marked with a *. The best result of each evaluation metric is in **bold**.

Model	infNDCG	infAP	MAP	R-Prec
BM25	0.2695	0.0736	0.1650	0.2198
$BM25 +$ $SEM_{d_{Para}-D^k_{PRF}}$	0.2980 + 10.58%*	0.0831 12.91%*	0.1758 +6.55%*	0.2345 + 6.69%*
$BM25 +$ $SEM_{d_{Sum}-D^k_{PRF}}$	**0.2986 +** **10.80%***	**0.0842 +** **14.40%***	**0.1791 +** **8.55%***	**0.2408 +** **9.55%***

Table 5. Evaluation results on the *Description* field on the TREC 2015 CDS task. The difference in percentage is measured against the baseline retrieval model *BM25*. A statistically significant difference is marked with a *. The best result of each evaluation metric is in **bold**.

Model	infNDCG	infAP	MAP	R-Prec
BM25	0.2724	0.0733	0.1641	0.2184
$BM25 +$ $SEM_{d_{Para}-D^k_{PRF}}$	0.2877 + 5.62%*	0.0837 14.19%*	0.1762 + 7.37%*	0.2325 + 6.46%*
$BM25 +$ $SEM_{d_{Sum}-D^k_{PRF}}$	**0.3016 +** **10.72%***	**0.0873 +** **19.10%***	**0.1806 +** **10.05%***	**0.2370 +** **8.52%***

2014 & 2015 CDS tasks are presented in Tables 8 and 9, respectively. According to the results, we have the observations as follows.

First, our proposed feedback-based CDS method has statistically significant improvements over the baseline retrieval model BM25 in most cases, which indicates the effectiveness of integrating semantic evidence into the frequency-based statistical models. Besides, according to Tables 8 and 9, our CDS method outscores the best automatic methods in both TREC 2014 and 2015 CDS tasks. This observation is promising in that a simple linear interpolation of the classical

Table 6. Comparison between our approach and $BM25 + Sim_{d_{Para}\text{-}Q}$ [28] on the TREC 2014 CDS task. The results are obtained based on the *Summary* field.

Method	infNDCG	infAP	MAP	R-Prec
$BM25 +$ $Sim_{d_{Para}\text{-}Q}$	0.2618	0.0763	0.1579	0.1518
$BM25 +$ $SEM_{d_{Para}\text{-}D^k_{PRF}}$	0.2698 + 3.06%	0.0935 + 22.54%*	0.1628 + 3.10%	0.2067 + 36.17%*
$BM25 +$ $SEM_{d_{Sum}\text{-}D^k_{PRF}}$	**0.2748 +** **4.97%***	**0.0953 +** **24.90%***	**0.1645 +** **4.18%**	**0.2083 +** **37.22%***

Table 7. Comparison between our approach and $BM25 + Sim_{d_{Para}\text{-}Q}$ [28] on the TREC 2015 CDS task. The results are obtained based on the *Summary* field.

Method	infNDCG	infAP	MAP	R-Prec
$BM25 +$ $Sim_{d_{Para}\text{-}Q}$	0.2742	0.0657	0.1642	0.1491
$BM25 +$ $SEM_{d_{Para}\text{-}D^k_{PRF}}$	0.2980 + 8.68%*	0.0831 + 26.48%*	0.1758 + 7.06%*	0.2345 + 57.28%*
$BM25 +$ $SEM_{d_{Sum}\text{-}D^k_{PRF}}$	**0.2986 +** **8.90%***	**0.0842 +** **28.16%***	**0.1791 +** **9.07%***	**0.2408 +** **61.50%***

Table 8. Comparison to *SNUMedinfo*, the best automatic run in the TREC 2014 CDS task. Results of *SNUMedinfo* are taken from those reported in [7]. $BM25 + SEM_{d\text{-}D^k_{PRF}}$ is the best result of our approach on this dataset, as in Table 3. No statistical test is conducted due to unavailability of the per-query result of *SNUMedinfo*.

Method	infNDCG	infAP
SNUMedinfo	0.2674	0.0659
$BM25 + SEM_{d\text{-}D^k_{PRF}}$	**0.2830**	**0.0911**

Table 9. Comparison to *WSU-IR*, the best automatic run in the TREC 2015 CDS task. Results of *WSU-IR* are taken from those reported in [3]. $BM25 + SEM_{d\text{-}D^k_{PRF}}$ is the best result of our approach on this dataset, as in Table 5. The difference between the two approaches is not statistically significant.

Method	infNDCG	infAP
WSU-IR	0.2939	0.0842
$BM25 + SEM_{d\text{-}D^k_{PRF}}$	**0.3016 + 2.62%**	**0.0873 + 3.68%**

BM25 model and our proposed semantic relevance score could have scored the best run in those tasks.

Second, according to Tables 6 and 7, our CDS method outperforms the method $BM25 + Sim_{d_{Para}\text{-}Q}$ proposed in [28], which integrates semantic evidence by measuring the cosine similarity between the embeddings of the patient record and biomedical article. As described in Sect. 1, patient records are much shorter than full-text biomedical articles, such that patient record is a weak indicator of relevance, thus our feedback-based CDS method is expected to outperform the method $BM25 + Sim_{d_{Para}\text{-}Q}$.

Third, comparing the two different ways of generating the article embeddings, *Term Summation* has a better performance than *Paragraph Embeddings* in most cases. As the full-text biomedical articles are usually very long, which contain large amount of irrelevant information, the mechanism of *Paragraph Embeddings* that considering the entire verbose texts while training embeddings may results in the sparse distribution of the semantic information in the embeddings of articles, such that the embeddings of articles generated by *Paragraph Embeddings* is not suitable to represent semantic relevance for long texts. In contrast, *Term Summation* generates embeddings of biomedical articles by only considering the top-k most informative words in the articles, which effectively reduces irrelevant information in the embeddings of biomedical articles.

Finally, comparing between the evaluation results obtained by using the *Summary* and *Description* fields, although using the *Description* field as queries obtained worse baseline retrieval results, the final performance of using *Description* field by integrating semantic evidence is better than using the *Summary* field in most cases. One possible reason is that the *Description* field is much longer than the *Summary* field, such that the relevant biomedical articles are returned by content-based retrieval models with relatively low ranking. By integrating the semantic evidence of relevance, the lowly ranked relevant documents are promoted in the ranking list which leads to improved retrieval performance.

6   Application of the Semantic Relevance Score to Other State-of-the-Art Methods

In this section, we use the best TREC run in 2015, WSU-IR, as the baseline to examine if our proposed method can still improve over the strongest baseline as far as we are aware of. We do not conduct the same comparison to *SNUMedinfo*, the best TREC CDS run in 2014, due to unavailability of per-query results. In addition, the latent Dirichlet allocation (LDA) model [6] is applied to generate the distributed representations of biomedical articles for comparison with the neural embedding model Word2Vec in our study.

Tables 10 and 11 present the evaluation results based on the automatic and manual runs submitted by WSU-IR [3] in the TREC 2015 CDS Task A, respectively. *wsuirdaa* and *wsuirdma* in Tables 10 and 11 are the submitted automatic and manual runs respectively, and are used as our strong baselines. *wsuirdaa* $+ SEM_{d_{Para}\text{-}D^k_{PRF}}$ and *wsuirdaa* $+ SEM_{d_{Sum}\text{-}D^k_{PRF}}$ correspond to applying *Paragraph Embeddings* and *Term Summation* respectively, the same to *wsuirdma* $+ SEM_{d_{Para}\text{-}D^k_{PRF}}$ and *wsuirdma* $+ SEM_{d_{Sum}\text{-}D^k_{PRF}}$. According to the results, we can see that there are still statistically significant improvements over the strong baselines *wsuirdaa* and *wsuirdma* in most cases when applying

Table 10. The evaluation results on the TREC 2015 CDS task A - automatic. The difference in percentage is measured against the baseline retrieval model *wsuirdaa* [3]. A statistically significant difference is marked with a *. The best result of each evaluation metric is in **bold**.

Method	infNDCG	infAP	MAP	R-Prec
wsuirdaa	0.2939	0.0842	0.1864	0.2306
wsuirdaa + $SEM_{d_{Para}\text{-}D^k_{PRF}}$	0.3130 + 6.50%*	0.0896 + 6.41%*	0.1905 +2.20%	0.2396 +3.90%
wsuirdaa + $SEM_{d_{Sum}\text{-}D^k_{PRF}}$	**0.3157 + 7.42%***	**0.0898 + 6.65%***	**0.1926 + 3.33%**	**0.2469 + 7.07%***

Table 11. The evaluation results on the TREC 2015 CDS task A - manual. The difference in percentage is measured against the baseline retrieval model *wsuirdma* [3]. A statistically significant difference is marked with a *. The best result of each evaluation metric is in **bold**.

Method	infNDCG	infAP	MAP	R-Prec
wsuirdma	0.3109	0.0880	0.1968	0.2493
wsuirdma + $SEM_{d_{Para}\text{-}D^k_{PRF}}$	0.3265 + 5.02%*	0.0940 +6.82%*	0.2015 + 2.39%	0.2605 + 4.49%
wsuirdma + $SEM_{d_{Sum}\text{-}D^k_{PRF}}$	**0.3335 + 7.27%***	**0.0963 + 9.43%***	**0.2054 + 4.37%**	**0.2643 + 6.02%***

Table 12. The evaluation results on the TREC 2015 CDS task A - automatic. The difference in percentage is measured against $wsuirdaa + SEM_{d_{LDA}\text{-}D^k_{PRF}}$. A statistically significant difference is marked with a *. The best result of each evaluation metric is in **bold**.

Method	infNDCG	infAP	MAP	R-Prec
wsuirdaa + $SEM_{d_{LDA}\text{-}D^k_{PRF}}$	0.2963	0.0853	0.1864	0.2306
wsuirdaa + $SEM_{d_{Para}\text{-}D^k_{PRF}}$	0.3130 + 5.64%*	0.0896 +5.04%*	0.1905 + 2.20%	0.2396 + 3.90%
wsuirdaa + $SEM_{d_{Sum}\text{-}D^k_{PRF}}$	**0.3157 + 6.55%***	**0.0898 + 5.28%***	**0.1926 + 3.33%**	**0.2469 + 7.07%***

both *Paragraph Embeddings* and *Term Summation*, indicating the effectiveness of our proposed semantic relevance score.

Tables 12 and 13 present the comparison between Word2Vec and LDA in generating the distributed representations of biomedical articles. The number of topics in LDA is set to 100 as used in [13]. $wsuirdaa + SEM_{d_{LDA}\text{-}D^k_{PRF}}$ and $wsuirdma + SEM_{d_{LDA}\text{-}D^k_{PRF}}$ in Tables 12 and 13 correspond to applying LDA model for generating the article representations, on top of the best TREC

Table 13. The evaluation results on the TREC 2015 CDS task A - manual. The difference in percentage is measured against $wsuirdma + SEM_{d_{LDA}\text{-}D^k_{PRF}}$. A statistically significant difference is marked with a *. The best result of each evaluation metric is in **bold**.

Method	infNDCG	infAP	MAP	R-Prec
$wsuirdma +$ $SEM_{d_{LDA}\text{-}D^k_{PRF}}$	0.3117	0.0887	0.1970	0.2494
$wsuirdma +$ $SEM_{d_{Para}\text{-}D^k_{PRF}}$	0.3265 + 4.75%*	0.0940 +5.98%*	0.2015 + 2.28%	0.2605 +4.45%
$wsuirdma +$ $SEM_{d_{Sum}\text{-}D^k_{PRF}}$	**0.3335 + 6.99%***	**0.0963 + 8.57%***	**0.2054 + 4.26%**	**0.2643 + 5.97%***

CDS runs in 2015. According to the results, there are statistically significant improvements over LDA in most cases when Word2Vec is utilized to generate the article embeddings. In fact, our experience indicates that the optimal value of hyper-parameter λ (See Eq. (10)) when applying LDA model is usually 1, such that the semantic relevance score does not work when LDA is used. Therefore, we may conclude that Word2Vec is more suitable than LDA for estimating the semantic similarity between biomedical articles.

7   Conclusions and Future Work

In this paper, we have proposed a novel feedback-based CDS method, which integrates the semantic similarity between a biomedical article and the corresponding pseudo relevance feedback set into frequency-based models. Experimental results show that integrating semantic evidence of relevance can indeed significantly improve the retrieval performance over the existing CDS approaches, including the best TREC results. In addition, a simple linear combination of the classical BM25 model with our proposed semantic relevance score ($BM25 + SEM_{d\text{-}D^k_{PRF}}$) would have achieved the best automatic runs on the TREC 2014 and 2015 CDS tasks. Compared to *Paragraph Embeddings*, *Term Summation* is more suitable to generate the embeddings of biomedical articles, due to the ability of reducing irrelevant information in the embeddings of biomedical articles. The comparison between Word2Vec and LDA shows that Word2Vec is more suitable than LDA for estimating the semantic similarity between biomedical articles.

In future research, we plan to utilize the semantic evidence for query expansion to further improve the performance of a CDS system.

Acknowledgments. This work is supported by the National Natural Science Foundation of China (61472391). We would like to thank the authors of [3] for kindly sharing their TREC runs with us.

References

1. Abacha, A., Khelifi, S.: LIST at TREC 2015 clinical decision support track: question analysis and unsupervised result fusion. In: TREC (2015)
2. Dai, A., Olah, C., Le, Q.: Document embedding with paragraph vectors. CoRR, abs/1507.07998 (2015)
3. Balaneshinkordan, S., Kotov, A., Xisto, R.: WSU-IR at TREC 2015 clinical decision support track: joint weighting of explicit and latent medical query concepts from diverse sources. In: TREC (2015)
4. Bengio, Y., Schwenk, H., Senécal, J., Morin, F., Gauvain, J.: Neural probabilistic language models. In: Holmes, D.E., Jain, L.C. (eds.) STUDFUZZ, vol. 194, pp. 137–186. Springer, Heidelberg (2006). doi:10.1007/3-540-33486-6_6
5. Bergstra, J., Bardenet, R., Kégl, B., Bengio, Y.: Algorithms for hyper-parameter optimization. In: Advances in Neural Information Processing Systems, pp. 2546–2554 (2011)
6. Blei, D., Ng, A., Jordan, M.: Latent Dirichlet allocation. J. Mach. Learn. Res. **3**, 993–1022 (2003)
7. Choi, S., Choi, J.: SNUMedinfo at TREC CDS Track 2014: Medical Case-based Retrieval Task. Technical report, DTIC Document (2014)
8. Chowdhury, G.: TREC: experiment and evaluation in information retrieval. Online Inf. Rev. **5** (2007)
9. Collobert, R., Weston, J., Bottou, L., Karlen, M., Kavukcuoglu, K., Kuksa, P.: Natural language processing (almost) from scratch. J. Mach. Learn. Res. **12**, 2493–2537 (2011)
10. Cummins, R.: Clinical decision support with the SPUD language model. In: TREC (2015
11. Cummins, R., Paik, J., Lv, Y.: A pólya urn document language model for improved information retrieval. ACM Trans. Inf. Syst. (TOIS) **33**(4), 21 (2015)
12. Goldberg, Y., Levy, O.: word2vec explained: deriving Mikolov et al'.s negative-sampling word-embedding method. CoRR, abs/1402.3722 (2014)
13. Goodwin, T., Harabagiu, S.: UTD at TREC 2014: query expansion for clinical decision support. Technical report, DTIC Document (2014)
14. Le, Q., Mikolov, T.: Distributed representations of sentences and documents. CoRR, abs/1405.4053 (2014)
15. Manning, C., Raghavan, P., Schütze, H.: Introduction to Information Retrieval. Cambridge University Press, New York (2008)
16. Metzler, D., Croft, W.: A Markov random field model for term dependencies. In: Proceedings of the 28th Annual International ACM SIGIR Conference on Research and Development in Information Retrieval, pp. 472–479. ACM (2005)
17. Mikolov, T., Yih, W., Zweig, G.: Linguistic regularities in continuous space word representations. In: HLT-NAACL (2013)
18. Mikolov, T., Chen, K., Corrado, G., Dean, J.: Efficient estimation of word representations in vector space. CoRR, abs/1301.3781 (2013)
19. Mnih, A., Hinton, G.: A scalable hierarchical distributed language model. In: Conference on Neural Information Processing Systems, Vancouver, British Columbia, Canada, pp. 1081–1088, December 2008
20. Palotti, J., Hanbury, A.: TUW @ TREC clinical decision support track 2015. In: TREC (2015)
21. Ponte, J., Croft, W.: A language modeling approach to information retrieval. In: Proceedings of the 21st Annual International ACM SIGIR Conference on Research and Development in Information Retrieval, pp. 275–281. ACM (1998)

22. Roberts, K., Simpson, M., Voorhees, E., Hersh, W.: Overview of the TREC 2015 clinical decision support track. In: TREC (2015)
23. Robertson, S., Walker, S., Beaulieu, M., Gatford, M., Payne, A.: Okapi at TREC-4. In: TREC, pp. 73–96 (1996)
24. Simpson, M., Voorhees, E., Hersh, W.: Overview of the TREC 2014 Clinical Decision Support Track. Technical report, DTIC Document (2014)
25. Song, Y., He, Y., Hu, Q., He, L.: ECNU at 2015 CDS track: two re-ranking methods in medical information retrieval. In: TREC (2015)
26. Turian, J., Ratinov, L., Bengio, Y.: Word representations: a simple and general method for semi-supervised learning. In: Proceedings of the 48th annual meeting of the association for computational linguistics, pp. 384–394. Association for Computational Linguistics (2010)
27. Vulić, I., Moens, M.: Monolingual and cross-lingual information retrieval models based on (bilingual) word embeddings. In: The International ACM SIGIR Conference, pp. 363–372 (2015)
28. Yang, C., He, B.: A novel semantics-based approach to medical literature search. In: 2016 IEEE International Conference on Bioinformatics and Biomedicine (BIBM), pp. 1616–1623. IEEE (2016)

Reordering Transaction Execution to Boost High Frequency Trading Applications

Ningnan Zhou[1,2], Xuan Zhou[3], Xiao Zhang[1,2(✉)], Xiaoyong Du[1,2], and Shan Wang[1,2]

[1] School of Information, Renmin University of China, Beijing, China
zhangxiao@ruc.edu.cn
[2] MOE Key Laboratory of DEKE, Renmin University of China, Beijing, China
[3] School of Data Science and Engineering, East China Normal University, Shanghai, China

Abstract. High frequency trading (HFT) has always been welcomed because it benefits not only personal interests but also the whole social welfare. While the recent advance of portfolio selection in HFT market generates more profit, it yields much contended OLTP workloads. Featuring in exploiting the abundant parallelism, transaction pipeline, the state-of-the-art concurrency control (CC) mechanism, however suffers from limited concurrency confronted with HFT workloads. Its variants that enable more parallel execution by leveraging find-grained contention information also take little effect. To solve this problem, we for the first time observe and formulate the source of restricted concurrency as *harmful ordering* of transaction statements. To resolve harmful ordering, we propose PARE, a pipeline-aware reordered execution, to improve application performance by rearranging statements in order of their degrees of contention. In concrete, two mechanisms are devised to ensure the correctness of statement rearrangement and identify the degrees of contention of statements respectively. Experiment results show that PARE can improve transaction throughput and reduce transaction latency on HFT applications by up to an order of magnitude than the state-of-the-art CC mechanism.

1 Introduction

The ever increasing CPU core counts and memory volume are witnessing a renaissance of concurrency control (CC) mechanisms in exploiting the abundant parallelism [12]. Transaction pipeline, the state-of-the-art CC mechanism, takes advantage over prior CC mechanisms, including two-phase locking (2PL), optimistic concurrency control (OCC), and multi-version concurrency control (MVCC), by allowing more parallel execution among conflicting operations [7,17]. However, it suffers from long time delays confronted with high frequency trading (HFT) applications.

HFT applications have been pervading the worldwide market since the last decade, ranging from individual investment, such as mutual fund management, to

© Springer International Publishing AG 2017
L. Chen et al. (Eds.): APWeb-WAIM 2017, Part II, LNCS 10367, pp. 169–184, 2017.
DOI: 10.1007/978-3-319-63564-4_14

social welfare, such as pension fund management. The recent advance in portfolio selection in HFT market encourages to compose each trade of both strong and weak investment signals in sake of risk management [16]. Because it is easier to capture strong investment signals, such as new production release, different portfolio selection algorithms tend to receive the same strong investment signals while differ at weak investment signals [4]. For example, Listing 1 describes the transactions produced by two portfolios which reflect the same strong investment signals to buy stocks from security Alphabet, Amazon and Twitter while share no weak investment signal in common. As a result, the HFT workloads interleave much contended operations with rarely contended ones, which reflect strong and weak investment signals respectively.

```
1  Transaction T1                          Transaction T2
2  update("Alphabet",    30);              update("Alphabet",    30);
3  update("Amazon",      30);              update("Twitter",     30);
4  update("Cisco",       10);              update("Facebook",    10);
5  update("Microsoft",   10);              update("Macy's",      10);
6  update("Tesla",       10);              update("Oracle",      10);
7  update("Twitter",     30);              update("Amazon",      30);
```

Listing 1. Two transactions generated by two portfolios, where the quantity 30 implies a strong investment signal and 10 a weak investment signal. Here update("t", v) represents that the transaction attempts to buy the security t with quantity v.

Figure 1(a) illustrates the execution of the two transactions in Listing 1 under the state-of-the-art CC mechanism transaction pipeline. The first conflicting operation, e.g., update on Alphabet, determines the serializable order of the two transactions, i.e., T_1 happens before T_2. Then, any operation in T_2 must follow this order; otherwise, the operation will be re-executed after a violation is detected. As Fig. 1(a) illustrates, this is equivalent to delaying the execution of any operation in T_2 till the completion of its corresponding conflicting operation in T_1. On the one hand, this allows T_2 to perform update(Alphabet) before T_1 completes. Thus transaction pipeline indeed outperforms all of 2PL, OCC and MVCC, which do not allow any overlapping execution because otherwise deadlock (2PL) or rollback (OCC and MVCC) will happen. On the other hand, the second update in T_2 on Twitter must delay till the completion of T_1. Compared to Fig. 1(b), which reorders the execution of T_1 and T_2 according to Listing 2, we can see that the delay in Listing 1 unnecessarily compromises throughput and increases transaction latency.

```
1  Transaction T1                          Transaction T2
2  update("Alphabet",    30);              update("Alphabet",    30);
3  update("Amazon",      30);              update("Amazon",      30);
4  update("Twitter",     30);              update("Twitter",     30);
5  update("Cisco",       10);              update("Facebook",    10);
6  update("Microsoft",   10);              update("Macy's",      10);
7  update("Tesla",       10);              update("Oracle",      10);
```

Listing 2. Reorder Transactions from Listing 1.

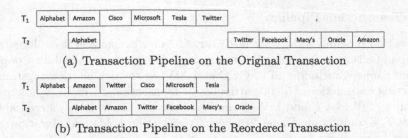

(a) Transaction Pipeline on the Original Transaction

(b) Transaction Pipeline on the Reordered Transaction

Fig. 1. Comparison of delay between original and reordered transactions.

Because delay is caused by conflicting operations only, much work focuses on extracting more fine-grained contention information from transaction semantics to benefit transaction pipeline [5,17,20,21]. However, Fig. 1(a) indicates that HFT applications benefit little from this kind of variants of transaction pipeline.

As a remedy, we present PARE, a pipeline-aware reordered execution of transactions, in this paper. To the best of our knowledge, it is the first time that reordering transaction execution is considered to benefit the CC mechanism of transaction pipeline. First, we observe and formulate harmful ordering of statements in transaction code and propose to eliminate harmful ordering by rearranging statements in decreasing order of the degree of contention. To this end, we devise two mechanisms. On the one hand, to preserve serializability after reordering, we devise a reordering block extraction algorithm. On the other hand, to measure the degree of contention, we devise a physical operator based counter. We evaluate the performance and practicality of PARE. The experiment results show that PARE improves transaction throughout and reduces transaction latency by up to an order of magnitude than the state-of-the-art CC mechanisms. In addition, the runtime overhead is limited.

The contributions of this paper are fourfold:

- We show the importance of reordering transaction execution for the state-of-the-art CC mechanism transaction pipeline under HFT applications.
- We observe and formulate harmful ordering under transaction pipeline and propose to rearrange statements in decreasing order of contention to eliminate harmful ordering.
- We propose two mechanisms to ensure the correctness of rearrangement of transaction statements and measure the degree of contention to enforce the elimination of harmful ordering.
- We conduct experiments to demonstrate the effectiveness and practicality of PARE.

2   Preliminary and Related Works

In this section, we first overview the CC mechanism of transaction pipeline and then review the related works. Readers familiar with transaction pipeline can pass Sect. 2.1.

2.1 Transaction Pipeline

Transaction pipeline is the state-of-the-art CC mechanism and it exploits much more parallelism than popular CC mechanisms. To ensure serializability, popular CC mechanisms including 2PL, OCC and MVCC restrict interleavings among conflicting transactions [7]. In particular, if transaction T_2 reads T_1's write to x, then all 2PL, OCC and MVCC produce schedules in which T_2's read always follows T_1's completion. Under 2PL, each transaction holds long-duration locks on records; any locks acquired by a transaction are only released at the end of its execution [6]. This long discipline constraints the execution of conflicting reads and writes; if transaction T_2 reads T_1's write to record x, and T_1 holds a write lock on x until it completes, T_2's read can only be processed after T_1 completes. Under OCC, transactions perform writes in a local buffer, and only copy these writes to the active database after validation [11]. Thus, a transaction's writes are only made visible at the very end of the transaction. Under MVCC, each write is associated with a timestamp when the transaction commits and any transaction can only read a record whose timestamp is less than the timestamp generated when the transaction begins [14]. Thus, MVCC similarly constrains conflicting transactions.

The emerging transactions from scientific computing, which for example splits a graph computing task into many ACID transactions [8–10], to daily life, which for example encodes HFT applications into ACID transactions, calls for more aggressive CC protocols. To cater such demands, transaction pipeline allows transactions to read uncommitted writes (dirty reads), and enforces a commit discipline on transactions that perform dirty reads [7,17]. In practice, if transaction T writes a record and later aborts, then any transaction that reads T's write also aborts. To ensure serializability, transaction pipeline determines the dependency relationship when the first conflicting operation is detected among conflicting transactions and later operations must not violate this relationship. For example, the first time T_2 reads T_1's write, it is determined that T_2 should happen after T_1. Then, T_2 may write another record y before T_1 attempts to read and write y. When T_1 accesses y, T_1 will detect the relationship violation and thus T_2 has to undo the work on record y and redo the execution after T_1 executes on record y. In this way, transaction pipeline in effect delays T_2's later conflicting operations till T_1's execution on these operations respectively.

2.2 Related Works

Our work is mainly related to the following two lines of research.

Variants of Transaction Pipeline are attracting the most attention in concurrency control research and have been built upon OCC and MVCC [5,7,17,20]. These methods chop transactions into pieces such that once each piece of transaction code begins to execute, it will not need to delay until the completion of the whole piece. Different static analysis techniques are designed to extend the scope of each piece to reduce runtime overhead. This is achieved by identifying more fine-grained contention information. In particular, IC3 [17] and

TheDB [20] assume fixed workload on simple Get/Put/Scan interfaces and thus cannot be applied to dynamic HFT workloads. MV3C [5] and PWV [7] annotate transactions by hand, which is also impractical in HFT applications. In contrast, our PARE operates on dynamically generated SQLs automatically and thus can be applied on fast changing transactions issued by HFT applications. In addition, these methods do not consider reordering transaction execution and thus suffer from inferior performance under HFT workloads.

Program analysis is widely adopted to improve database performance from the data application's perspective. Sloth [3] and Pyxis [2] are tools that use program analysis to reduce the network communication between the application and DBMS. I-confluence [1] determines whether a conflicting operation will preserve application invariants based on application dependent correctness criterion. QURO [21] tracks dependencies among transaction statements to reorder transaction execution for 2PL. DORA [15] also partitions transaction codes to exploit intra-transaction parallelism provided by multi-core server for 2PL. Majority of these techniques do not consider reordering transaction execution and thus cannot reduce delays for transaction pipeline. QURO seems the most related approach to PARE since both of them reorder transaction execution. However, two inappropriate design choices of QURO hinders transaction pipeline from boosting HFT applications: (1) QURO assumes fixed workload and requires to sample the degrees of contention for each statement in advance. Unfortunately, workload sampling is prohibitive because each transaction is dynamically generated according to the timely business information in HFT market. (2) QURO estimates the degree of contention at statement level and by the delayed time. As discussed in Sect. 3.3, this underestimates the degree of contention and repeatedly obtains inconsistent degree of contention. In contrast, PARE estimates at physical operator level and counts the occurrence frequency of contention.

3   Pipeline-Aware Reordered Execution

In this section, we first deduce the reordering strategy of PARE and then present two mechanisms to enable this strategy.

3.1   Reordering Strategy

We observe that there are two kinds of delays under transaction pipeline, which are the only source of inferior performance. The first kind of delays comes from the first conflicting operation, such as the update(Alphabet) of transaction T_2 in Fig. 1. Because it is undefined to write to the same object simultaneously, such delay is inevitable. Fortunately, this kind of delays lasts for only one operation and thus can be neglected in transaction execution. The second kind of delays can last indefinitely long time, such as the update(Twitter) of transaction T_2 in Fig. 1(a). We observe that such delay is formed when non-conflicting operations are encompassed by conflicting operations and formulate this observation as harmful ordering under transaction pipeline.

Definition 1. *Suppose that two transactions consisting of n and m statements are denoted by $T_1 = [s_1, s_2, \ldots, s_n]$ and $T_2 = [s'_1, s'_2, \ldots, s'_m]$, a harmful ordering exists if there exist $1 \leq i < j \leq n$ and $1 \leq p < q \leq m$ such that the following three conditions hold:*

1. s_i *conflicts with* s'_p *and* s_j *conflicts with* s'_q
2. $\forall i < x < j, p < y < q$, s_x *does not conflict with* s_y
3. $j - i \neq q - p$

For example, in Listing 1, where $n = m = 6$, a harmful ordering exists by setting $i = 1$, $j = 2$, $p = 1$ and $q = 6$. Condition 1 holds because `update(Alphabet)` and `update(Twitter)` of T_1 and T_2 are two conflicting operations. Condition 2 and 3 hold because there is no operations between `update(Alphabet)` and `update(Twitter)` in T_1 while there are other operations between `update(Alphabet)` and `update(Twitter)` in T_2.

Theorem 1. *Any delay after the first conflicting operation is caused by harmful ordering.*

Proof. Assume that there is no harmful ordering, there are three possibilities: (1) condition 1 does not hold. Then, there is no conflicting operations or only one confliction operation. In the first case, transactions can execute without any delay. In the second case, the transactions delay at the first conflicting operation. (2) condition 1 holds while condition 2 is violated. In this case, $\forall i < x < j$, $p < y < q$ such that s_x conflicts with s'_y, we have $x - i = y - p$. In this way, if s_i is first executed, when T_2 is about to execute s'_y, s_x must have been completed and thus there is no delay. Otherwise if $x - i \neq y - p$, we can construct another harmful ordering which contradicts with our assumption. (3) condition 1 and 2 hold while condition 3 is violated. In this case, if $s_i(s'_p)$ is first executed, when it is about to execute $s'_q(s_j)$, $s_j(s'_q)$ must have been completed and thus there is no delay. In summary, if there is no harmful ordering, no delay after the first conflicting will occur.

To eliminate harmful ordering, it works to rearrange statements in increasing or decreasing order of contention. In this way, the Condition 2 in Definition 1 will not hold in overall cases. In this paper, we make an arbitrary choice of reordering in decreasing order of contention and experiments show that these two strategies match each other. Applying this strategy to Listing 1, we obtain the reordered Listing 2 and Fig. 1(b) illustrates the execution and we can see that there is no delay longer than one operation.

To enforce this strategy, there are two obstacles: (1) reordering some statements may violate program semantics and (2) the degree of contention is fast changing and cannot be obtained in advance. Next, we show how to reorder statements safely and dynamically estimate the degree of contention.

3.2 Reordering Block Extraction

PARE manipulates on *reordering blocks*, which can be rearranged arbitrarily while preserving transaction semantics. These reordering blocks are extracted from transaction code automatically by leveraging data dependencies.

Definition 2 (Reordering Block). *Given each transaction in the form of a sequence of statements* $S = [s_1, s_2, \ldots, s_n]$, *a reordering block* B *is a subsequence* $B = [s_{i_1}, s_{i_2}, \ldots, s_{i_m}]$, *where* $1 \leq i_1 < i_2 < \cdots < i_m \leq n$, *such that every statement* s_{i_j} *in block* B *does not depend on any statement* s' *in other reordering blocks in terms of data flow. Formally,* $\forall s' \in S - B, \forall s_{i_j} \in B$, *the three types of dependencies should be eliminated:*

1. $readSet(s_{i_j}) \cap writeSet(s') = \emptyset$ *such that* s_{i_j} *does not depend on* s' *in terms of* read-after-write *dependency.*
2. $writeSet(s_{i_j}) \cap readSet(s') = \emptyset$ *such that* s_{i_j} *does not depend on* s' *in terms of* write-after-read *dependency.*
3. $writeSet(s_{i_j}) \cap writeSet(s') = \emptyset$ *such that* s' *and* s_{i_j} *do not depend on each other in terms of* write-after-write *dependency.*

In brief, each reordering block collects statements without the three types of dependencies with the remaining statements. If we partition the statements of a transaction into disjoint reordering blocks, it ensures serializability after arbitrary rearrangement without extra handling from existing database systems which enforce CC mechanism for serializable schedule.

Theorem 2. *Given a transaction S and a set of reordering blocks $\mathcal{B} = \{B\}$, where $\forall B, B' \in \mathcal{B}, B \cap B' = \emptyset$ and $\bigcup_{B \in \mathcal{B}} = S$, rearranging reordering blocks will not compromise serializability under serializable schedule.*

Proof. According to the definition of reordering blocks, the reordering blocks do not conflict with each other. Following the theory of Conflicting Serializability [18], these reordering blocks can be rearranged arbitrarily such that the behavior of the transaction will not be affected.

The three types of dependencies should be tracked on both program statements and SQL statements. To deal with program statements, it is straightforward for us to adopt the standard dataflow algorithm [13,21]. For simplicity, in this paper, we regard the if-statement and loop-statement as a whole statement, although sophisticated techniques such as loop fossil can make fine-grained dependency tracking [19,21]. As for SQL queries, we precisely model the dependency relationships among the query input and output:

1. Simple selection with no sub-queries: the read set encompasses all predicates in the where clause. Empty where clause leads to a universal read set. The write set is set to empty.
2. Other selection: the read set encompasses all predicates occurred in all where clauses. If empty where clause occurs in one of the sub-queries, the read set of the query is also a universal set. The write set is set to empty.
3. Simple update with no sub-queries: the read set is set as "simple selection with no sub-queries" does. The write set is the same as the read set.
4. Other update: the read set is set as "other selection" does. The write set is the same as the read set.

According to these rules, Algorithm 1 shows how to transform transaction code into reordering blocks. We iteratively check for each unused statement which statements cannot reorder with it (Line 4–9). All statements that cannot reorder each other will compose a reordering block (Line 11). The concatenation preserves the original execution order for statements in any reordering block (Line 9, where $\oplus$ denotes concatenation). After all statements assigned with certain reordering blocks, all reordering blocks are found (Line 4,12). For example, for Transaction T_1 in Listing 1, all statements are simple updates and the read set and write set of these statements are disjoint. In this way, each statement creates a reordering block. In other words, all updates can be reordered arbitrarily. In this paper, we focus on reordering selection and update. Dependency relationships of other SQL queries can be defined accordingly.

For example, for Transaction T_1 in Listing 1, all statements are simple updates and the read set and write set of these statements are disjoint. In this way, each statement creates a reordering block. In other words, all updates can be reordered arbitrarily. In this paper, we focus on reordering selection and update. Dependency relationships of other SQL queries can be defined accordingly.

Algorithm 1. Reordering block extraction

Input: A sequence of statements $S = [s_1, s_2, ..., s_n]$
Output: A set of Reordering Blocks $\mathcal{B}$

1 $\mathcal{B} = \emptyset$
2 **for** *each statement* $s \in S$ **do**
3 $\quad$ compute the **readSet** and **writeSet** of s
4 **for** $S \neq \emptyset$ **do**
5 $\quad B = [S[0]]$
6 $\quad$ **for** *each* $s' \in S, s' \notin B$ **do**
7 $\quad\quad$ **for** *each* $s \in B$ **do**
8 $\quad\quad\quad$ **if** $readSet(s) \cap writeSet(s') \neq \emptyset$ *or* $writeSet(s) \cap readSet(s') \neq \emptyset$ *or* $writeSet(s) \cap writeSet(s') \neq \emptyset$ **then**
9 $\quad\quad\quad\quad B = B \oplus s'$
10 $\quad S = S - B$
11 $\quad \mathcal{B} = \mathcal{B} \cup \{B\}$
12 **return** $\mathcal{B}$

3.3 Contention Estimation

In this section, we first show the design to measure the degree of contention and then show a memory-compact way to recycle counters.

To obtain the degree of contention of a reordering block, we first estimate the degree of contention of each statement thereof and use the maximum as the reordering block's degree of contention. In this way, we will not miss highly contended operations enclosed by reordering blocks with many less contended operations. Existing method uses waiting time and its variation on each statement to estimate the degree of contention [21]. This solution has two shortcomings: (1) the counter set for each unique statement underestimates the real degree of contention for the statement. For HFT application, each statement usually contains different parameters. For example, the SQL query for update(Alphabet, 30) is usually instantiated as update Stock set quantity -= 30, price = p where sid = Alphabet, where p is the latest price provided by user. Thus many statements updating the same quantity and price fields are counted in different counters. (2) estimating waiting time and its variance suffers from thrashing. This is because properly reordered highly contended operations will not wait for long time and thus are always categorized as less contended operations. Then, these operations will be placed to the end of the transaction and be considered as heavily contended operations again. To address these two issues, we propose an execution plan based estimation which collects the occurrence of each object in a time sliding window.

We observe that although conflicting operations may be represented in different SQLs, they must create the same physical operator referring to the same attributes. For example, although the update(Alphabet, 30) may update different prices, its corresponding physical operator will have the same type of update and indicate the same fields of quantity and price. The only difference is the parameters for the target values. Because contention only occurs at the operator placed at the bottom level of the execution plan in form of a tree structure, it is sufficient to allocate a counter recording the occurrence frequency for each physical operator at the bottom level of the execution plan. Algorithm 2 details the counter implementation. The time sliding window size WS is discretized into time slots with granularity of g so that when the time flows, the time slots are recycled to reflect the latest count. Once one physical operator is created at the bottom level of the execution plan, the Increment of the associated counter is invoked. Once we extract all reordering blocks, we invoke Get for each counter within each reordering block and pick up the maximum count as the degree of contention for each reordering block.

Algorithm 2. Time sliding window counter

Input: Time Sliding Window size WS, Granularity g
1 Initialize $C[0 : \lceil WS/g \rceil]$ to $[0, 0, ..., 0]$
2 **Function** Increment(t) // t is the current timestamp
3      $i = t \% (\lceil WS/g \rceil + 1)$;
4      $C[i] += 1$;
5      $i = (t - \lceil WS/g \rceil + 1) \% (\lceil WS/g \rceil + 1)$;
6      $C[i] = 0$;
7   return ;
8 **Function** Get():
9      $c = 0$;
10   **for** $i = 0$ **to** $\lceil WS/g \rceil$ **do**
11 | $c = c + C[i]$;
12   return c;

It may waste memory to allocate each physical operator with a counter, especially under HFT workloads where contended objects are fast changing so that the allocated counter may no longer be used in the next time period. To ease this concern, we maintain a counter pool and every time we need to allocate a new counter, we recycle the counter not used in the last time window.

Our counter mechanism has two parameters to set. From the perspective of the precision of the degree of contention, less g and moderate WS are welcomed. This is because larger g and WS will lead to stale degree of contention and the reordering will not eliminate harmful ordering. Too small WS compared to g will lost much contention information. From the perspective of memory overhead, large WS is not expected. However, too small g is prohibited by hardware. In this paper, experiments show that the setting of $WS = 1\,\mathrm{s}$ and $g = 100\,\mathrm{ms}$ works well under HFT workloads.

4   Experiment

In this section, we report experiment results on both workloads of HFT applications and typical OLTP workloads. We demonstrate the effectiveness and practicality of PARE.

4.1   Experimental Setting

In the experiments, we compare PARE against the original transaction pipeline implementation and a well-adopted popular CC mechanism MVCC. The evaluation is performed on a contended workloads representing the HFT application and a widely adopted OLTP benchmark TPC-C. The HFT workload runs on a simplified stock exchange market based on a single table: Stock (SecurityName, Price, Time, Quantity). Each item in the Stock table represents a security in the market with its name, latest price, latest trading time and the accumulated trading quantity. The HFT workload runs a unique type of transaction parameterized by a vector of security names. The n security names are selected by a portfolio selection algorithm. This design is motivated by the typical OLTP workload TPC-E, except that it allows various degrees of contention. In every f seconds, 5 out of n random securities are selected as strong investment signals. We mimic the dynamic nature and different degrees of contention of HFT applications by adjusting f and n respectively.

The experiments were carried out on HP workstation equipped with 4 Intel Xeon E7-4830 CPUs (with 32 cores and 64 physical threads in total) and a 56 GB RAM. The operating system was 64-bit Ubuntu 12.04. During the experiments, all databases were stored in an in-memory file system (tmpfs) instead of the hard disk, so that we could exclude the influence of I/O and focus on the efficiency of concurrency control.

4.2   Performance Comparison

In this section, we first compare the performance under HFT workloads to demonstrate the effectiveness of PARE. We present the results under three settings of extremely, highly and moderately contended workloads. Then, we compare the performance under typical OLTP workload, TPC-C, to demonstrate the broad scope of application of PARE.

Under extremely contended workloads, every transaction only contains the five conflicting operations ($n = 5$). As Fig. 2(a) illustrates, we can see that only PARE scales to 32 working threads and other methods crumble. What's more, Fig. 2(b) shows that PARE scales well without sacrificing transaction latency. On the other hand, the throughput of transaction pipeline increases by 25% from 1 worker to 32 workers and MVCC does not scale at all. This is because transaction pipeline always needs to delay due to harmful ordering. As for MVCC, every transaction will abort if its update operation is not the first to issue after the last transaction commits.

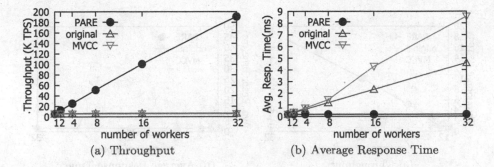

(a) Throughput   (b) Average Response Time

Fig. 2. Performance comparison under extremely contended HFT workloads.

Under highly contended workloads, the five conflicting operations are interleaved with five non-conflicting operations ($n = 10$). As Fig. 3(a) illustrates, PARE still outperforms the others. It is not surprising to observe PARE achieves only roughly half the throughput of that achieved under extremely contended HFT workloads. This time each transaction contains 10 updates, which doubles the workload than the extremely contended workload. Transaction pipeline this time boosts 66% from 1 worker to 32 workers and From Fig. 3(b), we can also see that given less than 8 workers, transaction pipeline scales better than MVCC, compared to Fig. 2(b). This is because the execution on non-conflicting operations makes use of the delay.

Under moderately contended workloads, the five conflicting operations are interleaved with forty five non-conflicting operations ($n = 50$). In this way, each transaction is ten times and five times heavier than that under extremely and highly contended workloads. As Fig. 4(a) illustrates, PARE still outperforms other methods significantly and it can be more obviously observed that transaction pipeline outperforms MVCC. In detail, transaction pipeline speeds up 4.8x

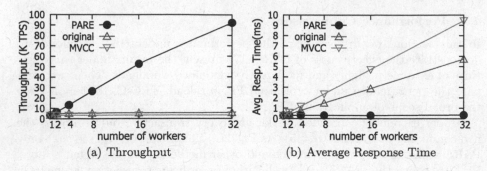

Fig. 3. Performance comparison under highly contended HFT workloads.

from 1 worker to 32 workers. This is also because given more non-conflicting operations, there are more opportunity for non-conflicting operations to use the delayed time that is wasted under previous workloads.

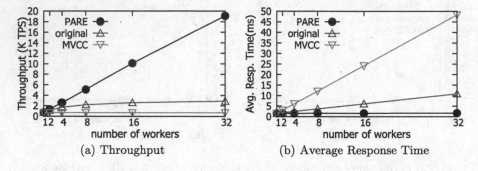

Fig. 4. Performance comparison under moderately contended HFT workloads.

In addition to the HFT workloads, we also compare the performance on the original TPC-C benchmark, the widely adopted OLTP workload. We populate the database with 10 warehouses. In Fig. 5(a), we can see that the performance of PARE and the original transaction pipeline matches and outperforms MVCC by up to 50%. Specifically, more workers result in larger performance gap between transaction pipeline and MVCC. This is because transaction under MVCC will abort due to the contended operations, i.e., update district table in New-Order transaction and consecutive updates on warehouse and district tables in Payment transaction. Rather, transaction pipeline and PARE are able to schedule the later update after the former update once the former update completes. This demonstrates that transaction pipeline indeed exploits more parallelism than MVCC. This also shows that the performance of transaction pipeline is not sensitive to single highly contended operation and two consecutive highly contended operations, which supports our defined "harmful ordering" because one contended operation and two consecutive contended operations do not form harmful ordering. Because there is no harmful ordering in TPC-C[1], the little performance margin between PARE and original transaction pipeline comes from the runtime overhead of PARE. We measure that the combination of extracting reordering blocks for each issued transaction and contention detection makes PARE less than 5% slower than the original transaction pipeline. This shows the PARE can also be applied to general workloads where harmful ordering may not exist.

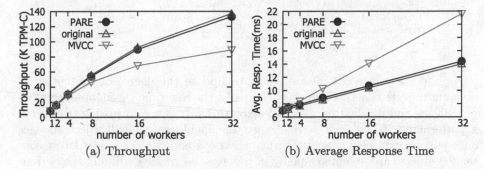

| (a) Throughput | (b) Average Response Time |

Fig. 5. Performance comparison under TPC-C.

In summary, PARE receives huge performance benefit under HFT applications when other concurrency control mechanisms plummet under such workloads. In addition, under rarely contended workloads such as TPC-C, PARE matches the performance of transaction pipeline due to its low runtime overhead.

4.3   Detailed Performance

In this section, we further demonstrate the practicality of PARE by evaluating the overhead of PARE in terms of reordering block extraction, contention identification and memory consumption.

It may be worried that extracting reordering blocks from each issued transaction program is time-consuming especially when harmful ordering cannot be guaranteed to appear in all workloads. To ease such concern, Table 1 profiles the amount of time spent on extracting reordering blocks and transaction execution under single thread respectively. We can see that for both typical OLTP workloads TPC-C and TPC-E, the overhead for the additional dependency analysis is very limited. This is due to our simple but efficient modeling of `readSet` and `writeSet` for SQL queries. Except for the Market-Feed transaction, which occupies 1% in TPC-E workload, the extraction of reordering blocks consumes less than 5% of the execution time. This demonstrates that the overhead of reordering block execution is limited and PARE is practical to typical OLTP workloads.

---

[1] The stock-level update routine in New-Order transaction exhibits a harmful ordering, but the conflicting operations are rarely contended due to the large item table.

Table 1. Execution time (ms) spent on reordering block extraction.

Workloads	Transaction type	Extraction time (ms)/Execution time (ms)	Ratio	Transaction type	Extraction time (ms)/Execution time (ms)	Ratio
TPC-C	New-Order	0.28/7.1	3.9%	Payment	0.26/8.4	3.1%
	Order-Status	0.12/7.6	1.6%	Delivery	0.07/6.2	1.1%
	Stock-Level	0.02/12	0.2%			
TPC-E	Broker-Volume	0.01/4.3	0.2%	Customer-Position	0.05/4.9	1.0%
	Market-Feed	0.21/3.8	**5.5%**	Market-Watch	0.17/9.6	1.8%
	Security-Detail	0.08/2.6	3.1%	Trade-Lookup	0.07/9.4	0.7%
	Trade-Order	0.36/7.4	4.8%	Trade-Result	0.41/8.3	4.9%
	Trade-Status	0.03/7.7	0.4%	Trade-Update	0.15/13.9	1.1%

It is performance-critical for PARE to update the degree of contention of different objects timely. Figure 6(a) compares the real-time throughput between $g = 100$ ms and $g = 500$ ms with WS fixed to 1 s. It is not surprising to observe a performance trough per second when the highly contended object changes once per second. This is because when stale contended objects and latest contended objects are weighted equally in counters, reordering will mix the current highly contended objects with non-conflicting objects and thus reordering will not take effect. Before the performance trough, it is expected that the performance is not sensitive to the misjudgement of highly contended objects. This is because at this time, statements are somewhat reordered in increasing rather than the desired decreasing order of degree of contention and thus still eliminates harmful ordering. This validates our argument that arbitrary choice of increasing or decreasing order of contention does not matter. Figure 6(b) compares the throughput among different settings of WS with g fixed to 100 ms. We can see that $WS = 200$ ms and $WS = 2$ s perform inferior to $WS = 1$ s. On the one hand, too short window size loses much contention information and harmful ordering occurs. On the other hand, too large window size covering many contention change will mix highly contended objects with lowly contended objects many times and thus harmful ordering occurs many times.

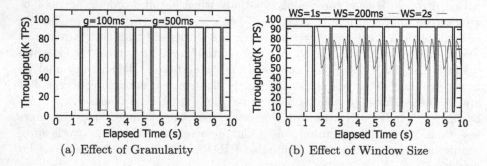

(a) Effect of Granularity   (b) Effect of Window Size

Fig. 6. Effect of counter setting.

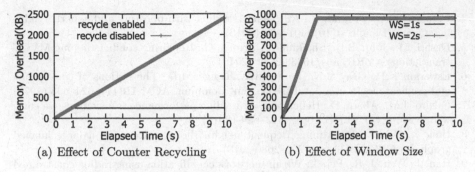

(a) Effect of Counter Recycling   (b) Effect of Window Size

Fig. 7. Overhead of memory consumption.

In the last set of experiments, we evaluate the memory consumption of PARE. Figure 7(a) shows that without recycling strategy, the memory overhead will continuously increase because more and more objects will be accessed. Figure 7(b) shows that less window size can save more memory because it saves the occupied memory of each counter and recycles counters more frequently.

5   Conclusion

This paper novelly proposes PARE to reorder transaction execution to improve the performance of the state-of-the-art concurrency control mechanism, transaction pipeline, under HFT applications. We for the first time observe and formulate the harmful ordering, which is brought with transaction pipeline and hidden under other concurrency control mechanisms. We deduce to reorder transaction execution in order of contention and further propose two mechanisms to extract reordering blocks and identify the degree of contention of objects. Experiments demonstrate that PARE outperforms the state-of-the-art concurrency control mechanism significantly on HFT applications in terms of both transaction throughput and latency and the overhead is limited on typical OLTP benchmarks.

Acknowledgement. This work is supported by Nature Science foundation of China Key Project No. 61432006 and The National Key Research and Development Program of China, No. 2016YFB1000702.

References

1. Bailis, P., Fekete, A., Franklin, M.J., Ghodsi, A., Hellerstein, J.M., Stoica, I.: Coordination avoidance in database systems. PVLDB **8**(3), 185–196 (2014)
2. Cheung, A., Madden, S., Arden, O., Myers, A.C.: Automatic partitioning of database applications. PVLDB **5**(11), 1471–1482 (2012)
3. Cheung, A., Madden, S., Solar-Lezama, A.: Sloth: being lazy is a virtue (when issuing database queries). In: SIGMOD 2014, pp. 931–942 (2014)

4. Creamer, G.G., Freund, Y.: Automated trading with boosting and expert weighting. Quant. Financ. **4**(10), 401–420 (2013)
5. Dashti, M., John, S.B., Shaikhha, A., Koch, C.: Repairing conflicts among MVCC transactions. CoRR, abs/1603.00542 (2016)
6. Eswaran, K.P., Gray, J.N., Lorie, R.A., Traiger, I.L.: The notions of consistency and predicate locks in a database system. Commun. ACM **19**(11), 624–633 (1976)
7. Faleiro, J.M., Abadi, D., Hellerstein, J.M.: High performance transactions via early write visibility. PVLDB **10**(5), 613–624 (2017)
8. Han, J., Wen, J.-R.: Mining frequent neighborhood patterns in a large labeled graph. In: CIKM 2013, pp. 259–268 (2013)
9. Han, J., Wen, J.-R., Pei, J.: Within-network classification using radius-constrained neighborhood patterns. In: CIKM 2014, pp. 1539–1548 (2014)
10. Han, J., Zheng, K., Sun, A., Shang, S., Wen, J.R.: Discovering neighborhood pattern queries by sample answers in knowledge base. In: ICDE 2016, pp. 1014–1025 (2016)
11. Kung, H.T., Robinson, J.T.: On optimistic methods for concurrency control. ACM Trans. Database Syst. **6**(2), 213–226 (1981)
12. Larson, P.-A., Blanas, S., Diaconu, C., Freedman, C., Patel, J.M., Zwilling, M.: High-performance concurrency control mechanisms for main-memory databases. PVLDB **5**(4), 298–309 (2011)
13. Marlowe, T.J., Ryder, B.G.: Properties of data flow frameworks: a unified model. Acta Inf. **28**(2), 121–163 (1990)
14. Neumann, T., Mühlbauer, T., Kemper, A.: Fast serializable multi-version concurrency control for main-memory database systems. In: SIGMOD 2015, pp. 677–689 (2015)
15. Pandis, I., Johnson, R., Hardavellas, N., Ailamaki, A.: Data-oriented transaction execution. PVLDB **3**(1–2), 928–939 (2010)
16. Shen, W., Wang, J., Jiang, Y.-G., Zha, H.: Portfolio choices with orthogonal bandit learning. In: IJCAI 2015, pp. 974–980 (2015)
17. Wang, Z., Mu, S., Cui, Y., Yi, H., Chen, H., Li, J.: Scaling multicore databases via constrained parallel execution. In: Proceedings of the 2016 International Conference on Management of Data, SIGMOD 2016, pp. 1643–1658 (2016)
18. Weikum, G., Vossen, G.: Transactional Information Systems: Theory, Algorithms, and the Practice of Concurrency Control and Recovery. Elsevier (2001)
19. Wolfe, M.J.: High Performance Compilers for Parallel Computing (1995)
20. Wu, Y., Chan, C.-Y., Tan, K.-L.: Transaction healing: scaling optimistic concurrency control on multicores. In: SIGMOD 2016, pp. 1689–1704 (2016)
21. Yan, C., Cheung, A.: Leveraging lock contention to improve oltp application performance. PVLDB **9**(5), 444–455 (2016)

Bus-OLAP: A Bus Journey Data Management Model for Non-on-time Events Query

Tinghai Pang[1], Lei Duan[1(✉)], Jyrki Nummenmaa[2,3], Jie Zuo[1],
and Peng Zhang[1]

[1] School of Computer Science, Sichuan University, Chengdu, China
pangtinghai@163.com, {leiduan,zuojie}@scu.edu.cn, zp_jy1993@163.com
[2] Faculty of Natural Sciences, University of Tampere, Tampere, Finland
Jyrki.Nummenmaa@staff.uta.fi
[3] Sino-Finnish Centre, Tongji University, Shanghai, China

Abstract. Increasing the on-time rate of bus service can prompt the people's willingness to travel by bus, which is an effective measure to mitigate the city traffic congestion. Performing queries on the bus arrival can be used to identify and analyze various kinds of non-on-time events that happened during the bus journey, which is helpful for detecting the factors of delaying events, and providing decision support for optimizing the bus schedules. We propose a data management model, called Bus-OLAP, for querying bus monitoring data, considering the characteristics of bus monitoring data and the scenarios of on-time analysis. While fulfilling typical requirements of bus monitoring data analysis, Bus-OLAP not only provides a flexible way to manage the data and to implement multiple granularity data query and update, but also supports distributed query and computation. The experiments on real-world bus monitoring data verify that Bus-OLAP is effective and efficient.

Keywords: Data management · OLAP · Parallel computing

1 Introduction

Public bus service plays an important and irreplaceable role in the traffic system of a city. On one hand, public bus is one of the most convenient ways for people to travel. On the other hand, public bus service is an effective way to reduce carbon dioxide emissions. Promoting the bus service can make the urban traffic condition better and provide convenience for people. *On-time bus* is an emerging bus running mode where the bus arrives at each bus stop according to the time table strictly, which is helpful for passengers to avoid wasting too much time on bus stop. There are several countries, e.g., the US, Finland, and Japan, where

This work was supported in part by NSFC 61572332, the Fundamental Research Funds for the Central Universities 2016SCU04A22, the China Postdoctoral Science Foundation 2016T90850, and the Academy of Finland Foundation 295694.

L. Chen et al. (Eds.): APWeb-WAIM 2017, Part II, LNCS 10367, pp. 185–200, 2017.
DOI: 10.1007/978-3-319-63564-4_15

Table 1. Instances of bus arrival record

Stop	Longitude	Latitude	Line	Date	Time	Delay	Weekday
1500	23.73550	61.49855	13	2016-08-03	16:40:15	210	Yes
3084	23.80464	61.47992	560	2016-08-04	17:20:09	610	Yes
3084	23.80464	61.47992	560	2016-08-05	00:09:48	−34	Yes
0098	23.73738	61.50003	28	2016-08-06	16:55:12	340	No

efforts have been made to implement on-time bus running mode to improve the bus service quality.

Clearly, a reasonable design of bus running route and time table is the key to carry out the on-time bus running mode. To this end, bus running data including the arrival information of a bus at every stop should be collected at first. Thanks to the development of sensors and Internet of Things, bus running data can be collected automatically. For example, the local government of Tampere, Finland, records and publishes the location information of each running bus every second since 2015. Table 1 lists some records of bus arrivals.

Example 1. In Table 1, "Stop" is the bus stop ID, "Line" is the bus line number, "Longitude" and "Latitude" indicate the location of a bus stop, "Date" and "Time" are the date and time of bus arrival, "Delay" is the difference between the arrival time and the scheduled time ('−' indicates ahead case), and "Weekday" indicates the day is weekday or not.

Bus arrival query, providing the statistics on the non-on-time arrivals of buses that happened under given conditions, is useful for the bus route design and the bus running schedule optimization. Typical bus arrival queries include: How many bus delaying events happened over a given time period? Which is the route where most delaying events happen? Do the bus delaying events aggregate and where?

Based on the query conditions, there are three kinds of bus arrival queries:

- Temporal queries, in which the query conditions are related to bus running time. For example, how many buses were delayed during 17:00–19:00?
- Spatial queries, in which the query conditions are related to bus stop locations. For example, given a bus stop ID "560", which is the nearest neighboring bus stop where delaying events happen?
- Spatial-temporal queries, in which the query conditions are related to both bus running time and bus stop locations. For example, which are the spatial-temporal zones with significant aggregation of delaying events?

To the best of our knowledge, there is no data management model dealing with the bus delay queries. We will review the related works in Sect. 2.

Since non-on-time analysis of bus monitoring data can be used to improve bus service, thus improving comfortable travel experience, we should provide an efficient model to manage bus journey data. However, there are some challenges that we need to address: (i) how to manage bus running data? (ii) how to fuse different source data? and (iii) how to query data efficiently? In this paper, to tackle these challenges of non-on-time query, we propose a model named Bus-OLAP to manage bus running data. The main contributions include: (i) building index of bus running data by bit vectors according to the application scenarios, such as the external factors (rush hour and holidays); (ii) implementing efficient index storage and updates; (iii) supporting distributed queries and computation based on Spark to accommodate the requirement of massive data processing and real-time response.

The rest of the paper is organized as follows. We review the related work in Sect. 2, and present the critical techniques of Bus-OLAP in Sect. 3. In Sect. 4, we report a systematic empirical study using real-world bus data and we conclude the paper in Sect. 5.

2   Related Work

The analysis on urban traffic plays an important role and attracts extensive attention from public. Traffic analysis helps to alleviate urban traffic congestion thereby improving environmentally-friendly traffic for people to travel. There are many researches about traffic analysis recently. Yuan *et al.* [19] analyze the history of taxis and passengers, and provide an easy way for passengers to pick taxis. Pang *et al.* [9] detect the anomalous behaviour of taxis in Beijing metropolitan area. Chen *et al.* [1] propose a model given the past trajectories and predict the next station of an object. Kong *et al.* [7] aim to deal with the problem of traffic congestion, and propose a method to predict the traffic congestion using the floating car trajectories data. Han *et al.* [5] focus on traffic dynamic prediction in urban transportation network, and find out the evolution of traffic states, which contributes to the adjustment of traffic management. Some other researches predict the urban traffic flow by the traffic state [3,11,14]. Wu *et al.* [15] detect the temporal-spatial aggregation of bus delay but do not consider the management of bus data. To the best of our knowledge, the existing works about traffic analysis are tackling the problems about vehicles with stochastic trajectories. However, the analysis of bus data is different from other vehicles. For example, the trajectory of a bus is fixed and the concern on bus monitoring data is non-on-time analysis, which are our main differences from other works.

Urban traffic data has spatial-temporal characteristics. The bus data management can improve the efficiency of storage, indexing and querying. Sistla *et al.* [10] propose a model, called MOST, to express the moving object by a time related function, which improves the efficiency of storage and querying moving object in database. Ting *et al.* [13] present a simplistic network model for moving objects. Some index structures such as R-tree [4] and B^+-tree [6] are also used to optimize spatial-temporal queries. Yu *et al.* [18] propose an algorithm YPK-CNN to monitor KNN queries. The works discussed above can improve the query efficiency by different ways. However, we want to provide a model which we can tailor to take into account efficiently the factors we choose, including external factors such as holidays and weather, which also affect bus running obviously.

Considering the analysis of large scale bus arrival data, parallel computing technologies should be applied to speed up efficiency of spatial-temporal analysis [8]. There are also many works that adapt the parallel technologies to cater the demand of large scale data queries. Xia et al. [16] design a method to conduct KNN queries based on Map-Reduce and apply it on real-time prediction of traffic flow. Eldawy et al. [2] build a system named GeoSpark based on Spark to improve the efficiency of spatial-temporal data analysis. Xie et al. [17] achieve parallel query based on Spark with R-tree index to deal with the efficiency of large scale data. In this work, we use Spark to build a parallel processing model based on the operation on bit vectors to conduct non-on-time queries.

3   Design of Bus-OLAP

In this section, we present our Bus-OLAP data management model for bus arrival queries. The framework of Bus-OLAP (Fig. 1) consists of data cleaning, transforming, loading, and query. Since the cleaning of bus monitoring data has been well studied in [12], so we leave it out due to limited space, and below we discuss the most important techniques used in Bus-OLAP: (i) data indexing, (ii) index operation, and (iii) efficient query and computation.

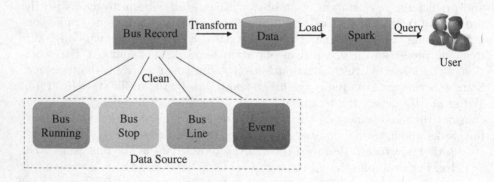

Fig. 1. Framework of Bus-OLAP

3.1   Data Indexing

To support efficient bus arrival queries, we implement an index based on attribute domain partition in Bus-OLAP. Specifically, we divide the domain of each attribute into several disjoint partitions initially. Then, for each attribute value of a record, we build a bit vector to indicate the partition it belongs to. Thus, we can apply bit operations to bus arrival queries, which improves the query efficiency.

Next, we introduce the details of the index building. For attribute A, the *domain* of A, denoted by $\mathcal{D}(A)$, is the set of all available values on attribute A. A *partition* of $\mathcal{D}(A)$, denoted by $\mathcal{P_D}(A)$, is a collection of non-empty subsets of $\mathcal{D}(A)$ such that (i) $\mathcal{D}(A) = \bigcup\limits_{d_i \in \mathcal{P_D}(A)} d_i$, and (ii) for $\forall d_i, d_j \in \mathcal{P_D}(A)$, $i \neq j$, $d_i \cap d_j = \emptyset$. Please note that the order of elements in $\mathcal{P_D}(A)$ is fixed in Bus-OLAP. We use $\mathcal{P_D}(A)_{[i]}$ to indicate the i-th element in $\mathcal{P_D}(A)$. Furthermore, there may be several different partitions for an attribute.

For a record r, we denote by $r.A$ the value of r on attribute A. Given a partition $\mathcal{P_D}(A)$ on attribute A, the index of $r.A$ is a *bit vector* $<b_1, b_2, ..., b_{|\mathcal{P_D}(A)|}>$ satisfying

$$b_i = \begin{cases} 1, & r.A \in \mathcal{P_D}(A)_{[i]} \\ 0, & r.A \notin \mathcal{P_D}(A)_{[i]} \end{cases} \tag{1}$$

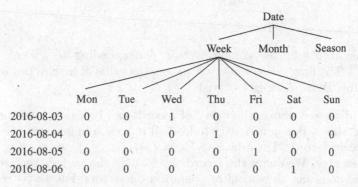

Fig. 2. An example of a partition on attribute "Date"

Example 2. A partition on attribute "Date" according to week is shown in Fig. 2. The domain "Date" is partitioned into seven subdomains, corresponding to the seven days of a week. The day "2016-08-03" in Table 1 represents 3rd Aug 2016, and it is Wednesday. Therefore, the index of first record in Table 1 is $<0, 0, 1, 0, 0, 0, 0>$ under partition $\mathcal{P_D}("Date")$.

Clearly, the space cost would be very high if the number of partitions ($|\mathcal{P_D}(A)|$) is large. In Bus-OLAP, the selection of attribute partition depends on: (i) the number of partitions; (ii) the requirement of query scenarios. Take an example of attribute "Date", divide the domain by different granularity such as week, month and season. Note that we do not divide the domain of "Date" by day because of the large space cost of partitions. Table 2 lists the partition criterion of main attributes domains in Table 1 by considering the application requirements.

For a partition $\mathcal{P_D}(A)$ on attribute A, the index of all records is denoted by a *bitmap* $V_A = \{v_1, v_2, ..., v_{|\mathcal{P_D}(A)|}\}$, where $v_i \in V_A$ is a column vector and corresponds to $\mathcal{P_D}(A)_{[i]}$, $1 \leq i \leq n$. If the value of k-th element of v_i is 1, the value of k-th record on attribute A belongs to $\mathcal{P_D}(A)_{[i]}$.

Table 2. Attribute domain partition criterion

Attributes	Criterion	Explanation
Latitude, Longitude	Distance	According to the geographic information, divide the map into grids
Stop	Location	The location of bus stop is fixed and the query of KNN stops and RKNN stops require the locations of stop
Date	Week, Month, Season	To query different granularity of time and partition Date by week (Mon, Tus, ...), month (Jan, Feb, ...) and Season (Spring, ...)
Time	Time interval	To query the on-time state of different period of time, we partition the time by time interval
Delay	Threshold	Partition delay by the threshold predefined by user

Example 3. In Fig. 2, the column vector corresponding to "Wed" is $v = (1, 0, 0, 0)^T$. The first element of v is 1 means the value of first record in Table 1 on attribute "Date" belongs to "Wed".

As we discussed above, the query of records can be easily implemented by vector operations. It is unnecessary to load all indexes into memory when querying the recent records. For a bitmap $V_A = \{v_1, v_2, ..., v_n\}$, we store $v_i \in V_A$ as a unit of storage. We query the records by loading the vectors that related to query conditions and then monitor operation on vectors. For a new record, we update the index efficiently by appending a bit to the end of vector.

3.2 Index Operation

We now introduce some operations between vectors. For a partition $\mathcal{P}_D(A)$, we define the *sub-partition* of $\mathcal{P}_D(A)$ as $\mathcal{P}_D(A)'$, where $\mathcal{P}_D(A)' \subseteq \mathcal{P}_D(A)$. The bitmap corresponding to $\mathcal{P}_D(A)'$ is $V_A' = \{v_1', v_2', ..., v_m'\}$. If there is an element $\mathcal{P}_D(A)'_{[i]}$ such that $r.A$ belongs to $\mathcal{P}_D(A)'_{[i]}$, then we say that record r satisfies sub-partition $\mathcal{P}_D(A)'$. We define the operation on V_A' as $OR(V_A') = v_1'|v_2'| \cdots |v_m'$, where $v_i' \in V_A'$ and "|" denotes the OR operation between vectors. Correspondingly, the result of $OR(V_A')$ is also a vector. The records satisfying sub-partition $\mathcal{P}_D(A)'$ are obtained by the bit operation on vectors.

Example 4. Consider Fig. 2, we want to find the records generated from Monday to Friday. The sub-partition $\mathcal{P}_D("Date") = \{Mon, Tue, Wed, Thu, Fri\}$, and the corresponding bitmap is $V_{Date} = \{v_1, v_2, v_3, v_4, v_5\}$, where $v_1 = (0, 0, 0, 0)^T$, $v_2 = (0, 0, 0, 0)^T$, $v_3 = (1, 0, 0, 0)^T$, $v_4 = (0, 1, 0, 0)^T$ and $v_5 = (0, 0, 1, 0)^T$. Then $OR(V_{Date}) = v_1|v_2|v_3|v_4|v_5 = (1, 1, 1, 0)^T$, and the records satisfying the query condition are the first three records in Table 1.

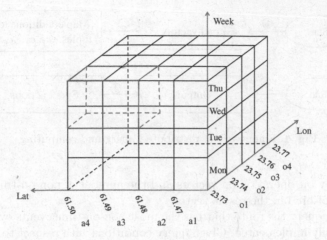

Fig. 3. An example of attribute combination query

Similarly, we define the AND operation between the vectors of different sub-partitions. For two sub-partitions $\mathcal{P}_\mathcal{D}(A_1)$ and $\mathcal{P}_\mathcal{D}(A_2)$, bitmap V_{A_1} corresponds to $\mathcal{P}_\mathcal{D}(A_1)$ and V_{A_2} corresponds to $\mathcal{P}_\mathcal{D}(A_2)$. The records that satisfy both these two sub-partitions can be obtained by AND operation $OR(V_{A_1})\&OR(V_{A_2})$.

Example 5. As shown in Fig. 3, consider querying the records whose longitude ranges from 23.73 to 23.74, latitude ranges from 61.48 to 61.50 and are generated on Wednesday and Thursday. The sub-partition of "Longitude" is $\{o_1\}$, "Latitude' is $\{a_3, a_4\}$ and "Date" is $\{Wed, Thu\}$. The result corresponds to the red area in Fig. 3 and is queried by bit operation $OR(\{a_3, a_4\})\&OR(\{o_1\})\&OR(\{Wed, Thu\})$.

3.3 Parallel Query and Computing

Non-on-time analysis of bus arrival is query sensitive, and requires quick response from the system. Figure 4 shows the framework of parallel querying and computing by vectors based on Spark. Next, we introduce the details of parallel query and computing by three typical kinds of queries in non-on-time analysis.

Temporal Queries. The typical temporal queries include: query the number of non-on-time events over a period of time; query the route that has the most non-on-time events over a period of time.

Consider to query the number of non-on-time events over a period of time, we denote the time interval as t, and the non-on-time condition z (ahead, on-time or delayed). The corresponding condition tuple in Fig. 4 is $<t, z>$. Note that there can be many time intervals, and there will be many condition tuples. We compute each condition tuple in parallel based on Spark. For each tuple, the query result

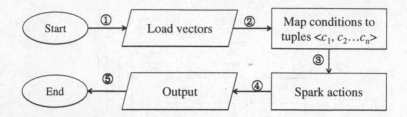

Fig. 4. Framework of distributed query and computing

is obtained by bit operation on vectors and the number of non-on-time events is the number of bit 1 in the result vector.

Similarly, query the route that has the most non-on-time events over a period of time is easily implemented. Given query conditions with respect to time interval t, non-on-time condition z and bus line l, the condition tuple is $<t, z, l>$. Each tuple computing returns the number of delaying events, and the query result is the line that has the maximum delaying events over all tuple results.

Spatial Queries. There are some typical spatial queries such as KNN and RKNN queries of a given bus stop. In non-on-time analysis, if there is a bus stop p that has many delaying events, the analyst may want to know the bus stops nearing to p and analyze the reason that causes this delay phenomenon.

After partitioning the location information "Longitude" and "Latitude", the map is divided into grids. For a bus stop q, the main steps of querying the k nearest neighbor bus stops are: (i) extend the search space by spreading a layer of grids based on q; (ii) when there is a rectangular area that includes at least k stops, find the k-th stop p that away from q in distance, and calculate the distance r between q and p; (iii) continue to extend the search space under the limit of radius r and find k nearest neighbor stops.

Example 6. Considering an example in Fig. 5, search five bus stops that nearest neighboring to bus stop q. The area R_1 includes five bus stops, and p_5 is the 5-th bus stop away from q by distance. The maximum search space is area R_2 that does not exceed radius r. And the search result is the nearest five bus stops in R_2 away from q.

Please note that the query of KNN bus stops is also obtained by the operation of vectors. Consider the area R_1 in Fig. 5, the vector operation on longitude is $op_{lon} = OR(\{o_2, o_3, o_4, o_5, o_6\})$ and latitude is $op_{lat} = OR(\{a_2, a_3, a_4, a_5, a_6\})$. When extending the rectangular area from R_1 to R_2, the new result on Longitude is $op_{lon}|OR(\{o_1, o_7\})$ and Latitude is $op_{lat}|OR(\{a_1, a_7\})$. Obviously, every time we extend the search area with a layer, we just need to conduct the bit operation on vectors iteratively.

The query of RKNN is to find out all bus stops whose KNN stops include a given bus stop q. It is implemented using parallel computation on Spark. The condition tuple is $<q, p, k>$, where q is the given bus stop and each tuple in parallel process returns the result that whether p is one of the KNN stops of q.

Spatial-Temporal Queries. Detecting the spatial-temporal zone of bus delay aggregation can be done by querying the zone where delay aggregation occurs in a period of time. The significant of spatial-temporal aggregation delay is measured by log-likelihood ratio. Given a zone S and a time interval T, the likelihood of a bus delay happening is denoted as

$$
\mathcal{L}(S,T) = \begin{cases} \mathcal{D}(S,T) \times log(r_1) + (\mathcal{D}(\widetilde{S}, \widetilde{T}) - \mathcal{D}(S,T)) \times log(r_2) \\ -\mathcal{D}(\widetilde{S}, \widetilde{T}) \times log\frac{\mathcal{D}(\widetilde{S}, \widetilde{T})}{\mathcal{N}(\widetilde{S}, \widetilde{T})} & , r_1 > r_2 \\ 0 & , r_1 \le r_2 \end{cases}
$$

$$(2)$$

where,

$$
r_1 = \frac{\mathcal{D}(S,T)}{\mathcal{N}(S,T)}, r_2 = \frac{\mathcal{D}(\widetilde{S}, \widetilde{T}) - \mathcal{D}(S,T)}{\mathcal{N}(\widetilde{S}, \widetilde{T}) - \mathcal{N}(S,T)}
$$

where, $\mathcal{N}(S,T)$ is the total number of events that buses arrive bus stops in zone S and time interval T, and $\mathcal{D}(S,T)$ is the number of delaying events. $\widetilde{S}$ and $\widetilde{T}$ are the maximal zone and maximal time interval. S and T are the observed zone and time interval.

We aim to query the zone with a time interval that has maximal LLR in our delay aggregation analysis. Figure 6 shows the process of delay aggregation query with Spark. Step 1 in Fig. 6 corresponds to step 2 in Fig. 4, step 2 and 3 in Fig. 6 correspond to step 3 in Fig. 4. Step 1 joins the query conditions into condition tuples $<o, a, d, t, de>$, where the symbols denote "Longitude", "Latitude", "Date", "Time" and "Delay" respectively. Step 2 partitions tuples into every node of Spark and gets the new tuples of LLR. The last step searches the tuple that has the maximal LLR, which is the result of our query.

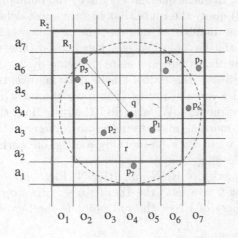

Fig. 5. An example of KNN query

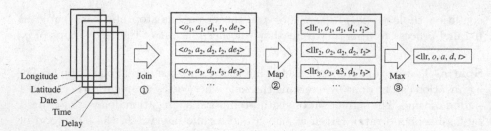

Fig. 6. Spark process of delay aggregation query

The related queries discussed above are the typical application scenarios of Bus-OLAP. It is easy to perform the queries by vector operations with Spark.

4   Empirical Evaluation

In this section, we report a systematic empirical study on real-world bus dataset from Tampere[1]. The dataset includes the bus stop information of Tampere, 43 bus routes, 75 bus stops of a line on average and 6,297,520 records generated from all workdays during 3rd August 2015 and 30th October 2015.

All algorithms are implemented in Java and compiled by JDK 7. The distributed environment is built by Spark 1.6.2 and includes eight nodes. Each node is a computer with an Intel Core i7-6700 3.40 GHz CPU, and 64 GB main memory, running Ubuntu 14.04 operating system.

4.1   Effectiveness

We verify the effectiveness of Bus-OLAP by three kinds of typical queries (temporal queries, spatial queries and spatial-temporal queries). For the temporal queries, there are two frequent queries: (1) query the number of delaying events in every workday; (2) query the route that has the most delaying events.

Figure 7 shows the number of non-on-time arrival events on workdays of August, September and October. We set the non-on-time threshold as three minutes and find out that there are more delaying events than ahead events of bus arrival. Table 3 lists the routes where delaying events occur frequently. The Lines 13, 3 and 8 are the routes that have the most delaying events on afternoon of all three months. We also find that the lines that occur delaying events frequently on morning and afternoon are different. The possible reason is that many people go to work on morning and go off work on afternoon and causes different traffic congestion of bus.

For the spatial queries of KNN and RKNN, Fig. 8 presents an example of spatial query. We set $k = 8$, and the ID of target bus stop is "560". Figure 8 (a) illustrates the query of KNN, the red point is the bus stop "560" and the

---

[1] http://trafficdata.sis.uta.fi.

Table 3. Query on lines with the most delaying events

Month	Time	Bus line	# of arrival events	# of delaying events	Rank
Aug	Morning	13	55198	8981	1
		28	35443	5766	2
		29	35320	3832	3
	Afternoon	13	93636	21929	1
		3	102989	15549	2
		8	74464	12773	3
Sep	Morning	13	58212	8515	1
		28	31081	7134	2
		3	63086	6046	3
	Afternoon	13	91410	22647	1
		3	103046	15742	2
		8	76246	14825	3
Oct	Morning	3	68771	6902	1
		28	24959	5908	2
		8	61284	5776	3
	Afternoon	8	20124	85300	1
		3	114396	20039	2
		13	81781	18105	3

blue points are the KNN bus stops that nearest neighboring to "560". Similarly, Fig. 8 (b) shows the RKNN query of bus stop "560" and there are eight bus stops whose 8-NN bus stops include "560".

Detection of bus delay aggregation is a typical spatial-temporal query. We partition the domain of longitude and latitude by the same distance interval and the search space becomes to a $2000 * 1000$ grids. We constraint the maximal search range to $100 * 100$ grids. Let's analyze several situations with different

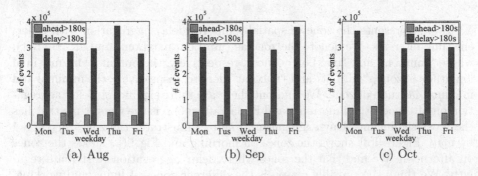

Fig. 7. Number of non-on-time events

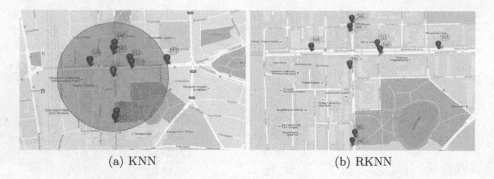

(a) KNN   (b) RKNN

Fig. 8. Spatial query ($k = 8$, $q = 560$) (Color figure online)

time intervals: (1) on which weekday the most significant delay aggregations take place; (2) over which time interval do the most significant delay aggregation occur; (3) which day and time interval has most significant delay aggregation.

Table 4. Zone of spatial-temporal delay aggregation of every day in a week

Weekday	Time	Zone (minLon, maxLon, minLat, maxLat)	LLR
Mon	Morning	(23.810540, 23.844397, 61.495706, 61.510013)	121.74
	Afternoon	(23.595092, 23.633565, 61.502209, 61.518467)	723.79
Tue	Morning	(23.809386, 23.844781, 61.490666, 61.505623)	163.04
	Afternoon	(23.597015, 23.633949, 61.515053, 61.530661)	638.80
Wed	Morning	(23.812079, 23.844397, 61.496194, 61.509850)	130.92
	Afternoon	(23.595092, 23.633565, 61.502209, 61.518467)	641.86
Thu	Morning	(23.810540, 23.844397, 61.495869, 61.510176)	130.64
	Afternoon	(23.593168, 23.631641, 61.496356, 61.512614)	737.03
Fri	Morning	(23.812079, 23.844397, 61.496194, 61.509850)	76.64
	Afternoon	(23.623562, 23.661265, 61.489691, 61.505623)	1111.67

Table 4 presents the zone of spatial-temporal delay aggregation with different time intervals. We denote the zone as (minLon, maxLon, minLat, maxLat), where "minLon" and "maxLon" denote, respectively, the minimal and maximal longitude of zone, "minLat" and "maxLat" denote, respectively, the minimal and maximal latitude of zone. We find out there are more significant delay aggregation on afternoon than morning, and Friday afternoon is the worst time that has the most significant delay aggregation. Figure 9 illustrates the zones in Table 4 by map. Figure 9(a) shows the zones on morning and Fig. 9(b) shows the zones on afternoon. We find that the zones where delay aggregations distribute regularly. We think the possible reason is the different zones of living and working. The query results also reflect the moving traffic flow. Besides, we also conduct

(a) Morning   (b) Afternoon

Fig. 9. Illustration of delay aggregation zones in Table 4

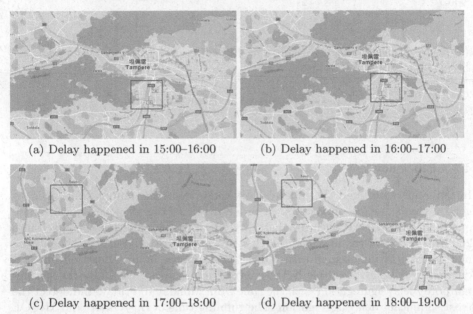

(a) Delay happened in 15:00–16:00   (b) Delay happened in 16:00–17:00

(c) Delay happened in 17:00–18:00   (d) Delay happened in 18:00–19:00

Fig. 10. Results of delay aggregation detection in different time interval

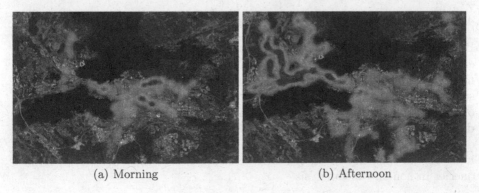

(a) Morning   (b) Afternoon

Fig. 11. Bus delay heat map

the experiments on different time interval of a day. As shown in Fig. 10, we can observe the moving of traffic flow from 15:00 to 19:00. And Fig. 11 shows the heat map of bus delay, which verifies the effectiveness of delay aggregation detection by Bus-OLAP.

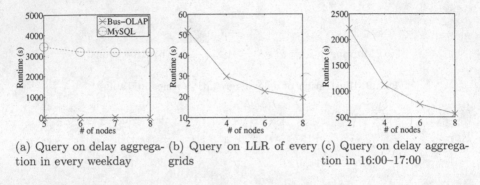

(a) Query on delay aggregation in every weekday   (b) Query on LLR of every grids   (c) Query on delay aggregation in 16:00–17:00

Fig. 12. Runtime w.r.t. number of nodes

4.2 Efficiency

In our paper, we build the index by bit vector, and the non-on-time query is got by the operation on vectors. In order to verify the efficiency, we compare our model Bus-OLAP with database MySQL 5.5. We partition the space into 200*100 grids and the maximal search space is one grid. In our experiment, the number of delaying events in Bus-OLAP is computed by bit operation but it is queried by MySQL database in the contrast experiment. We query the spatial-temporal delay aggregation with the time from Monday to Friday, and detect the most significant delay aggregation. The running time of Bus-OLAP and MySQL is shown in Fig. 12(a) and Bus-OLAP performs better than MySQL.

4.3 Scalability

Bus-OLAP should adapt to large scale query and computing under the complex environment of data analysis. We apply Bus-OLAP to verify the scalability of model. We partition the space into 2000*1000 grids and the maximal search space is 100*100 grids. Figure 12(b) shows the running time of querying on LLR of every grid; Fig. 12(c) shows the running time of querying delay aggregation during a time interval from 16:00 to 17:00. The running time of Bus-OLAP decreases when increasing the number of nodes shown in Fig. 12. Bus-OLAP can improve the efficiency of queries by increasing nodes when there are complex queries in non-on-time analysis.

5   Conclusions

In this paper, we design a model named Bus-OLAP supporting non-on-time queries to monitor bus data. The Bus-OLAP supports flexible data indexing by bit vector, index storage and update. Besides, Bus-OLAP supports parallel queries and computing based on Spark. Our experiments verify the effectiveness and efficiency of Bus-OLAP. In our future work, we will consider more scenario application of on-time analysis and fuse more implicit factors that affect the running time of bus. Besides, we also provide the result of analysis to analyst, and realize the flexible management of urban traffic.

References

1. Chen, M., Liu, Y., Yu, X.: Predicting next locations with object clustering and trajectory clustering. In: Cao, T., Lim, E.-P., Zhou, Z.-H., Ho, T.-B., Cheung, D., Motoda, H. (eds.) PAKDD 2015. LNCS, vol. 9078, pp. 344–356. Springer, Cham (2015). doi:10.1007/978-3-319-18032-8_27
2. Eldawy, A., Mokbel, M.F.: A demonstration of spatialhadoop: an efficient mapreduce framework for spatial data. Proc. VLDB Endow. 6(12), 1230–1233 (2013)
3. Ghosh, B., Basu, B., O'Mahony, M.: Multivariate short-term traffic flow forecasting using time-series analysis. IEEE Trans. Intell. Transp. Syst. 10(2), 246–254 (2009)
4. Guttman, A.: R-trees: a dynamic index structure for spatial searching. In: Proceedings of Annual Meeting on SIGMOD 1984, pp. 47–57 (1984)
5. Han, Y., Moutarde, F.: Analysis of large-scale traffic dynamics in an urban transportation network using non-negative tensor factorization. Int. J. Intell. Transp. Syst. Res. 14(1), 36–49 (2016)
6. Jagadish, H.V., Ooi, B.C., Tan, K.L., Yu, C., Zhang, R.: iDistance: an adaptive B^+-tree based indexing method for nearest neighbor search. ACM Trans. Database Syst. 30(2), 364–397 (2005)
7. Kong, X., Xu, Z., Shen, G., Wang, J., Yang, Q., Zhang, B.: Urban traffic congestion estimation and prediction based on floating car trajectory data. Future Gener. Comput. Syst. 61, 97–107 (2016)
8. Liu, D., Chen, H., Qi, H., Yang, B.: Advances in spatiotemporal data mining. J. Comput. Res. Dev. 50(2), 225–239 (2013)
9. Pang, L.X., Chawla, S., Liu, W., Zheng, Y.: On mining anomalous patterns in road traffic streams. In: Tang, J., King, I., Chen, L., Wang, J. (eds.) ADMA 2011. LNCS, vol. 7121, pp. 237–251. Springer, Heidelberg (2011). doi:10.1007/978-3-642-25856-5_18
10. Sistla, A.P., Wolfson, O., Chamberlain, S., Dao, S.: Modeling and querying moving objects. In: Proceedings of the 13th International Conference on Data Engineering, pp. 422–432 (1997)
11. Stathopoulos, A., Karlaftis, M.G.: A multivariate state space approach for urban traffic flow modeling and prediction. Transp. Res. Part C Emerg. Technol. 11(2), 121–135 (2003)
12. Syrjärinne, P., Nummenmaa, J.: Improving usability of open public transportation data. In: 22nd ITS World Congress, pp. 5–9 (2015)
13. Ting, R.H., De Almeida, T., Ding, Z.: Modeling and querying moving objects in networks. Int. J. VLDB 15(2), 165–190 (2006)

14. Wang, Y., Papageorgiou, M., Messmer, A.: Real-time freeway traffic state estimation based on extended kalman filter: a case study. Transp. Sci. **41**(2), 167–181 (2007)
15. Wu, X., Duan, L., Pang, T., Nummenmaa, J.: Detection of statistically significant bus delay aggregation by spatial-temporal scanning. APWeb 2016. LNCS, vol. 9865, pp. 277–288. Springer, Cham (2016). doi:10.1007/978-3-319-45835-9_24
16. Xia, D., Li, H., Wang, B., Li, Y.: A map reduce-based nearest neighbor approach for big-data-driven traffic flow prediction. IEEE Access **4**, 2920–2934 (2016)
17. Xie, X., Xiong, Z., Hu, X., Zhou, G., Ni, J.: On massive spatial data retrieval based on spark. In: Chen, Y., et al. (eds.) WAIM 2014. LNCS, vol. 8597, pp. 200–208. Springer, Cham (2014). doi:10.1007/978-3-319-11538-2_19
18. Yu, X., Pu, K.Q., Koudas, N.: Monitoring k-nearest neighbor queries over moving objects. In: Proceedings of the 21st International Conference on Data Engineering, pp. 631–642 (2005)
19. Yuan, J., Zheng, Y., Zhang, L., Xie, X., Sun, G.: Where to find my next passenger. In: Proceedings of the 13th International Conference on Ubiquitous Computing, pp. 109–118 (2011)

Distributed Data Mining for Root Causes of KPI Faults in Wireless Networks

Shiliang Fan[1], Yubin Yang[1(✉)], Wenyang Lu[1], and Ping Song[2]

[1] State Key Laboratory for Novel Software Technology,
Nanjing University, Nanjing, China
`yangyubin@nju.edu.cn`
[2] Huawei Technologies Co. Ltd., Shanghai, China
`songping@huawei.com`

Abstract. In the field of wireless network optimization, with the enlargement of network size and the complication of network structure, traditional processing methods cannot effectively identify the causes of network faults in the face of increasing network data. In this paper, we propose a root-cause-analysis method based on distributed data mining (DRCA). Firstly, we put forward an improved decision tree, where the selection of the best split-feature is based on the feature's purity-gain, and then we skillfully convert the problem of root-cause-analysis into modeling of an improved decision tree and interpretation of the tree model. In order to solve the problem of memory and efficiency associated with large-scale data, we parallelize the algorithm and distribute the tasks to multiple computers. The experiments show that DRCA is an effective, efficient, and scalable method.

Keywords: KPI faults · Root-cause-analysis · Improved decision tree · Distributed data mining · Parallelization algorithm

1 Introduction

Wireless network optimization is an important part of the network operation process [2]. In general, the monitoring of network performance relies on a predefined set of key performance indicators (KPIs). Faults in these KPIs always represent the deterioration of network performance [2].

The network system outputs numerous KPI reports daily. Engineers traditionally analyze these reports to identify the specific causes of KPI faults based on accumulated experience and data processing tools [1,7,8]. However, the enlargement of network size and the complication of network structure leads to inaccuracies and inefficiencies with the traditional method. There are two issues pertaining to KPI faults in wireless networks: firstly, the KPI reports for analysis lack root causes labels, so the problems cannot be transformed into a multi-classification of the root causes; secondly, with the limitation of RAM capacity and CPU speed [11], it is difficult to obtain timely and effective results when dealing with large-scale KPI reports.

© Springer International Publishing AG 2017
L. Chen et al. (Eds.): APWeb-WAIM 2017, Part II, LNCS 10367, pp. 201–209, 2017.
DOI: 10.1007/978-3-319-63564-4_16

2　Problem Definition and Analysis

Each row of the report represents one sample. Assuming the KPI report has $N(N$ is very large) samples in total, expressed as $x_i(1 \leqslant i \leqslant N)$, each column represents a feature of x_i. The structure of the report is as shown in Table 1.

Table 1. The structure of the KPI report

Index	Start time	Period	Cell	Indicator$_1$	$\cdots$	**Indicator$_K$(KPI)**	Indicator$_M$
1							
$\cdots$							
N							

We define a set S containing all *Indicators* from *Indicator$_1$* to *Indicator$_N$*, with the exception of **Indicator$_K$(*KPI*)**. To determine the set of $\{Root\_Cause_i\}$ $(Root\_Cause_i \in S)$ leading to the KPI fault, and the corresponding set of $\{Weight_i\}$ (*Weight$_i$* represents the influence of *Root_Cause$_i$* on the KPI fault), we take a series of indicators in the set S as features, and the occurrence of KPI($Indicator_K$) fault as labels to train the binary classification model. Then, we interpret the model to obtain each root cause and its weight. Based on the considerations of interpretability and parallelizability, we have selected the decision tree [10] as the classification model.

3　Method Introduction

3.1　Data Preparation

We regard whether the KPI fault occurs as the training target (label), so it is necessary to extract the *KPI* of each x_i and compare the value with the industry threshold. If the KPI fault occurs, then $label_i = -1$ otherwise $label_i = 1$.

3.2　Model Training

Theoretical Principle. Based on the idea of the decision tree [9,10] we view "*KPI fault occurrence*" as the classification target. In the growth of the decision tree, we continuously select $Indicator_i(Indicator_i \in S)$ as the decision-making basis. If $Indicator_i$ can provide more information for classification, it is more likely to be a root cause.

However, it is clear that our real intention of the training decision tree is not to classify, but to obtain the extent of effective information brought by each indicator for classification. As shown in Fig. 1(a) and (b), there is a non-linear relationship between information entropy and the distribution of a variable X in binary classification. Therefore, there will be some deviation if we use entropy-gain to quantify the influence of each feature on classification.

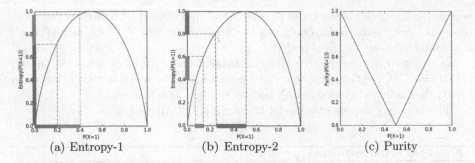

(a) Entropy-1   (b) Entropy-2   (c) Purity

Fig. 1. The relationship between the distribution and entropy/purity.

Therefore, we propose another criterion to select features when training the decision tree. Assuming the proportion of negative samples whose $label = -1$ in data-set D is p^-, we define the **purity** of D as follows:

$$Purity(D) = |2 * p^- - 1|. \qquad (1)$$

As shown in Fig. 1(c), there is linear relationship between purity and the distribution of the variable X in binary classification. Since all indicators are continuous, we use the dichotomy to process these continuous features. Suppose that continuous feature a has L different values on D, recording as $\{a_1, a_2, \ldots, a_L\}$ after ranking in ascending order. The data-set D can be divided into subsets D_t^- and D_t^+ by split-point t, where D_t^- contains the samples with values on feature a not greater than t, and D_t^+ the samples with values greater than t. We adopt directly each $a_i(1 \leqslant i \leqslant L)$ forming the set of alternative split-points:

$$T_a = \{a_i | 1 \leqslant i \leqslant L\}. \qquad (2)$$

Based on the definition of purity, we define the **purity-gain** similarly to the information entropy-gain:

$$
\begin{aligned}
Gain(D, a) &= \max_{t \in T_a} Gain(D, a, t) \\
&= \max_{t \in T_a} \sum_{\lambda \in \{-,+\}} \frac{|D_t^\lambda|}{|D|} Purity(D_t^\lambda) - Purity(D).
\end{aligned} \qquad (3)
$$

Concrete Implementation. To calculate the purity-gain of each feature, we gather their information in parallel using MapReduce. The breadth-first strategy is used to generate the decision tree layer by layer so that with a single execution, the statistical information of all tree-nodes on the same layer can be obtained simultaneously.

In order to record the position of each tree-node and to divide the data-set into the tree-nodes expediently, we design the following data structure:

```
Node :
    Path : Map<indicator_id, Pair<indicator_value, comparison>>
    Label : String
```

where *Path* represents the series of conditions that should be met from the root-node to the current *Node*. *Label* can be *"inside"* or *"leaf"* which means *Node* is an internal node or a leaf node.

In order to avoid excessive growth of the decision tree, we set a parameter called *limit*. When the purity-gain of the best split-feature on a node is less than *limit*, the node will stop splitting and be marked as a leaf node.

The training process of a model is described as Algorithm 1.

Algorithm 1. DecisionTreeDriver

Input: $DPath$: file path of the data-set after data preparation
Output: $Tree$: the trained model of decision tree
1 $currentQueue \leftarrow$ new LinkedList$<Node>$;
2 $rootNode \leftarrow$ new $Node()$;
3 Insert $rootNode$ to $currentQueue$;
4 **while** $currentQueue$ *is not empty* **do**
5      $newQueue \leftarrow$ new LinkedList$<Node>$;
6   write $currentQueue$ to HDFS ;
7      $ProduceStatisticalInfo(DPath)$; //run MapReduce Job to do statistics on data-set
8   **for** each node $\in currentQueue$ **do**
9   read the statistical information from HDFS ;
10         $bestPG, bestFeatureID, bestSplitPoint \leftarrow SelectBestSplitFea(node)$;
11   **if** $bestPG < limit$ **then**
12             $node.Label \leftarrow$ "leaf"
13   **else**
14             $node.Label \leftarrow$ "inside";
15             $leftPath \leftarrow node.Path.add(bestFeatureID,Pair<bestSplitPoint,0>)$;
16             $leftNode \leftarrow$ new $Node(leftPath)$;
17             $rightPath \leftarrow node.Path.add(bestFeatureID,Pair<bestSplitPoint,1>)$;
18             $rightNode \leftarrow$ new $Node(rightPath)$;
19   insert $leftNode,rightNode$ to $newQueue$;
20   add $node$ to $Tree$
21      $currentQueue \leftarrow newQueue$;
22 return $Tree$

The function $ProduceStatisticalInfo$ represents statistics on large-scale samples with MapReduce. In the phase of Map, for each $Node_j$, if the sample x_i meets the *Path* of $Node_j$ then a key-value pair recording as $<key, value>$ will be output. The Reducers receive the pairs of $<key, value>$ output by Mappers and add the values of the same key up recording as $value\_sum$. The $value\_sum$ is actually the number of samples which meet the *key*. Then write all pairs of $<key, value\_sum>$ to HDFS.

$SelectBestSplitFea$, as the name suggests, is the function to select the best split-feature of each node according to the output above produced by MapReduce. With the help of the characteristic that all pairs of $<key, value\_sum>$ are sorted by *key* and no *key* is repeated after the Reduce function, we only need to traverse the output once to determine the purity-gain of each feature.

3.3 Reverse Interpretation

After training the model, suppose that the decision tree has T nodes numbered $1, 2, \ldots, T$. We use an array named $node\_fea$ to record the best split-feature of each node while another named *node_pg* records the purity-gain of the above-mentioned split-feature. We make use of MapReduce to divide all samples into the corresponding paths of the decision tree in parallel, and then record the number of negative samples passing through each tree-node in the array *statistic*.

In the first instance, we acquire the indexes of tree-nodes with the same best split-features, for each $Indicator_i(Indicator_i \in S)$ expressed as ind_i:

$$Indexes(ind_i) = \{index | node\_fea[index] = ind_i(ind_i \in S)\}. \tag{4}$$

Since the same split-feature may appear in different nodes, we define the weighting purity-gain for each feature as follows:

$$P\_Gain(ind_i) = \sum_{j \in Index(ind_i)} \frac{statistic[j] \times node\_pg[j]}{\sum_{k \in Index(ind_i)} statistic[k]}. \tag{5}$$

Considering that the number of samples divided by different split-features is also different while training the model, the absolute weight of ind_i is defined as:

$$Abs\_Weight(ind_i) = \frac{P\_Gain(ind_i) \times \sum_{j \in Index(ind_i)} statistic[j]}{\sum_{k=1}^{T} statistic[k]} \times 100\%. \tag{6}$$

Then the relative weight of ind_i is obtained:

$$Rela\_Weight(ind_i) = \frac{Abs\_Weight(ind_i)}{\sum_{ind_i \in S} Abs\_Weight(ind_i)} \times 100\%. \tag{7}$$

4   Experiment and Analysis

In order to verify the effect of **DRCA** in mining the root cause of KPI faults, we conduct a series of relevant experiments and validate the effectiveness and efficiency of this method.

4.1   Experimental Environment

The data-set used in this study is provided by a company in the field of communication and sampled from a realistic communication network. The data-set contains 2203740 samples and the composition of each sample has been explained previously. Excluding *Index*, *StartTime*, *Period*, and *Cell* there are 57 indicators in the data-set, including 5 KPIs. The remaining 52 indicators may possibly lead to KPI faults. Furthermore, we take a Hadoop cluster with 1 master and 4 slaves to do experiments.

4.2   Criteria

Regarding the problem raised in this paper, it's difficult to do an experiment compared with other conventional methods. On one hand, those methods can't output the weight of each root-cause; On the other hand, due to the lack of sufficient labeled data as a criterion, we can't quantify the accuracy of the result mined by each method. To verify the effectiveness of the model, we propose the following reliability criteria, relying on the analysis of the problem combined with the real situation:

(1) The root-causes obtained should be relatively stable and concentrated;
(2) The root-causes obtained should coincide partly with the manual experience.

Additionally, the efficiency is evaluated by comparing with the serial algorithm.

4.3   Effectiveness Experiment

In the experiment, we focus on the KPI called *"RRC Setup Success Rate"*. According to the industry standard, this KPI breaks down when its value is less than 99.5%. We mine the possible root-causes which lead to the fault of it from the set S. When training the decision tree model, we change the feature selection criterion from entropy-gain (Method 1) to purity-gain (Method 2). Then, we compare the results of the two methods under the different values of *limit*.

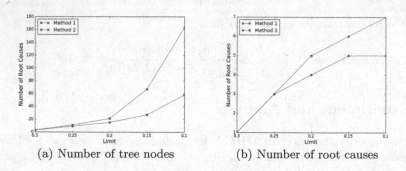

(a) Number of tree nodes   (b) Number of root causes

Fig. 2. The mining results of two different methods.

It can be seen from Fig. 2(a) that, with the limit decreasing, the growth rate of the decision tree trained by Method 1 is significantly higher than that of Method 2. As shown in Fig. 2(b), there is not a significant difference in the number of root causes between the two methods. However, further analysis of Table 2 shows that the root causes resulting from Method 1 are not stable and the weight of fractional root causes changes significantly. By comparison, the root causes obtained by Method 2 are considered to be stable.

Drawing on the experience provided by the above company, the main reasons for the fault of *"RRC Setup Success Rate"* are the following: weak coverage, strong interference, and channel congestion. The root causes mined by Method 2 mostly point to the *"Physical Down link Control Channel"* (PDCCH), which is perceptibly consistent with the manual experience — channel congestion.

Table 2. The root-causes and their weight of two methods

Limit	Method 1		Method 2	
	Root cause	Weight	Root cause	Weight
0.3	L.ChMeas.PDCCH.AggLvl8Num	100.00%	L.ChMeas.PDCCH.AggLvl8Num	100.00%
0.25	L.ChMeas.PDCCH.SymNum.3	34.93%	L.ChMeas.PDCCH.SymNum.3	44.27%
	L.ChMeas.PDCCH.AggLvl8Num	31.46%	L.ChMeas.PDCCH.AggLvl8Num	32.15%
	L.ChMeas.PDCCH.SymNum.2	27.67%	L.ChMeas.PDCCH.SymNum.2	23.58%
0.15	L.ChMeas.PDCCH.SymNum.2	22.74%	L.ChMeas.PDCCH.SymNum.3	30.71%
	L.ChMeas.PDCCH.SymNum.3	19.25%	L.ChMeas.PDCCH.AggLvl8Num	22.31%
	L.ChMeas.PDCCH.AggLvl8Num	16.21%	L.ChMeas.PDCCH.SymNum.2	14.33%
	L.ChMeas.CCE.Avail	14.69%	L.ChMeas.CCE.Avail	16.22%
	L.ChMeas.PDCCH.SymNum.1	13.57%	RRC ConnReq attempt	11.06%
	RRC ConnReq attempt	10.34%		

Considering the stability, concentration, and consistency with manual experience of the root causes resulting from Method 2, it can be proved that the method we propose is effective.

4.4   Efficiency Experiment

Simultaneously, we achieve the serial version of our method and compare it with DRCA. The following experiments deal with the KPI named "*RRC Setup Success Rate*", and are conducted on the basis of $limit = 0.15$.

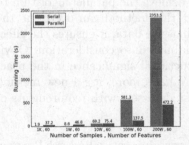 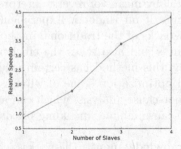

Fig. 3. The running time of two methods.   **Fig. 4.** The relative speedup of DRCA.

The 1 K, 10 K, 100 K, 1 M, and 2 M samples are selected from the original data-set as the experimental data, and the running time of serial algorithm is compared with that of the DRCA on different data. As can be seen from Fig. 3, in terms of small-scale data, the parallel algorithm is actually not as efficient as the serial. With the increase in the number of samples, the efficiency of the stand-alone serial algorithm decreases while DRCA performs well, which shows that DRCA can greatly improve the efficiency in the face of large-scale data.

Then, we test the relative speedup of DRCA with the 1 M samples by changing the number of slave nodes in the cluster. As shown in Fig. 4, with the increase in the number of slaves in the cluster, the computing capability of DRCA is enhanced, which proves that the algorithm is well scalable.

5   Related Work

In root-cause-analysis and fault detection, there are many researches based on machine learning, including supervised models such as logistic regression [5] and k-nearest neighbor [6], as well as unsupervised models such as clustering [3]. However, these models [3–6] are not suitable for our research, as we require not only the root-causes of KPI faults, but also the weight of each root-cause. Thus the model must have strong interpretability in order to interpret the answer from the trained model, while the models mentioned above do not meet these requirements. Further, facing large-scale data, the stand-alone models mentioned above cannot better solve the bottlenecks of memory and computing speed.

6   Conclusions

In this study, first we analyze the difficulties in identifying the root-causes of KPI faults with the manual method. Then, we propose a DRCA, which draws the concept of purity and purity-gain from the main idea of the decision tree, and constructs an improved decision tree using purity-gain as the selection criterion of the best split-feature. Finally, we obtain the suspicious root-causes and their weight by means of reverse interpretation. We achieve the parallel algorithm to execute it on Hadoop. Experiments show that this method can overcome the drawbacks of the traditional methods. Facing massive data, it ensures the effectiveness and also takes the efficiency and scalability into consideration. Obviously, this method has certain value and instructional significance in the practical optimization scene of wireless networks. What's more, it's a new method of root-cause-analysis, which can adopt to most data-sets with requirements of root-cause mining, including big data.

Acknowledgement. This work is sponsored by Huawei Innovation Research Program (HIRP Grant No. YB2015090007), the Natural Science Foundation of China (Grant Nos. 61673204, 61321491), the Program for Distinguished Talents of Jiangsu Province, China (Grant No. 2013-XXRJ-018), and the Fundamental Research Funds for the Central Universities (Grant No. 020214380026).

References

1. Method and system for optimising the performance of a network. https://www.google.com/patents/US20040266442
2. Ye Ouyang, M., Fallah, H.: A performance analysis for UMTS packet switched network based on multivariate KPIS. In: IJNGN, pp. 80–92 (2010)

3. Julisch, K.: Clustering intrusion detection alarms to support root cause analysis. ACM Trans. Inf. Syst. Secur. **6**(4), 443–471 (2003)
4. Chen, M., Zheng, A.X., Lloyd, J., Jordan, M.I., Brewer, E.: Failure diagnosis using decision trees. In: ICAC, pp. 36–43 (2004)
5. Lpez-Soto, D., Yacout, S., Angel-Bello, F.: Root cause analysis of familiarity biases in classification of inventory items based on logical patterns recognition. Comput. Ind. Eng. **93**, 121–130 (2016)
6. Bauer, M., Cox, J.W., Caveness, M.H., Downs, J.J., Nina, F.: Nearest neighbors methods for root cause analysis of plantwide disturbances. In: Industrial and Engineering Chemistry Research, pp. 5977–5984 (2007)
7. Doggett, M.: A statistical comparison of three root cause analysis tools. J. Ind. Technol. **20**(2), 2–9 (2004)
8. Doggett, M.: Root cause analysis: a framework for tool selection. Qual. Manag. J. **12**(4), 34–45 (2005)
9. Kwad, N., Choi, C.H.: Input feature selection for classification problem. IEEE Trans. Neural Netw. **13**(1), 143–159 (2002)
10. Quinlan, J.R.: Induction of decision trees. Mach. Learn. **1**, 81–106 (1984)
11. Narlikar, G.J.: A parallel, multithreaded decision tree builder. Technical report, Carnegie Mellon University (1998)

Precise Data Access on Distributed Log-Structured Merge-Tree

Tao Zhu[1], Huiqi Hu[1(✉)], Weining Qian[1], Aoying Zhou[1], Mengzhan Liu[2], and Qiong Zhao[2]

[1] School of Data Science and Engineering, East China Normal University, Shanghai, China
tzhu@stu.ecnu.edu.cn, {hqhu,wnqian,ayzhou}@dase.ecnu.edu.cn
[2] Bank of Communications, Shanghai, China
{liumengzhan,qiongzhao}@bankcomm.com

Abstract. Log-structured merge tree decomposes a large database into multiple parts: an in-writing part and several read-only ones. It achieves high write throughput as well as low read latency. However, read requests have to go through multiple structures to find the required data. In a distributed database system, different parts of the LSM-tree are stored distributedly. Data access issues extra network communications for a server in the query layer to pull entries from the underlying storage layer. This work proposes the precise data access strategy. A Bloom filter-based structure is designed to test whether an element exists in the in-writing part of the LSM-tree. A lease-based synchronization strategy is used to maintain consistent copies of the Bloom filter on remote query servers. Experiments show that the solution has 6× throughput improvement over existing methods.

Keywords: Data access · Distributed system · Consistency

1 Introduction

Log-Structured merge tree [5] organizes all data entries in multiple components: a Memtable and several SSTables, following the notations used in [8]. The Memtable is a memory-based structure, optimizing for high write throughput. The SSTable is a disk-based structure, offering large storage capacity and servicing read requests only. Data in the Memtable are migrated into a SSTable in batch. It has been widely adopted by distributed storage systems such as BigTable [8], where the Memtable and SSTables are kept in the main memory and distributed file system (e.g. GFS [7]) respectively.

Systems using log-structured storage offer excellent read/write performance but lack some important features. Thus, some database systems (e.g. Megastore [10] and Percolator [9]) choose to build a query layer upon these storage systems to add SQL interface or transaction support. A node in the query layer interacts with the underlying storage layer through network communication. A problem is

© Springer International Publishing AG 2017
L. Chen et al. (Eds.): APWeb-WAIM 2017, Part II, LNCS 10367, pp. 210–218, 2017.
DOI: 10.1007/978-3-319-63564-4_17

that data access on a distributed LSM-tree issues many useless communications. A read operation has to iterate over the Memtable and all SSTable until locate the required data item. The dedicated item only exists in one structure and accessing all other structures is useless.

This work targets at the distributed database system where Memtable, SSTables and query processing nodes (noted as p-node in the following) are deployed on different servers and proposes an effective way to precisely locate the storage structure for accessing. Before processing a read request, a p-node is able to identify the right structure for reading without contacting the storage layer. In summary, we make the following contributions: (1) a Bloom filter with low maintaining and synchronization overhead is designed to encode data existence for the Memtable; (2) a lease-based strategy is designed for p-nodes to maintain a copy of the Bloom filter of Memtable and ensure read consistency when using the copy; (3) A data access algorithm is designed to support a p-node in determining the right structure for data access.

2   Relate Works

[5] proposes the log-structured merge tree. The author exploits a multi-level structure for large database storage. BigTable [8] extends such mechanism in distributed system. It keeps the write-optimized index in main memory while the read-only files in GFS [7]. Percolator [9] and Megastore [10] build their query servers directly on the BigTable. Our work acts as a data access optimizations between the query layer of a database system and the underlying storage layer.

Some other optimizations are also designed for log-structured storage. [8] relies on data compaction to merge multiple SSTables together and reduce the number of SSTable to be visited. Muhammad [12] does performance analysis on the overhead of data compaction and proposes some improvements. bLSM-tree [11] uses the Bloom filter [2] to reduce SSTable access, which is adopted in [8]. In a different, our work is able to filter access to the Memtable. Besides, our technique is designed for distributed system to reduce the network communications between the application servers and the storage layer.

3   Preliminary

Storage Model. A typical structure of a distributed log-structured storage system is illustrated in Fig. 1, with a Memtable, several SSTables and p-nodes.

Memtable is the in-memory structure which services for data reads and writes. To ensure durability, redo log entries [4] are forced into durable storage for recovery purpose. Each write operation firstly flushes its redo log entries into the disk, after which, its modifications are applied into the Memtable. Group commit [3] is used to improve the disk utilization by combining multiple redo log flushing in a disk write, because existing disk device only offers limited IOPS.

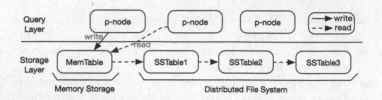

Fig. 1. Data storage and access on distributed LSM-tree

SSTable is the immutable structure where data is stored in lexicographic order based on their primary key. The SSTable is generated by freezing an active Memtable. The frozen Memtable is transferred into distributed file system and becomes the SSTable. A new Memtable replaces the old one for servicing further writes. As time goes by, there are several SSTables generated as illustrated in Fig. 1, where *SSTable*1 is the latest one and *SStable*3 is the oldest ones.

In Fig. 1, to read an entry with key k, a p-node has to go through the 1st Memtable, 2nd SSTable, 3rd SSTable... until seeing the data with k as its key. In addition, Memtable and SSTables are distributedly stored. Hence, a p-node has to issue many remote data access via the network.

Precise Data Access is to let a p-node determine visiting either the Memtable or one SSTable without contacting the underlying storage servers. Let Memtable be m and its owned key set be $\mathcal{K}_m = \{k_1^m, k_2^m, \dots\}$; SSTable be s and its owned key set be $\mathcal{K}_s = \{k_1^s, k_2^s, \dots\}$. Given a query key k, a p-node is required to answer whether $k \in \mathcal{K}_m$ or $k \in \mathcal{K}_s$ stands. It is easy to answer whether $k \in \mathcal{K}_s$ by caching the Bloom filter of SSTable on p-nodes (as discussed in Sect. 2). But, answering whether $k \in \mathcal{K}_m$ is of much more difficulties. Hence, we aim at determining whether Memtable access is necessary for a read operation. There is no essential difference between one SSTable or multiple. We assume there only one SSTable in the following.

The kernel problem is to answer whether a remote evolving set contains a typical element or not. An intuitive solution is to maintain a Bloom filter for the Memtable as well, and synchronize the structure to multiple p-nodes. However, there are two difficulties here. Firstly, since the Memtable is stored in *remote*, its Bloom filter has to be synchronized to p-nodes through network. But the Bloom filter is of large size. Direct synchronization tends to exhaust the network bandwidth. Secondly, as the Memtable services data writes, it is evolving over the time, a copy of its Bloom filter on a p-node may not remain the same with the source one after synchronization. Potential difference between the primary Bloom filter and its copies tends to lead to inconsistency read. Understanding these difficulties, we present solutions in the following sections.

4   Entry Existence

Figure 2 illustrates how to maintain and synchronize a Bloom filter ($\mathcal{B}_m$) for the Memtable. When data writes happen on the Memtable, redo log entries

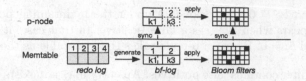

Fig. 2. The Bloom filter maintenance and synchronization

are prepared. Before flushing a group of redo entries into disk, updates on $\mathcal{B}_m$ are generated from redo entries based on the policy in Sect. 4. These updates act as the modification log entries for the $\mathcal{B}_m$ (short for bf-logs). To reduce synchronization cost, $\mathcal{B}_m$ is not directly sent to p-nodes, instead bf-logs are transported to p-nodes and a copy of $\mathcal{B}_m$ on a p-node catches up with the source by applying (replaying) the identical bf-logs.

Maintenance. The Memtable can be probably changed by the following types of operations: *insert*, *update* and *delete*. Considering an entry e with key k:

1. Update operation modifies an existing record entry e. (i) If $k \notin \mathcal{K}_m$, then e is newly created in the Memtable. A bf-log is generated for k, which adds existence of k into $\mathcal{B}_m$; (ii) If $k \in \mathcal{K}_m$, then a previous operation has added k into $\mathcal{K}_m$ and handled its update on $\mathcal{B}_m$, the current one does nothing.

2. Delete operation is treated as special *update*, which adds an deleted flag for k. A read operation checks whether the entry is deleted via the flag.

3. Insert operation writes an non-existing record entry e into the Memtable. (i) If $k \notin \mathcal{K}_m$, then $k \notin \mathcal{K}_s$ must also stand. To read e, a p-node can easily find $k \notin \mathcal{K}_s$ by checking $\mathcal{B}_s$ and e can only be found on the Memtable. The p-node can infer the fact without querying $\mathcal{B}_m$. Thus, there is no necessity in modifying $\mathcal{B}_m$ when inserting e into Memtable. (ii) If $k \in \mathcal{K}_m$, this means the entry must be tagged with a deleted flag, its bits in $\mathcal{B}_m$ should be already processed by one previous *delete* operation. Therefore, we do not need to modify $\mathcal{B}_m$ again.

In summary, $\mathcal{B}_m$ is only modified when an entry is newly created on the Memtable by a *update* or *delete* operation.

Data Access based on $\mathcal{B}_m$. Considering the query $q(k)$ in Definition 1, we denote b_m (b_s) is 1 when all hashing bits of the key k are **1** in the (copy of) $\mathcal{B}_m$ $(\mathcal{B}_s)$ respectively. Temporarily, we assume there is no false positive in the Bloom filter. Based on the maintaining policy of $\mathcal{B}_m$, we locate an entry using the following rules.

(1) If $b_m = 0$ and $b_s = 0$, then e is either non-existing or newly inserted into Memtable. A p-node will access Memtable.
(2) If $b_m = 0$ and $b_s = 1$, then e never receives any modification after it is written into the SSTable. A p-node will directly access the SSTable.
(3) If $b_m = 1$ and $b_s = 1$, e is stored on SSTable at first and then get modified. A p-node should visit the Memtable to read the entry.

(4) $b_m = 1$ and $b_s = 0$ is not possible under the maintaining policy for $\mathcal{B}_m$. It only appears when $\mathcal{B}_m$ sees a false positive. In that case, it actually has $b_m = 0$ and $b_s = 0$. Hence, a p-node takes the Memtable as the destination.

As the Bloom filter contains false positives. An entry may not exists in Memtable or SSTable even if $\mathcal{B}_m$ or $\mathcal{B}_s$ confirms its existence respectively. When the entry is not returned by the first access, a p-node access the rest structure to handle any potential false positive.

Lightweight Synchronization. A p-node synchronize the remote $\mathcal{B}_m$ to its local by pulling bf-logs from the Memtable server and replaying these entries. As discussed above, only a small part of operations generate bf-logs, the number of bf-logs is much smaller than that of redo log entries. It means that the bf-log synchronization has lower network overhead than the log replication [6].

Each bf-log is indexed by a monotonically increasing serial number. The Memtable server keeps the newest bf-logs in a circular buffer. A p-node pulls bf-logs from the remote by sending the largest serial number N ever received. The Memtable server replies with all bf-logs whose serial number is bigger than N. As the circular buffer has limited memory, new bf-logs may overwrite the oldest ones. The $\mathcal{B}_m$ is sent to a p-node when some required bf-logs are missing.

5   Consistence

A copy of $\mathcal{B}_m$ on a p-node may fall behind the primary one. As a result, a p-node may miss some newly committed entries and suffer from inconsistent read. We present a lease-based solution and use Fig. 3 to explain the design.

Group Commit. The Memtable commits write operations with following steps: (1) Generation. Redo entries are buffered in memory. They are flushed into the disk in a fixed period, called the *group interval*, e.g. *from $t_s(g_1)$ to $t_s(g_2)$*. (2) Start phase begins at a time $t_s(g_x)$ with a group of redo entries formed. Then, bf-logs are generated and applied into $\mathcal{B}_m$, e.g. *from $t_s(g_1)$ to $t_w(g_1)$*. (3) Write phase begins at a time $t_w(g_x)$. The write thread is writing redo entries into the disk, e.g. *from $t_w(g_1)$ to $t_p(g_1)$*, which generally several milliseconds to finish under the hard disk driver. (4) Publish phase begins at a time $t_p(g_x)$ after the write thread has finished disk writing. Data modifications of the group are applied into the Memtable, e.g. *from $t_p(g_1)$ to $t_e(g_1)$*. After that, the group ends at a time $t_e(g_x)$.

Invariance. Both $\mathcal{B}_m$ and the Memtable keep invariant during a period. Considering two successive groups g_1 and g_2, $\mathcal{B}_m$ is invariant from $t_w(g_1)$ to $t_s(g_2)$ and Memtable is invariant from $t_e(g_1)$ to $t_p(g_2)$. With the temporary invariance of Memtable and $\mathcal{B}_m$, a lease-based mechanism can be designed to ensure the read consistency when a p-node uses a copy of $\mathcal{B}_m$ in data access.

Lease Definition. A lease L_x is a contract given by the Memtable server and held by each p-node. It contains an invariant Bloom filter $\mathcal{B}'_m$ (a version of $\mathcal{B}_m$ at some time) and a expiration time t_x, and guarantees that for each entry in the Memtable, its bits are correctly set in $\mathcal{B}'_m$ based on the maintaining policy before t_x is reached. It is safe for a p-node to use $\mathcal{B}'_m$ before t_x is reached.

Lease Design. A lease can begin after a group has updated $\mathcal{B}_m$, i.e. $t_w(g_x)$, and end before the next group begin to publish, i.e. $t_p(g_{x+1})$. For example, L_1 can last from $t_w(g_1)$ to $t_p(g_2)$ and $\mathcal{B}'_m$ is the version of $\mathcal{B}_m$ at $t_w(g_1)$.

Correctness. Between $t_w(g_1)$ and $t_p(g_2)$, the Memtable has two versions while $\mathcal{B}'_m$ includes all bf-logs from g_1 and all previous ended groups. (1) Before $t_p(g_1)$, the Memtable m_0 contains data entries committed by all groups end in prior to g_1. It is safe to use $\mathcal{B}'_m$ because all bf-logs generated by these groups have been applied in $\mathcal{B}'_m$. On the other hand, $\mathcal{B}'_m$ also contains bf-logs from g_1. Though data entries created by g_1 are not included in m_0 at present, a p-node would still be directed to the Memtable when reading them. Such reading can be viewed as a false positive and does not lead to consistency problem. (2) After $t_p(g_1)$, the Memtable m_1 contains data entries committed by g_1 and all previous ended groups. Now $\mathcal{B}'_m$ is the exact structure for m_1.

Two successive leases have overlap in the time-line. A new lease is available for acquisition before the in-using one is going to be expired. In the overlap, both their $\mathcal{B}_m$ work correctly in accessing Memtable. The proof is straightforward.

Lease Implementation. A lease L_x is generated at the time $t_w(g_x)$, containing the current largest bf-log serial number N and the expiration time t_x. Its $\mathcal{B}'_m$ is created by replaying all bf-logs whose serial number is small than N. The t_x can be any time before the next group publishes, i.e. $t_p(g_{x+1})$. However, the timestamp is not known in advance, but can be inferred by adding the *current time*, the *group interval* and *disk writing time* together (e.g. L_1 in Fig. 3). Local processing time, e.g. from $t_s(g_x)$ to $t_w(g_x)$, is ignored as it is very short. Group interval can be given by system configuration. Disk write time can be estimated from the time used for previous groups.

Commit Wait. Since t_x is inferred, it can be smaller or bigger than $t_p(g_{x+1})$. (1) If $t_x > t_p(g_{x+1})$, the Memtable should not allow g_{x+1} to publish its content. Otherwise, inconsistent read may happen since L_x is not expired now. As a result, the publish phase of g_{x+1} is blocked until t_x is reached. It is called as *commit wait*. To avoid *commit wait*, we prefer to use the lower bound of the estimated disk write time in determining the t_x.

Acquisition. In each synchronization, a p-node pulls a lease and bf-logs whose serial numbers are in $(N_1, N_2]$ from the Memtable server (N_1 the largest serial

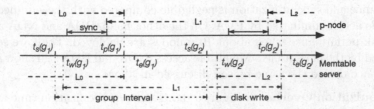

Fig. 3. Group commit and lease management

number ever received, N_2 is the one specified by the lease). Synchronization is required when the lease is going to expire soon. Typically, a p-node tries to acquire a new lease when the in-using one will be expired in 400 us.

A p-node checks whether the Bloom filter is usable by confirming that the *current time* is smaller than the expiration time of the lease. A problem is the time deviation between servers. PTP [1] can be used to synchronize server clocks, which achieves less than 50 us under a local area network. A p-node infers the time of a remote server by adding its local time with the largest deviation.

6   Experiment

The experiments use 15 servers, equipped with two 2.00 GHz 6-Core processors, 192 GB DRAM, connected by 1 GB switch. The Memtable is stored on a server, the SSTable is shareded over 3 servers. The rest deploy p-nodes. All experiments use the YCSB benchmark with 1 million records in the database. 95% records are stored in the SSTable, and records are accessed in uniform distribution. The workload contains unlimited read requests and 10 K writes per second. Three methods are evaluated and compared. (1) *NDA* is the basic access method in LSM-tree. (2) *BDA* maintains Bloom filter of SSTable to prone useless SSTable access. (3) *PDA* is the method presented in this work. Performance are evaluated by read operations processed per second (ops).

Concurrency. Figure 4 shows the performance of different methods by varying the number of clients connected with the system. Overall, PDA has the best performance under all cases. It reaches about 1100 k ops when 450 clients are used, which is about 6 times that of the NDA or BDA. The performance of NDA and BDA increases with more clients are simulated, but stabilizes once the Memtable server is overloaded. They easily make the Memtable server be performance bottleneck since they have to access the Memtable for every request. On the other hand, performance of PDA improves all the time and does not witness bottleneck from Memtable access. Secondly, NDA and BDA share similar performance because the SSTable is well merged and cached on each p-node. Reducing SSTable access does not contributes to performance.

Scalability. Figure 5 evaluates performance by varying the number of p-nodes connected with storage servers. By deploying more p-nodes, the synchronization overhead of PDA is increased. But PDA still shows linear scalability with respect to the number of p-nodes used. The overhead introduced by Bloom filter maintenance and synchronization is negligible compared with those unnecessary Memtable access eliminated by PDA. On the other hand, BDA and NDA achieve their peak performance when about 10 p-nodes are deployed. They are severely influenced by the mass useless Memtable access. NDA and BDA still show similar performance due to the same reason discussed in above.

Synchronization Overhead. Figure 6 shows the synchronization time and frequency by varying the group interval. It always takes about 200 us for a p-node to extend a new lease. The time used is relatively very short compared with the

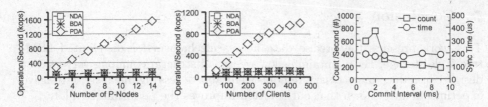

Fig. 4. Scalability.   Fig. 5. Concurrency.   Fig. 6. Synchronization.

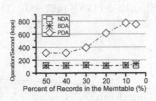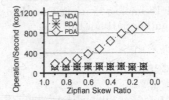

Fig. 7. Storage.   Fig. 8. Skewness.

group interval. Secondly, when Memtable flushes one group of redo entires per 2 ms, each p-nodes issues about 700 synchronizations per second. Synchronization frequency decreases because a p-node gets a longer lease. An exception is when 1 ms group interval is used. When using a short group interval, many small groups are formed. Writing small groups increases the average disk write time because HDD favors large sequential writes. As a result, the disk write time is increased, making each p-node receive a longer lease again.

Storage Distribution. Figure 7 shows the performance by varying the percentage of records stored in the Memtable. When about 50% records should be read from Memtable, PDA achieves about 300 k ops. With the percentage goes down, the performance keeps increasing. In comparison, both NDA and BDA are not sensitive to the parameter. Given a record who has its lasted version in the Memtable, PDA process in the same with the others. Thus, when the percentage of these records increases, the performance of PDA get closer to that of NDA/BDA. But it still shows about 200% improvement even when 50% records should be read from Memtable. In real deployment, Memtable does not contain a large percent of records.

Skewed Access Distribution. Figure 8 shows the performance under a skewed access distribution. In YCSB, request parameters are generated under a Zipfian distribution, which uses θ to adjust the skewness. When $\theta = 0.9$, PDA achieves about 187 k ops while NDA/BDA is about 128 k ops. PDA has about 1.46$\times$ improvements. It is because most records read are also get updated under a very skewed workload. With θ goes down, performance of PDA increases.

7   Conclusion

This work presents the precise data access mechanism for distributed LSM-tree style storage. By maintaining low overhead structures among servers, our design can reduce unnecessary remote Memtable access significantly. Extensive experiments have shown that our solution improves the performance a lot.

Acknowledgements. This is work is partially supported by National Hightech R&D Program (863 Program) under grant number 2015AA015307, National Science Foundation of China under grant numbers 61332006, 61432006 and 61672232, and the Youth Science and Technology- "Yang Fan" Program of Shanghai (17YF1427800). The corresponding author is Huiqi Hu.

References

1. PTP. https://en.wikipedia.org/wiki/Precision_Time_Protocol
2. Bloom, B.: Space/time trade-offs in hash coding with allowable errors. CACM **13**, 422–426 (1970)
3. DeWitt, D., Katz, R., et al.: Implementation techniques for main memory database systems. In: SIGMOD, pp. 1–8 (1984)
4. Mohan, C., Haderle, D., et al.: ARIES: a transaction recovery method supporting fine-granularity locking and partial rollbacks using write-ahead logging. TODS **17**, 94–162 (1992)
5. O'Neil, P., Cheng, E., Gawlick, D., O'Neil, E.: The log-structured merge-tree (LSM-tree). Acta Informatica **33**, 351–385 (1996)
6. Gray, J., Helland, P., ONeil, P., Shasha, D.: The dangers of replication and a solution. In: SIGMOD, pp. 173–182 (1996)
7. Ghemawat, S., Gobioff, H., Leung, S.T.: The Google file system. In: SOSP, pp. 29–43 (2003)
8. Chang, F., Dean, J., et al.: Bigtable: a distributed storage system for structured data. In: OSDI, pp. 4:1–4:26 (2008)
9. Peng, D., Dabek, F.: Large-scale incremental processing using distributed transactions and notifications. In: OSDI, pp. 1–15 (2010)
10. Baker, J., Bond, C., et al.: Megastore: providing scalable, highly available storage for interactive services. In: CIDR, pp. 223–234 (2011)
11. Sears, R., Ramakrishnan, R.: bLSM: a general purpose log structured merge tree. In: SIGMOD, pp. 217–228 (2012)
12. Ahmad, M., Kemme, B.: Compaction management in distributed key-value datastores. In: PVLDB, pp. 850–861 (2015)

Cuttle: Enabling Cross-Column Compression in Distributed Column Stores

Hao Liu[1], Jiang Xiao[2(✉)], Xianjun Guo[3], Haoyu Tan[1],
Qiong Luo[1], and Lionel M. Ni[4]

[1] Department of Computer Science and Engineering, HKUST, Kowloon, Hong Kong
{hliuag,hytan,luo}@cse.ust.hk
[2] Huazhong University of Science and Technology, Wuhan, China
jiangxiao@hust.edu.cn
[3] Deepera Inc., Ocean Coast City, Shenzhen, China
xjguo@deepera.com
[4] University of Macau, Zhuhai, China
ni@umac.mo

Abstract. We observe that, in real-world distributed data warehouse systems, data columns from different sources often exhibit redundancy. Even though these systems can employ both general and column-oriented compression schemes to reduce the data storage pressure, such cross-column redundancy (CCR) is not recognized or exploited effectively. Therefore, we propose Cuttle, a column storage system that enables cross-column compression to reduce CCR. Specifically, we identify three kinds of CCR and develop a referential transformation encoding (RTE) scheme to compress multiple columns of data with CCR. Furthermore, we address the CCR selection problem and propose a greedy algorithm to generate cross-column compression schemes. Our experiments on real-world datasets show that Cuttle can further reduce data size by half after applying both the column-oriented and general compression schemes, and that the query processing performance with Cuttle is improved by 20% without any change to the application programs.

Keywords: Big data · Compression · Storage optimization · Data management

1 Introduction

Nowadays, many emerging big data applications try to capture subtle correlations between different data sets to make better predictions or other forms of analytical results. These applications often use scalable distributed data warehouse systems (e.g., Hive [9] and Vertica [3]) gathering and integrating as much as possible data from as many as possible sources. As the data size grows so fast, the storage capacity (or budget) quickly becomes the bottleneck. In such situations, it is mandatory to apply data compression techniques to save costly storage resources.

© Springer International Publishing AG 2017
L. Chen et al. (Eds.): APWeb-WAIM 2017, Part II, LNCS 10367, pp. 219–226, 2017.
DOI: 10.1007/978-3-319-63564-4_18

Year	Month	Day	Quarter	FlightDate	AirportID	AirportSeqID	State	StateName	CDepTime	DepTime	DelayMin	WheelsOff	DepTimeBlk
2012	12	05	4	2012-12-05	11433	1143301	MI	Michigan	2140	2140	0	2151	2100-2200
2012	12	12	4	2012-12-12	11042	1104202	OH	Ohio	0552	0554	2	0609	0500-0600
2012	12	20	4	2012-12-22	11042	1104202	OH	Ohio	0557	0630	34	0638	0600-0700
2012	12	23	4	2012-12-23	13871	1387102	NB	Nebraska	0645	0647	2	0702	0600-0700
2012	12	31	4	2012-12-31	14783	1478302	MO	Missouri	0802	0802	0	0811	0800-0900

Fig. 1. A flight information table extracted from United States Department of Transportation data warehouse. Each tuple records a flight departing from a place (State, StateName) at time (DepTime) and arriving to another place at some time.

Traditional compression techniques can effectively eliminate most of the data redundancies in the scope of a single column. However, we observe there are a variety of data correlations *across* columns. The data redundancies caused by such correlations, which we refer to as *cross column redundancy* (CCR), cannot be eliminated by traditional techniques. For example, in the table shown in Fig. 1, the value of AirportID in each row is a fixed-length prefix of the corresponding AirportSeqID. Another example is DelayMin roughly equal to the difference between DepTime and CDepTime. We have detected more than 60 CCRs in the full Flight dataset in Fig. 1. Compressing them by our proposed approach can further save more than 50% storage space and the average performance of typical queries is improved by 10%. In contrast, the storage space saved by traditional compression techniques usually comes at the cost of degrading the average query performance by 20% to 50%.

The observations of CCR motivate us to design Cuttle, which is, to the best of our knowledge, the first system that exploits CCR for data compression. Cuttle achieves three major objectives. First, Cuttle compresses CCR at the physical layer without changing the logical schema. It is fully compatible with most existing compression techniques (summarized in Sect. 3). Second, it can automatically generate compression plans and encode data into compact storage layouts. Third, compared with using existing compression techniques only, Cuttle significantly improves the data compression ratio with negligible processing overhead. In fact, in many cases, the performance of common queries can be improved.

The main contributions of this paper are summarized as follows.

- We identify three types of CCRs that are ubiquitous in real-world data warehouse and propose a unified cross-column encoding scheme to compress them.
- We study the CCR selection problem and propose an automatic compression planning algorithm that can automatically choose a subset of CCRs for compressing the table.
- We evaluate Cuttle with three real-world datasets. The results show that, when CCR compression is enabled, the compression ratio is normally improved by 1.5× to 2× and the query performance can be improved by up to 22%.

The remainder of our paper is structured as follows. We summarize the state of the art in Sect. 2. We formally define CCR and introduce the CCR encoding scheme in Sect. 3. The CCR selection problem is discussed in Sect. 4. Section 5 presents the experiments. Finally, we conclude our work in Sect. 6.

2 Related Work

Traditional database compression schemes can be divided into two categories: heavyweight and lightweight [5,10]. The first category is the traditional general data compression techniques, including GZip [1], Zlib [4], and LZO [2]. These heavyweight compression schemes can achieve a better compression ratio than the lightweight ones. Its major drawback is the excessive time taken for decompression. Also, querying partial data requires to decompress the entire table. The second category is encoding schemes that are aware of the structure or the semantics of data. The category of lightweight compression includes Null supression, RLE and dictionary encoding [7,9]. Its primary advantage is fast compression/decompression time and the ability of partial decompression. These techniques are particularly effective in the context of column-stores where data of a single table are physically stored together. Compared with the methods in the first category, they require much less processing overhead when querying the compressed data. In most cases, heavyweight schemes and lightweight schemes are orthogonal techniques. Therefore, it is common practice to jointly use them to further enhance the compression ratio. Existing database systems such as MonetDB [6], CStore [8], and Hive [9] all implement these traditional database compression schemes.

3 Cross Column Redundancy

In this section, we first define cross-column redundancy (CCR) and then propose a lightweight compression scheme that addresses CCRs.

3.1 CCR Definition

Let T be a table, $R = \{r_1, r_2, \ldots, r_n\}$ and $C = \{c_1, c_2, \ldots, c_m\}$ be the n rows and m columns in T. $r[c]$ denote the element at row r and column c. A CCR in table T is a triplet $\varphi = \langle X, A, f \rangle$ where X is a subset of columns for C ($X \subseteq C$), $A \in C$, and f is a transformation function such that we can derive A based on X and f. In addition, we refer to any column $c \in X$ as an *lhs* column and A as the *rhs* column. We define three types of CCRs:

- Exact CCR. In the first type of CCR, the *rhs* column is determined by the *lhs* columns. In other words, we can compute each element of the *rhs* column based on the corresponding element in the *lhs* columns via the transformation function f. For example, FlightDate is the concatenation of Year, Month and Day padded with '-'. Similarly, StateName $= f(\text{State})$, where f is a mapping function.

- **Approximate CCR.** Given a table of n records and a threshold λ, an **approximate CCR** stipulates that over $\lambda \cdot n$ elements in the *rhs* column are determined by the corresponding elements in the *lhs* columns, i.e., we can compute at least $\lceil \lambda \cdot n \rceil$ elements in the *rhs* column based on the *lhs* columns via the transformation function f. For example, $\langle \{\texttt{DepTime}\}, \texttt{DepTimeBlk}, f \rangle$ is an approximate CCR as there are 4 of 5 rows satisfying $\texttt{DepTime} = f(\texttt{DepTimeBlk})$ except the first row.
- **Similar CCR.** The elements in the *rhs* column cannot be directly computed from those in the *lhs* columns via the transformation function f. However, the elements on both sides are similar. In this case, f is a similarity function (e.g., edit distance for strings). For example, $\texttt{WheelsOff} = f(\texttt{DepTime})$, where f is a distance function that computes the time difference. In the first row, the time difference between $\texttt{WheelsOff}$ and $\texttt{DepTime}$ is 11.

3.2   Referential Transformation Encoding

According to the definition of CCR, values in an *rhs* column can be derived from its *lhs* columns. In this section, we propose Referential Transformation Encoding (RTE), a lightweight compression scheme that reduces CCR.

Given a CCR $\varphi = \langle X, A, f \rangle$, RTE encodes the *rhs* column A into a $\langle f, I, D \rangle$ triple, where f is the transformation function in φ, I is the indicator vector, and D is the differential value vector:

- **Transformation function.** The transformation function f can be applied to each element of the *lhs* columns to obtain the corresponding element in A: $r[A] = f(r[X])$.
- **Indicator vector.** The indicator vector is for an **approximate CCR**. Recall that in an **approximate CCR**, we can compute most elements in *rhs* based on *lhs*, but not all. Therefore, RTE only encodes those *rhs* elements that can be derived from *lhs* and leaves the remaining elements unencoded. As such, RTE introduces an indicator vector I to mark if an element is encoded.
- **Differential vector.** The differential vector is for a **similar CCR**, where the transformed elements from *rhs* column are not equal, but similar to those in *lhs*. We can not remove those values in *lhs* from the table. As the differences are typically smaller than the original values, RTE stores differences for a similar CCR.

For decompression, RTE decodes each element of *rhs* via the triple (f, I, D) and the *lhs* columns. As RTE is orthogonal to intra-column compression techniques, e.g., dictionary encoding, we can further improve the compression ratio.

4   CCR Selection Problem

Given a table T of m columns, and a set of CCRs Φ, one column may act as the *rhs* column in multiple CCRs. However, for efficiency and simplicity, one column can only be encoded once in Cuttle. Therefore, given the compressed data size (or reduced size) of each CCR, the CCR selection problem is to choose a subset of CCRs such that (1) each column acts as the rhs column at most once, (2) the overall compression ratio of the table is maximized, and optionally (3) the decompression speed is within a given time constraint. Given k CCRs and the estimated compressed size of each CCR, there are 2^k possible CCR combinations. We prove that the CCR selection problem is NP-hard. It can be reduced to the well-known dominating set problem. We omit the proof in this paper due to page limit.

Since finding the optimal compression plan in polynomial time is impossible, Cuttle employs a fast greedy algorithm to provide a near optimal compression plan. Specifically, we add one CCR at a time to the optimal compression plan set Φ^o such that the added CCR φ can maximize the space saving. Each time after a new φ added into Φ^o, we remove the added CCR φ and the other CCRs $\varphi' \in \Phi$ that have the same rhs column as φ. The algorithm terminates when Φ is empty or no CCRs in Φ can be added to reduce the storage cost further. The running time of the greedy algorithm is $O(k^2 n)$, where k is the size of Φ and n is the number of rows for computing the compressed column size.

5 Experiments

In this section, we present our experimental results on Cuttle. First, we evaluate the compression ratio and query performance of Cuttle without heavyweight compression techniques in a column store. Second, we integrate existing compression techniques into Cuttle and compare its compression ratio and query performance with existing heavyweight compression technique (we use Zlib in this paper).

5.1 Experimental Setup

Environment. We evaluate Cuttle in distributed data warehouse environments. The Cuttle prototype extends Hive and runs on top of Hadoop. We run Cuttle on an 8 node Amazon EC2 cluster. Each node has an Intel Xeon E5-2670v2 (2.5 GHz/4-core) CPU, and 30.5 GB 1600 MHz DDR3 main memory. The data warehouse system we use in the experiment is Hive 2.1.0.

Datasets. In practice, existing data warehouse systems contain many CCRs. We perform our analysis on the following three real-world datasets: (1) the Bankcard[1] dataset is a global financial company's bankcard transaction dataset; (2) the Flight[2] dataset is the flight route data from United States Department of Transportation; (3) the NFL2015[3] dataset contains all the plays from the 2015–2016 National Football League (NFL) season.

---

[1] http://en.unionpay.com/.
[2] https://www.transtats.bts.gov/.
[3] https://www.kaggle.com/maxhorowitz/nflplaybyplay2015.

Implementation. We integrate RTE into Hive via a set of UDF functions, leveraging the standard UDF interface in modern database systems and mapping the original table to an implicit view maintained inside Cuttle. In other words, it can be deployed on existing data warehouse systems without modification.

5.2   Storage Performance

Let us first study the compression power of Cuttle. We compare the storage performance of Cuttle with the baseline ORC file, which is used in Hive, in storage size and compression ratio. Table 1 shows the results with/without a heavyweight compression technique (i.e., Zlib). The compression ratio is computed by *original size/compressed size*.

RTE. When no heavyweight compression technique is applied, the compression ratios of Cuttle on three datasets are over 12, which is 1.5× of the baseline ORC file. Especially, Cuttle results in 2.68 GB only on the $NFL2015$ dataset, 1.7× more compact than the baseline.

RTE + Zlib. When the heavyweight compression scheme (Zlib in this paper) is applied, Cuttle achieves a 33 compression ratio whereas the baseline ORC file only achieves 17.83 to 30.1. The higher compression is because we can apply Zlib after RTE, but the down side is that Zlib compressed data cannot be partially queried on.

In summary, Cuttle achieves a higher compression ratio than the baseline ORC file, and further improves the compression ratio by another 2.8× with heavyweight compression techniques.

Table 1. Size and compression ratio of each dataset

	Original	Bankcard	Flight	NFL2015
		120.14 GB	60.81 GB	39.86 GB
Without Zlib	ORC	14.26 GB	6.71 GB	4.55 GB
	Cuttle	9.41 GB	4.39 GB	2.68 GB
With Zlib	ORC	5.19 GB	3.41 GB	2.02 GB
	Cuttle	3.57 GB	1.7 GB	1.4 GB

5.3   Query Performance

Second, to study the impact of RTE compression on query performance, we compare the query execution time of the Cuttle table with the ORC file format in Hive. Specifically, we execute aggregation and scan queries on each table since they are the most common in a data warehouse environment. For each dataset, the first aggregation query accesses the columns that are not encoded, and the second one accesses columns that are encoded by RTE. The scan operations are performed in the same fashion. We first execute queries on the baseline ORC file in Hive and Cuttle table with RTE only and then compare the query performance on compressed tables with both RTE and Zlib.

Querying RTE compressed tables. Figure 2(c) shows the running time of aggregation and scan queries on RTE compressed tables. On the Bankcard dataset, we observe that Cuttle is 11% faster than the baseline ORC file in aggregation and 23% faster in the scan query. We also observe that query performance improvement on the Bankcard dataset is greater than both on Flight and on NFL. This is because the transformation function on Flight and NFL is more complex, which requires more decompression time. In general, Cuttle is faster than the baseline ORC file disk I/O reduction. Since the encoded table is smaller, queries on the encoded table requires less I/O and memory.

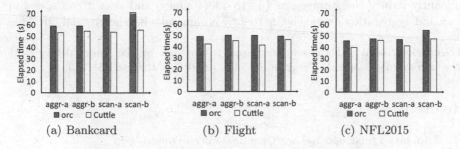

(a) Bankcard	(b) Flight

(c) NFL2015

Fig. 2. Query performance of Cuttle with RTE only

Querying RTE and Zlib compressed tables. Figure 3(c) shows the running time of aggregation and scan queries on tables compressed by both RTE and Zlib. Interestingly, the query performance improvement of Cuttle reduces with the integration with Zlib. This is because tables in Cuttle are already compressed by RTE before applying Zlib, and as a result querying requires two decompression operations. On unencoded columns, the query performance of Cuttle is similar or slightly faster than the baseline ORC file. On encoded columns, the query performance of Cuttle is similar or slightly slower than the baseline ORC file.

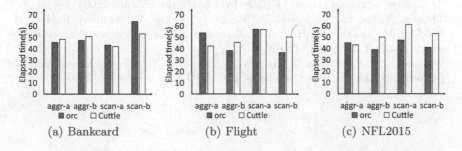

(a) Bankcard   (b) Flight   (c) NFL2015

Fig. 3. Query performance of Cuttle with both RTE and Zlib

Overall, Cuttle achieves a better query performance after RTE compression, and the query performance is close to that on the original ORC file after subsequent Zlib compression.

6   Conclusion

In this paper, we have presented Cuttle, a cross-column compression system that can automatically eliminate CCRs. We give the formal definition of CCR, and propose a lightweight compression scheme, RTE. RTE encodes CCRs into column storage and can work together with heavyweight compression schemes. We study the CCR selection problem and propose a greedy algorithm to generate the compression scheme. We evaluate the performance of Cuttle on three real-world datasets. The experimental results demonstrate that Cuttle can significantly reduce the storage size (up to 43.44 times) and that it can speed up scan and aggregation queries by up to 23% compared with the original ORC file.

Acknowledgment. This research is supported in part by the National Key Basic Research and Development Program of China (973) Grant 2014CB340303.

References

1. Gzip. http://man.openbsd.org/OpenBSD-current/man1/gzip
2. Lz. https://www.lzop.org
3. Vertica. https://www.vertica.com/
4. Zlib. http://www.zlib.net/
5. Abadi, D., Madden, S., Ferreira, M.: Integrating compression and execution in column-oriented database systems. In: Proceedings of the 2006 ACM SIGMOD International Conference on Management of Data, pp. 671–682. ACM (2006)
6. Idreos, S., Groffen, F., Nes, N., Manegold, S., Mullender, S., Kersten, M., et al.: MonetDB: two decades of research in column-oriented database architectures. Bull. IEEE Comput. Soc. Tech. Committee Data Eng. **35**(1), 40–45 (2012)
7. Roth, M.A., Van Horn, S.J.: Database compression. SIGMOD Rec. **22**, 31–39 (1993)
8. Stonebraker, M., Abadi, D.J., Batkin, A., Chen, X., Cherniack, M., Ferreira, M., Lau, E., Lin, A., Madden, S., O'Neil, E., O'Neil, P., Rasin, A., Tran, N., Zdonik, S.: C-store: a column-oriented DBMS. In: VLDB, pp. 553–564 (2005)
9. Thusoo, A., Sarma, J.S., Jain, N., Shao, Z., Chakka, P., Anthony, S., Liu, H., Wyckoff, P., Murthy, R.: Hive: a warehousing solution over a map-reduce framework. VLDB **2**, 1626–1629 (2009)
10. Westmann, T., Kossmann, D., Helmer, S., Moerkotte, G.: The implementation and performance of compressed databases. SIGMOD Rec. **29**, 55–67 (2000)

Machine Learning and Optimization

Optimizing Window Aggregate Functions via Random Sampling

Guangxuan Song, Wenwen Qu, Yilin Wang, and Xiaoling Wang(✉)

Shanghai Key Laboratory of Trustworthy Computing,
International Research Center of Trustworthy Software,
East China Normal University, Shanghai, China
{guangxuan_song,vinny_qu,ly-wrx.1993}@163.com, xlwang@sei.ecnu.edu.cn

Abstract. Window functions have been a part of the SQL standard since 2003 and have been well studied during the past decade. As the demand increases in analytics tools, window functions have seen an increasing amount of potential applications. Although the current mainstream commercial databases support window functions, the existing implementation strategies are inefficient for the real-time processing of big data. Recently, some algorithms based on sampling (e.g., online aggregation) have been proposed to deal with large and complex data in relational databases, which offer us a flexible tradeoff between accuracy and efficiency. However, sampling techniques have not been considered for window functions in databases. In this paper, we first propose two algorithms to deal with window functions based on two sampling techniques, Naive Random Sampling and Incremental Random Sampling. The proposed algorithms are highly efficient and are general enough to aggregate other existing algorithms of window functions. In particular, we evaluated our algorithms in the latest version of PostgreSQL, which demonstrated superior performance over the TPC-H benchmark.

Keywords: Window function · Query optimization · Sample

1 Introduction

Window functions, as a feature of relational database, were introduced in SQL:1999 as extended documents and formally specified in SQL:2003. Window functions allows aggregation to no longer be limited to a single tuple but over a set of tuples of a particular range. Unlike aggregation with the group-by clause which only outputs one tuple for each group, the OVER() clause associates each tuple in a window with the aggregated value over the window. As a result, window functions greatly facilitates the formulation of business intelligence queries by using the aggregation functions such as ranking, percentiles, moving averages and cumulative totals and overcome the shortcomings of alternatives such as grouped queries, correlated subqueries and self-joins [1,2]. Moreover, the syntax of a window function is so user-friendly that database users without professional programming skills may deal with business intelligence queries gracefully. Due

© Springer International Publishing AG 2017
L. Chen et al. (Eds.): APWeb-WAIM 2017, Part II, LNCS 10367, pp. 229–244, 2017.
DOI: 10.1007/978-3-319-63564-4_19

to these desirable features, window functions are widely used in current main-stream commercial databases in recent years, such as Oracle, DB2, SQL Server, Teradata, (Pivotal's) Greenplum [3, 21, 22].

However, the existing implementation strategies of window functions are too expensive to process big data. While some pioneering works have been proposed to speed up the evaluation of window functions [4–6, 10, 11], the speed-up cannot keep pace with the ever-growing data size generated by modern applications, and none of them can meet the requirement of real-time processing which is vital to online services.

Researchers have realized the necessity of tackling queries over big data using sampling since 1990s. The motivation of sampling is that analytical queries do not always require precise results. It is often more desirable to return an approximate result with an acceptable margin of error much more quickly. Database users can specify confidence intervals in advance and stop the query processing as long as desired accuracy is reached.

Online aggregation is a classic sampling technique first proposed in 1997 [12]. It improves the interactive behavior of database systems processing expensive analytical queries: a user gets estimates of an aggregate query online as soon as the query is issued. For example, if the final answer is 1000, after k seconds, the user gets the estimates in form of a confidence interval like [990, 1020] with 95% probability. This confidence keeps on shrinking as the system gets more and more samples. And online aggregation has been used in standard SQL queries, such as join [14] and group by [13]. However few systems fully support online aggregation. Besides, we are not aware of any work studying online aggregation for window functions.

This work develops several different sampling approaches, based on the sampling technologies in online aggregation, for evaluating window functions. The proposed sampling algorithms speed up query evaluation by reducing the number of tuples used in window aggregation functions.

The contributions of this paper are summarized as follows:

- We designed the Naive Random Sampling (NRS) and Incremental Random Sampling (IRS) algorithms for the quick evaluation of window functions in relational databases. NRS randomly samples tuples in a window for aggregation, and IRS improves upon NRS by avoiding repeated sampling.
- We prove that our algorithms generate unbiased estimators for various aggregate functions and provide an approach to calculate the confidence intervals. We also explain how to adjust parameters (e.g., sampling rate) according to the unique characteristics of the window functions, to satisfy various user requirements.
- We implemented our algorithms in the latest version of PostgreSQL and used the TPC-H benchmark to test the performance of our methods.

The rest of this paper is organized as follows. Section 2 discusses the background of window functions and sampling, and introduces the related work.

Section 3 describes our algorithms and presents the formulae to calculate confidence intervals. We show the experimental results in Sect. 4. Finally, this paper is concluded in Sect. 5 with a vision towards the future work.

2   Background and Related Work

Window functions. Window functions are part of the SQL:2003 standard in order to accommodate complex analytical queries. The general form of a window aggregate function is as follows:

```
SELECT *, Agg(expression) OVER(
    [PARTITION BY partition_list]
    [ORDER BY order_list [frame_clause]])
FROM table_list;
```

where *Agg* can be any of the standard aggregation functions such as *SUM*, *AVG*, *COUNT* and *expression* is the attribute to be aggregated upon. The *OVER* clause defines the aggregation window, inside which *PARTITION BY*, *ORDER BY* and *frame_clause* determine the properties of the window jointly (In fact, a frame is just a window). Examples include *MAX* (velocity) *OVER* (*PARTITION BY* roadid, carid *ORDER BY* time_t *ROWS BETWEEN 10 PRECEDING AND 10 FOLLOWING*), which means we monitor the maximal speed of the car within 20 min of each time in a car monitoring application.

Figure 1 illustrates the process of determining the range of a window. Firstly, the *PARTITION BY* clause divides the tuples into disjoint groups according to the attribute columns in *partition_list*. Nextly, attributes in *order_list* are used to sort tuples in each partition. Finally, we confirm the bounds of the frame, i.e. the range of tuples to aggregate upon, for the current tuple by *frame_clause* and invoke the related aggregate function to compute the aggregation result. The first two processes are called *reordering*, and the last one is called *sequencing*.

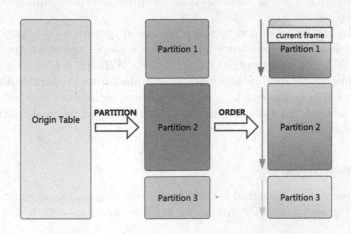

Fig. 1. Determine the window of current tuple

For each row in a partition, a window frame identifies two bounds in the window and only the rows between these two bounds are included for aggregate calculation. There are two frame models, *ROWS* and *RANGE*. In the *ROWS* model, the boundaries of a frame can be expressed as how many rows before or after the current row belong to the frame; in the *RANGE* model, the boundaries are specified as a range of values with respect to the value in the current row [10].

Like traditional aggregate functions, a window aggregate function is composed of a state transition function *transfun* and an optional final calculation function *finalfun*. To compute the final result of an aggregate function, a transition value is maintained for the current internal state of aggregation, which is continuously updated by *transfun* for each new tuple seen. After all input tuples in the window have been processed, the final aggregate result is computed using *finalfun*. Then we determine the frame boundaries for the next tuple in the partition and do the same work again.

Sampling. Sampling has been widely used in various fields of Computer Science, such as statistics, model training, query processing. With the advent of Big Data era, sampling is gaining unprecedented popularity in various applications, since it is prohibitively expensive to compute an exact answer for each query, especially in an interactive system. Sampling techniques provide an efficient alternative in applications that can tolerate small errors, in which case exact evaluation becomes an overkill.

Online aggregation [12], as a classic approach based on sampling, provides users with continuous updates of aggregation results as the query evaluation proceeds. It is attractive since estimates can be timely return to users, and the accuracy improves with time as the evaluation proceeds.

There are four components in a typical online aggregation system. The first one is the approximate result created by the system. The second and third parts consist of reliability and confidence interval which reflect the accuracy and feasibility of the result. The progress bar is the last part that reports the schedule information for the current task. All values improve over time until they meet a user's needs.

Since the results generated by sampling are approximate, it is important to calculate an associated confidence interval. For a simple aggregate query on one table, like *SELECT op(expression) FROM table WHERE predicate*, it's easy to achieve online aggregation by continuously sampling from the target table. Standard statistical formulas can help us get unbiased estimators and estimate the confidence interval. A lot of previous work [13–16] have made great contributions on this problem.

2.1 Related Work

A window function is evaluated in two phases: resorting and sequencing. Accordingly, existing approaches for evaluating window functions mainly fall into two categories.

Cao et al. [4] improved the full sort technique for evaluating window functions by proposing more competitive hash-sort and segment-sort algorithms. Then Cao et al. [5] proposed a novel method for the multi-sorting operations in one query. They found that most of the computation time is spent in sorting and partitioning when window size is small. Cao et al. proved that it is an NP-hard problem to find the optimal sequence to avoid repeated sorting for multiple window functions in one query and provided a heuristic method. Other earlier studies, such as [6–9], made a lot of contributions with regard to the ORDER BY clause and GROUP BY clause and they proposed optimization algorithms based on either the function dependencies or the reuse of the intermediate results.

Besides, there are several recent works about window operators and their functions. Leis et al. [10] proposed a general execution framework of window functions which is more suitable for memory-based systems. A novel data structure named Segment Tree is proposed to store aggregates for sub-ranges of an entire group, which helps reduce redundant calculations. However, the approach is only suitable for distributive and algebraic aggregates. Wesley et al. [11] proposed an incremental method for three holistic windowed aggregates. However, all the above works are towards exact evaluation and they face performance bottlenecks when processing massive data.

Sampling has been used in approximate query evaluation [17,18], data stream processing [25], statistical data analysis [26] and other fields [27,28]. Moreover, as the classic application of sampling, Hellerstein et al. [12] first proposed the concept of online aggregation. Since then, research on online aggregation has been actively pursued. Xu et al. [13] studied online aggregation with *group by* clause and Wu et al. [16] proposed a continuous sampling algorithm for online aggregation over multiple queries. Qin and Rusu [20] extended online aggregate to distributed and parallel environments. Li et al. [14] studied online aggregation on the queries with join clauses. These techniques based on sampling give us an opportunity to calculate the aggregate easily. In general, sampling is required to be uniform and independent in order to get reasonable results, but both [14,19] proved that a non-uniform or a non-independent sample can be used to estimate the result of an aggregate.

There are many other valuable works [23,24] on online aggregation and query sampling, but none of them deals with the problem of window functions. In this paper, we study how to evaluate online aggregate for queries involving.

3   Window Function Execution

Definition 1 (*window W*). *A window consists of a set of rows and is denoted by a triple $W_i(h, t, V)$, where i indicates that it is the i-th window, h is the window head and t is the window tail, and V is the transition value for the window.*

As mentioned in Sect. 2, window functions in commercial database systems are generally evaluated over a windowed table in a two-phase manner. In the first phase, the windowed table is reordered into a set of physical window partitions, each of which has a distinct value of the *PARTITION BY* key and is sorted on the *ORDER BY* key. The generated window partitions are then pipelined into the second phase, where the window function is sequentially invoked for each row over its window frame within each window partition.

Algorithm 1 shows the details of second phase. Firstly, for each window W_i in the *partition*, we initialize the parameters of W_i and place the read pointer at the start of the window (Line 5). Then determine whether the start position of the current window is equal to that of the previous one. If the condition is true, we can reuse the previous result and assign it to $W_i.V$ (Line 7). Nextly, we place the read pointer at the end of the previous window and traverse all tuples from $W_{i-1}.t$ to $W_i.t$. If it is false, we just traverse all tuples from $W_i.h$ to $W_i.t$. Finally, return the result and prepare for the next window.

Algorithm 1. Window Function Execution

---

1   **Input:** the whole table T after partitioning and reordering
2   **Output:** aggregate result for each tuple in the table
3   **for** *each partition in the table T* **do**
4   **for** *each window W_i in the partition* **do**
5   init $W_i.h, W_i.t, W_i.V$
6   **if** $W_i.h == W_{i-1}.h$ **then**
7            $W_i.V \leftarrow W_{i-1}.V$;
8   **for** *each row r in $(W_{i-1}.t, W_i.t]$* **do**
9               $W_i.V \leftarrow$ transfunc$(W_i.V, r)$;
10   **else**
11   **for** *each row r in $[W_i.h, W_i.t]$* **do**
12              $W_i.V \leftarrow$ transfunc$(W_i.V, r)$;
13   **return** $W_i.V$;

An example of the processing of window functions is shown in Fig. 2, where the aggregate function is summation and the frame model adopted is *ROW*. The entire table is partitioned and sorted according to *attr_2* and makes summation on *attr_1*, and red brackets in Fig. 2 represent windows. Then we find that W_2 begins with the same tuple as W_1 and the result of W_2 is incrementally updated. However, the start tuple is different between W_1 and W_2, we have to traverse all the tuples to evaluate a result.

For a table with N tuples, if we use a general window function query, like *SUM(attr_1) OVER (PARTITION BY attr_2 ORDER BY attr_2 ROWS BETWEEN S/2 PRECEDING AND S/2 FOLLOWING)*. In most cases, we must use S tuples to get one result for each tuple. And the total execution cost is $\Theta(NS)$ without considering partitioning and sorting.

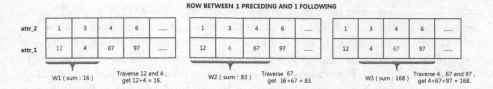

Fig. 2. An example of window functions (Color figure online)

4   The Sampling Algorithms

The evaluation cost of the window functions is proportional to the number of tuples in the table and in each window. Thus, the execution time increases sharply as the data size grows. Since the table size N is a constant, we propose to reduce the number of tuples checked in each window, S, by sampling.

Definition 2 (*Sampling window SW*). *The sampling window contains a subset of rows in window W_i and is defined by a triple $SW_i(sh, st, indexarr)$, where i represents the i-th sampling window, indexarr is an array which contains all the indexes of sampled tuples, sh is the head of sampling window, i.e., the first sampled tuple's position in indexarr and st is the tail of sampling window.*

4.1   Naïve Sampling

Our first algorithm is named as NRS (Naïve Random Sampling) and it randomly samples tuples in each window. Our algorithm generates a new temporary sampling window based on the current window by random sampling. This new sampling window is then used to estimate the aggregate results. The sampling process is independent and random and each tuple in the current window is selected with equal probability. Hence, the new sampling window is a subset of the current window.

Algorithm 2. Naïve Sampling

---

1 **Input:** the whole table T after partition and order
2 **Output:** approximate aggregation result for each tuple in the table
3 **for** *each partition in the table T* **do**
4   **for** *each window W_i in the partition* **do**
5   initialize W_i and compute the boundaries of W_i;
6        $SW_i \leftarrow$ sample($W_i.h$, $W_i.t$, rate);
7   **for** *each row r in $SW_i.indexarr$* **do**
8           $W_i.V \leftarrow$ adjust_transfunc<agg>($W_i.V$, r);
9   **return** $W_i.V$;

The detailed algorithm is given by Algorithm 2. Firstly, we determine the boundaries of the current window (Line 5). Then we generate a new sampling window SW_i by randomly sampling tuples from the current window W_i with a given sampling *rate*. And only indexes of tuples are stored in the SW_i (Line 6). Finally, we traverse all tuples in SW_i to estimate the result of aggregation function. To make the estimate unbiased, we use a new transition function *adjust_transfunc* to calculate intermediate aggregate results and it varies depending on the aggregation function (Line 8). For example, if the aggregate function is SUM, we multiply the attribute value of each sampled tuple r by the total number of tuples in the current window, and the adjusted values of all sampled tuples are averaged as the final aggregate result.

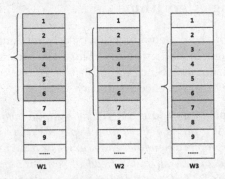

Fig. 3. An example of NRS algorithm

Figure 3 illustrates how the NRS algorithm works. There are three windows: W_1, W_2, W_3. Each time, we sample a subset of tuples from the whole window. In W_1, tuple 1, tuple 2 and tuple 3 are selected. Hence, the number of tuples processed is reduced from 6 to 3. However, we have to resample each window as the window slides, and there is no sample reuse between any two consecutive windows.

4.2   Incremental Sampling

It is inefficient to sample every window in Algorithm 2, since a large number of common tuples are shared by adjacent windows. This provides an opportunity for us to improve the sampling strategy to reuse the previously sampled tuples. This leads to an incremental sampling algorithm named IRS (Incremental Random Sampling).

Algorithm 3. Incremental Sampling

---

1 **Input:** the whole table T after partition and order
2 **Output:** approximate aggregation result for each tuple in the table
3 **for** *each partition in the table T* **do**
4   **for** *each window W_i in the partition* **do**
5   initialize W_i and compute the boundaries of W_i;
6        $SW_i \leftarrow SW_{i-1}$;
7   **while** $SW_i.indexarr[SW_i.sh] < W_i.h$ **do**
8           $\lfloor$ remove_head(SW_i);
9   create a empty SW structure named T_SW;
10   **if** $SW_i.indexarr[SW_i.st] < W_i.t$ **then**
11         $\lfloor$ $T\_SW_i \leftarrow$ sample($SW_i.indexarr[SW_i.st]$, $W_i.t$, *rate*);
12   append(SW_i, T_SW);
13   **for** *each row r in $SW_i.indexarr$* **do**
14         $\lfloor$ $W_i.V \leftarrow$ adjust_transfunc<agg>($W_i.V$, r);
15   **return** $W_i.V$;

Algorithm 3 shows the procedure of incremental sampling. Firstly, we reuse the previous structure SW_{i-1} (Line 6). Then tuples, which are beyond the range of current window, are removed from SW_i (Line 8). Function *remove_head* is used to remove the first item in $SW_i.indexarr$ and update $SW_i.sh$ depending on how many tuples are removed. Then we use a temporary structure T_SW to store the indexes of newly sampled tuples in the range from $SW_i.indexarr[SW_i.st]$ to $W_i.t$ (Line 9–11). Next, we append the indexes in $T\_SW.indexarr$ to SW_i and update $SW_i.st$ (Line 12). Finally, we aggregate the tuples in SW_i to estimate the final result (Line 13–14).

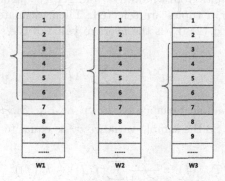

Fig. 4. An example of IRS algorithm (Color figure online)

Figure 4 shows a simple example of the IRS algorithm. The curly bracket on the left of a partition represents the range of the current window W_i and each small entry represents a tuple in table T. White entries are the tuples that fall outside the range of the current window, while yellow and gray rectangles represent the tuples in the current window. Among the tuples in the current window, yellow ones are the sampled tuples and are those stored in SW.

In Fig. 4, W_1 contains 6 tuples, and the 3 tuples marked in yellow are the sampled ones. Then tuple 1, tuple 2 and tuple 5 are stored in $SW_1.indexarr$ and are used to estimate the result of W_1. Next, when current window slides to W_2, we firstly remove the indexes in $SW_1.indexarr$ that are out of the range. In this example, tuple 1 is removed, but tuple 2 and tuple 5 are still available. At the same time, we have to update parameters of SW_1 when tuple 1 is removed. After that, we copy SW_1 to SW_2 and process the new tuple (tuple 7) in W_2. Tuple 7 is not sampled and the result of W2 is estimated by tuple 2 and tuple 5. The process of W_3 is similar. We find that tuple 2 needs to be removed and the new tuple 8 is added to $SW_3.indexarr$. So W3 can be estimated by tuple 5 and tuple 8.

4.3 Estimator and Confidence Interval

Confidence interval is an important component of approximation algorithms and indicates the proximity of each approximate result to extra result. Here we give the formulae related to the estimator and confidence interval of SUM.

Consider a window W containing m tuples, denoted by $t_1, t_2, ..., t_m$. Suppose $v(i)$ is an auxiliary function whose result is equal to m times the value of t_i when applied to t_i. Then we define a new function $F(f)$, and the extra result α of sum function is equal to $F(v)$.

$$F(f) = \frac{1}{m} \sum_{i=1}^{m} f(i) \tag{1}$$

Now we sample tuples from window W and get a sequence of tuples, denoted as $T_1, T_2, ..., T_n$, after n tuples are selected. Then estimated result $\overline{Y_n}$ can be calculated by using Eq. (2). $\overline{Y_n}$ is the average value of the samples and is equal to $T_n(v)$.

$$T_n(f) = \frac{1}{n} \sum_{i=1}^{n} f(T_i) \tag{2}$$

The formula of sample variance is shown as Eq. (3). $T_n(f)$ is the average value of the samples. So the sample variance $\widetilde{\sigma^2}$ is equal to $T_{n,2}(v)$.

$$T_{n,q}(f) = \frac{1}{n-1} \sum_{i=1}^{n} (f(T_i) - T_n(f))^q \tag{3}$$

According to central limit theorem, we can conclude that the left-hand side follows normal distribution with mean 0 and variance 1.

$$\frac{\sqrt{n}(\overline{Y_n} - \alpha)}{\sqrt{T_{n,2}(v)}} \Rightarrow N(0,1) \tag{4}$$

After we get $\overline{Y_n}$ and $\widetilde{\sigma^2}$, we have

$$P\left\{|\overline{Y_n} - \alpha| \leq \varepsilon\right\} \approx 2\Psi\left(\frac{\varepsilon\sqrt{n}}{\sqrt{T_{n,2}(v)}}\right) - 1 \tag{5}$$

where Ψ is the cumulative distribution function of the $\mathcal{N}(0,1)$ random variable. Finally, we define z_p as the $\frac{p+1}{2}$-quantile of the normal distribution function and set Eq. 5 equal to p. Then a confidence interval is computed as

$$\varepsilon_n = \frac{z_p\widetilde{\sigma}}{\sqrt{n}} \tag{6}$$

Other aggregation functions, such as COUNT, AVG, VAR, have the similar proof process. And [14,15] give more details about the proofs of a few common aggregation functions. It should be noted that incremental sampling algorithm(IRS) is non-independent and uniform. But P. J. Haas and J. M. Hellerstein [19] proved that we could estimate the aggregate by using a non-independent and uniform sample method.

5   Experiments

Our proposed techniques were implemented in the latest version PostgreSQL 9.6.1, and we compared them with PostgreSQL's native window function implementation over the TPC-H benchmark.

5.1   Experimental Setup

We run PostgreSQL on a computer of Lenovo with an Intel(R) Core(TM) i5-4460M CPU at 3.20 GHz. The system has 24 GB 1600 MHz DDR3 of RAM.

5.1.1   Dataset
We use the TPC-H DBGEN instruction to generate two "order" tables, with 1.5 million rows and 15 million rows respectively. There are 9 attribute columns in Table "order" and it covers a wide variety of data types, such as INTEGER, DOUBLE, VARCHAR, DATE, et al. More information can be obtained from TPC's official website.

5.1.2   Comparison Algorithm
- **PG:** The default implementation of the PostgreSQL itself.
- **NRS:** The naive random sampling method as shown in Algorithm 2.
- **IRS:** The incremental random sampling method as shown in Algorithm 3.

5.2   Experimental Results

In this section, we use the following query template in our experiments to demonstrate our implementation:

```
SELECT o_orderkey, aggf(o_totalprice) OVER
(ORDER BY col Rows BETWEEN pre PRECEDING AND fel FOLLOWING)
FROM orders;
```

where aggf are aggregation functions, such as sum, avg, count and so on, frame model is *ROW*, and the bounds of window are limited by *pre* and *fel*.

5.2.1   Comparison with PostgreSQL

Figure 5 shows the average execution time of SUM and AVG aggregate functions based on three algorithms: default algorithm in PostgreSQL, Native Random Sampling (NRS) and Incremental Random Sampling (IRS). The sampling rates of NRS and IRS are set to 0.3. As expected, our algorithms are far better than the default algorithm in PostgreSQL, especially dealing with large windows. In other words, NRS and IRS are more insensitive to window size. Besides, IRS is slightly better than NRS in terms of efficiency, because IRS avoids repeated sampling when calculating the window functions.

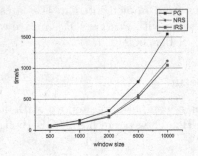

Fig. 5. The execution time of different algorithms

5.2.2   Effect of Sampling Rate

Sampling rate is a very important parameter in our methods. We use different functions on different data sets to explore the impact of sampling rate. Since Naive Random Sampling (NRS) and Incremental Random Sampling (IRS) are similar in this respect, all the pictures in Fig. 6 show the results of IRS algorithm. Figure 6a and b are the results of AVG function on the table with 1.5 M tuples. The results show that execution time is negatively correlated with sampling rate and error is positively correlated with it in any window size condition. And we can get the same conclusions with Fig. 6c and d whose results are based on the bigger table with 15 M tuples and aggregate function is SUM.

What's more, Fig. 6a and c also show that the slopes of the three curves are not similar and the black curve is the smoothest. It means the growth trend of

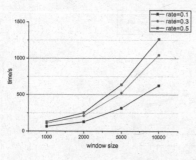

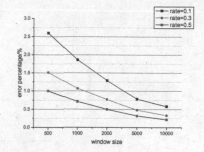

(a) The execution time of different sampling rates on the 1.5MB tuples with AVG

(b) The error of different sampling rates on the 1.5M tuples table with AVG

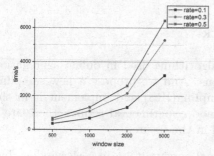

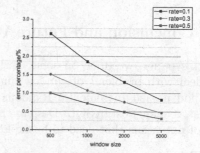

(c) The execution time of different sampling rates on the 15M tuples table with SUM

(d) The error of different sampling rates on the 15M tuples table with SUM

Fig. 6. Execution time and error on different data sets

time consumption with various rates are different as the size of window increases. In more details, IRS with low sampling rate is more insensitive to window size than that with high rate. This is the result of sample data management and the influence becomes larger as the sampling tuples become more.

Then from Fig. 6b and d, we can find that with fixed sample rate, the error becomes smaller when window size becomes bigger. It can be explained by the law of large numbers. The larger the volume of sample data is, the closer we are to the real value; the smaller the volume of sample data is, the higher the randomness is. And Fig. 6b and d also show that the error difference among various sampling rates becomes smaller as the window becomes larger. This conclusion is consistent with what is shown in Fig. 7. In Fig. 7, we can find error of big window is more stable and is more insensitive to sampling rate than that of small window. We can also use the law of large numbers to explain these conclusions.

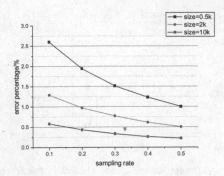

Fig. 7. Effect of sampling rate on error

6   Conclusion and Future Work

In this paper, we draw on some of the algorithms used in online aggregation to deal with problems of window functions. And we propose a new sampling method named Incremental Random Sampling(IRS) and implement it in the latest version PostgreSQL. Despite the good experimental results, there are a lot of potentials that need to be exploited in the future:

- IRS algorithm reduces the computational consumption in each window, but does not decrement the number of windows over the whole table. And it's time consuming that the table needs to be reordered into different window partitions in the first phase. So when the amount of data becomes larger, it's very useful if we could find a pioneered sampling algorithm based on the whole data set and give the formulation of confidence intervals.
- The system we have implemented is not an online system and it needs users to give the sampling rate. In the future, we want our system to be more intelligent and become a real online system.

Acknowledgement. This work was supported by NSFC grants (No. 61532021 and 61472141); Shanghai Knowledge Service Platform Project (No. ZF1213), Shanghai Leading Academic Discipline Project (Project Number: B412) and Shanghai Agriculture Applied Technology Development Program (Grant No. G20160201).

References

1. Zuzarte, C., Pirahesh, H., Ma, W., Cheng, Q., Liu, L., Wong, V.: Winmagic: subquery elimination using window aggregation. In: Proceedings of the 2003 ACM SIGMOD International Conference on Management of Data (SIGMOD 2003), pp. 652–656. ACM (2003)
2. Bellamkonda, S., Ahmed, R., Witkowski, A., Amor, A., Zait, M., Lin, C.-C.: Enhanced subquery optimizations in oracle. Proc. VLDB Endow. **2**(2), 1366–1377 (2009)

3. Ben-Gan, I.: Microsoft SQL Server 2012 High-Performance T-SQL Using Window Functions. Microsoft Press (2012)
4. Cao, Y., Bramandia, R., Chan, C.Y., Tan, K.-L.: Optimized query evaluation using cooperative sorts. In: Proceedings of the 26th International Conference on Data Engineering, ICDE, Long Beach, California, USA, pp. 601–612. IEEE (2010)
5. Cao, Y., Chan, C.-Y., Li, J., Tan, K.-L.: Optimization of analytic window functions. Proc. VLDB Endow. 5(11), 1244–1255 (2012)
6. Cao, Y., Bramandia, R., Chan, C.-Y., Tan, K.-L.: Sort-sharing-aware query processing. VLDB J. 21(3), 411–436 (2012)
7. Neumann, T., Moerkotte, G.: A combined framework for grouping and order optimization. In: Proceedings of the 30th International Conference on Very Large Data Bases (VLDB 2004), pp. 960–971. VLDB Endowment (2004)
8. Simmen, D., Shekita, E., Malkemus, T.: Fundamental techniques for order optimization. In: Proceedings of the 1996 ACM SIGMOD International Conference on Management of Data (SIGMOD 1996), pp. 57–67. ACM (1996)
9. Wang, X.Y., Chernicak, M.: Avoiding sorting and grouping in processing queries. In: Proceedings of the 29th International Conference on Very Large Data Bases (VLDB 2003), pp. 826–837. VLDB Endowment (2003)
10. Leis, V., Kan, K., Kemper, A., et al.: Efficient processing of window functions in analytical SQL queries. Proc. VLDB Endow. 8(10), 1058–1069 (2015)
11. Wesley, R., Xu, F.: Incremental computation of common windowed holistic aggregates. Proc. VLDB Endow. 9(12), 1221–1232 (2016)
12. Hellerstein, J.M., Haas, P.J., Wang, H.J.: Online aggregation. ACM SIGMOD Rec. 26(2), 171–182 (1997)
13. Xu, F., Jermaine, C.M., Dobra, A.: Confidence bounds for sampling-based group by estimates. ACM Trans. Database Syst. 33(3), 16 (2015)
14. Li, F., Wu, B., Yi, K., Join, W., et al.: Online aggregation via random walks. In: Proceedings of 35th ACM SIGMOD International Conference on Management of Data (SIGMOD 2016), pp. 615–629. ACM (2016)
15. Haas, P.J.: Large-sample and deterministic confidence intervals for online aggregation. In: Proceedings of International Conference on Scientific and Statistical Database Management, pp. 51–63. IEEE (1997)
16. Wu, S., Ooi, B.C., Tan, K.: Continuous sampling for online aggregation over multiple queries. In: SIGMOD, pp. 651–662. ACM (2010)
17. Olken, F.: Random sampling from databases. Ph.D. thesis, University of California at Berkeley (1993)
18. Wang, L., Christensen, R., Li, F., et al.: Spatial online sampling and aggregation. Proc. VLDB Endow. 9(3), 84–95 (2016)
19. Haas, P.J., Hellerstein, J.M.: Ripple joins fro online aggregation. In: Proceedings of ACM SIGMOD International Conference on Management of Data, pp. 287–298. ACM (1999)
20. Qin, C., Rusu, F.: PF-OLA: a high-performance framework for parallel online aggregation. Distrib. Parallel Databases 32(3), 337–375 (2014)
21. Window functions for postgreSQL desgin overview (2008). http://www.umitanuki.net/pgsql/wfv08/design.html
22. Bellamkonda, S., Bozkaya, T., Ghosh, B., Gupta, A., Haydu, J., Subramanian, S., Witkowski, A.: Analytic functions in oracle 8i. Technical report (2000)
23. Jin, R., Glimcher, L., Jermaine, C., et al.: New sampling-based estimators for OLAP queries. In: International Conference on Data Engineering, p. 18. IEEE (2016)

24. Joshi, S., Jermaine, C.M.: Sampling-based estimators for subset-based queries. VLDB J. **18**(1), 181–202 (2009)
25. Vitter, J.S.: Random sampling with a reservoir. ACM Trans. Math. Softw. **11**(1), 37–57 (1985)
26. Murgai, S.R.: Reference Use Statistics: Statistical Sampling Method Works (University of Tennessee at Chattanooga), p. 54. Southeastern Librarian (2006)
27. Adcock, B., Hansen, A.C.: Generalized sampling and infinite-dimensional compressed sensing. Found. Comput. Math. **16**(5), 1263–1323 (2016)
28. Bengio, S., Vinyals, O., Jaitly, N., et al.: Scheduled Sampling for Sequence Prediction with Recurrent Neural Networks. Computer Science (2015)

Fast Log Replication in Highly Available Data Store

Donghui Wang[1], Peng Cai[1(✉)], Weining Qian[1,2], Aoying Zhou[1], Tianze Pang[2], and Jing Jiang[2]

[1] Institute for Data Science and Engineering, East China Normal University,
Shanghai 200062, People's Republic of China
donghuiwang@stu.ecnu.edu.cn, {pcai,wnqian,ayzhou}@sei.ecnu.edu.cn
[2] Software Development Center, Bank of Communications, Shanghai 201201,
People's Republic of China
{pangtz,jiangj5}@bankcomm.com

Abstract. Modern large-scale data stores widely adopt consensus protocols to achieve high availability and throughput. The recently proposed Raft algorithm has better understandability and widely implemented in large amount of open source projects. In these consensus algorithms including Raft, log replication is a common and frequently used operation which has significant impact on the system performance. Especially, since the commit latency is capped by the slowest follower out of the majority followers responded to the leader, it's important to design a fast scheme to process the replicated logs by follower nodes. Based on the analysis on how the follower node handles the received log entries in Raft algorithm, we figure out the main factors influencing the duration time from when the follower receives the log and to when it acknowledges the leader this log was received. In terms of these factors we propose an effective log replication scheme to optimize the process of flushing logs to disk and replaying them, referred to as Raft with Fast Followers (FRaft). Finally, we compare the performance of Raft and FRaft using YCSB benchmark and Sysbench test tools, and experimental results demonstrate FRaft has lower latency and higher throughput than the Raft only using straightforward pipeline and batch optimization for log replication.

Keywords: Log replication · High availability · Consensus system · Raft

1 Introduction

Today's modern applications often require the back-end data store not only to provide the acceptable system performance but also to meet the high availability requirements. State machine replication is regarded as the most general approach to implementing a highly available data store where the data is replicated across a set of servers and consensus protocols are used to guarantee the consistency among different copies. Consensus protocols, including Paxos or its

© Springer International Publishing AG 2017
L. Chen et al. (Eds.): APWeb-WAIM 2017, Part II, LNCS 10367, pp. 245–259, 2017.
DOI: 10.1007/978-3-319-63564-4_20

variants [11,17,18], Viewstamped Replication [19] and Zab [15], reach an agreement on each operation and ensure all replicas execute the operation in the same order. Consensus is the fundamental problem in distributed systems and these protocols have became the key component of large-scale and fault-tolerant data store [6,10,21].

In contrast to the famous Paxos protocol, the recently proposed Raft algorithm has better understandability and widely implemented in large amount of open source projects [2,20]. During the execution of these consensus protocols including the recently proposed Raft, log replication is a common and frequently used operation which has significant impact on the system performance. In Raft, a transaction can be committed if its log has been replicated on the majority of followers. However, log replication algorithm also comes with inevitable performance problems because of the latency caused by network and processing time in followers (mainly from disk latency).

Raft achieves consensus among a group of members via an elected leader. Only leader can accept new request from clients, and then replicates log entries to followers. When the leader accept the acknowledgment from the majority of followers, it commits the transaction and both leader and followers can safely apply log entries to their replicas. In the naive implementation of Raft, the leader propagates one request at a time. In general, this is highly ineffective because multiple network transmissions increase the delay of each request. There are two optimizations widely used in the implementation of consensus protocols [5,7, 12,13]: batching and pipeline. Batching is to pack several requests into a single *AppendEntries* RPC, which spread the overhead on a set of requests. Pipeline allows the leader to propagate a new *AppendEntries* RPC to followers before the previous ones are acknowledged [16]. Pipeline can effectively improve the throughput especially in the WAN network with high latency.

Although batching and pipeline can improve the performance of Raft consensus protocol, the follower still needs to wait for flushing a batch to disk before processing the next batch in the task queue which holds the many batches sent by the leader. On the other hand, the strategy of replaying logs after they are committed incurs a large amount of expensive memory copy operation (see the details in the problem analysis section). To address these challenges, we redesign the log replication scheme for Raft protocol. The basic idea is to separate flushing a batch log from the log processing flow. Instead of directly writing the received batch logs to disk by followers, the batch is immediately moved from the task queue to a batch buffer. By this way, the next batch can be handled without any blocking. A single thread is used to monitor the batch buffer, and asynchronously flush a group of batch to disk in order to reduce disk IO overhead. Furthermore, in order to decrease the operations of memory copy, the received logs are also replayed immediately but the applied results are invisible until the corresponding transaction are committed.

The time consuming on processing the logs by follower has significant impact on the throughput and the end-to-end transaction response time as perceived by the user. In this paper, we optimize the log flushing and replay in the follower, referred to as Raft with Fast Followers (FRaft). Since FRaft has the advantage of reducing the follower servers' processing time, it has lower latency and higher throughput than the Raft only using straightforward pipeline and batch optimization for log replication.

We summarize our contributions of this paper as follows:

- According to the analysis of processing log by followers from engineering perspective, we claim that follower servers' processing time has became one of obstacles for performance promotion in Raft protocol.
- We propose a new log replication approach based on Raft, namely FRaft, that the follower is designed to more effectively handle the received logs and pre-replay the logs to avoid additional memory copy. Accordingly, we also present the corresponding recovery strategy.
- We implemented Raft and FRaft based on the open source database Ocean-Base developed by Alibaba.
- We conduct extensive experiments to evaluate Raft and FRaft under different benchmarks with YCSB and Sysbench test tools.

The paper is organized as follows. In Sect. 2, we present the analysis of log replication in Raft and introduce a motivation example that demonstrates the bottleneck of log processing in the follower. Section 3 present our proposed log replication approach FRaft and its strategy of relaying logs. In Sect. 4, we list the recovery approach under different anomalies. In Sect. 5, we conduct an experimental study to compare the performance of different approaches. In Sect. 6, we introduce the related works, and conclude this paper in Sect. 7.

2 Problem Analysis

2.1 Log Replication in Raft

Figure 1 demonstrates the basic log replication scheme without any optimizations in Raft [20]. Both leader and follower maintain a *commitIndex* which means log entries before it have been applied. Firstly, when receiving a new request from a client, the leader appends corresponding log entries to disk and replicates it to all followers by broadcasting *AppendEntries* RPCs encapsulated log entries and *commitIndex*. Secondly, each follower appends these log entries to disk and pushes it into *commitqueue* waiting for being applied to memtable (which is often implemented with B+ Tree or SkipList by in-memory DBMS or Key-Value data store) in the next round, and then sends acknowledgment to leader. When the majority of followers return success, the leader updates its *commitIndex* and applies log entries to leader's replicas. Finally, the leader will broadcast new *commitIndex* in the next *AppendEntries* RPC. Once follower learns *commitIndex* of leader, it applies log entries whose index is between its own *commintIndex* to leader's *commitIndex* to memtable. This means the log entries received and flushed in this round will be reread for applying in the next round, which will incur additional memory copy and thus increase transaction latency. In addition, the approach of writing log to disk directly in followers causes high disk I/O and increases the waiting time of next log being processed.

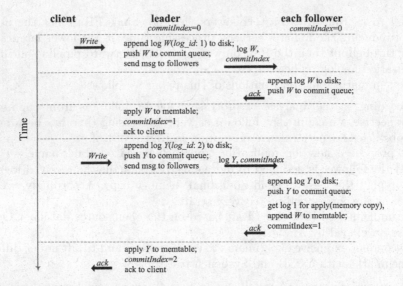

Fig. 1. The basic log replication scheme in Raft without any optimizations.

2.2 A Motivation Example

Figure 2 presents the Raft with batch and pipeline optimization. Similar to standard Raft, follower appends the received log entries from leader to the disk sequentially for durability while these uncommitted log entries cannot be directly applied to memtable. Both leader and follower keep track of the *commitIndex*. Once a follower knows *commitIndex* of leader, it will apply the log entries whose log index between *commitIndex* of its own to *commitIndex* of leader to memtable.

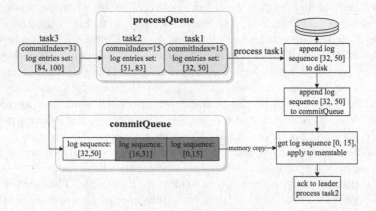

Fig. 2. A motivation example of follower's processing on Raft, with batch and pipeline optimization.

However, in practice, follower should reread the log entries from disk and apply them to memtable which causes highly disk I/O usage during log replication. An optimized way is caching these uncertain log entries into buffer to avoid disk I/O, but still invokes additional memory copy. As shown in Fig. 2, follower appends a batch log entries to the *commitQueue* which holds the uncertain log entries. Then, reading log entries for applying from *commitQueue* will produce a memory copy. Moreover, as follower appends the received log entries to disk directly which requires a disk I/O for each batch log entries, which increase the overall I/O cost of processing logs. Even worse, the next batch log entries is blocked in the *processQueue* and can not be processed until the prior one finished.

3   Fast Log Replication Approach

FRaft optimizes the process of appending log entries to disk and applying log entries to memtable. We append a batch of log entries to a in-memory buffer at first, and use a designated thread to flushing several batches via one disk I/O; Also, log entries are applied ahead of time to avoid additional memory copy.

3.1   Process of the Follower

Figure 3 shows the fast scheme to process the replicated log entries by follower nodes in FRaft. The *processQueue* is used for receiving tasks sent by leader. Firstly, we resolve the log entries of the head task of *processQueue* and append them to a specified buffer W. This approach decrease the stall time of each batch in the *processQueue*. Secondly, this task is pushed into *waitQueue* to wait for responding to client when log entries of this task has been appended to disk. Since more batches are flushed to disks only once by a single thread,

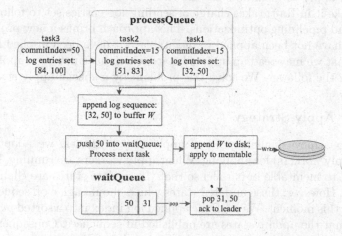

Fig. 3. Processing logs by the follower in FRaft

Algorithm 1. Follower processing algorithm on FRaft

```
 1: procedure FOLLOWERPROCESS(processQueue)
 2:     while TRUE do
 3:         task ← processQueue.pop()
 4:         resolove task
 5:         log_sequence ← task.logSequence
 6:         log_end_index ← task.log_end_index
 7:         append log_sequence to W
 8:     •   waitQueue.push(log_end_index)
 9:     end while
10: end procedure
 1: procedure REPLAY(W)
 2:     while TRUE do
 3:         append W to disk
 4:         for all log_sequence ∈ W do
 5:             log_end_index ← log_sequence.log_end_index
 6:             apply log_sequence to memtable
 7:             if waitQueue.head <= log_end_index then
 8:                 waitQueue.pop()
 9:             end if
10:             send response to client
11:         end for
12:     end while
13: end procedure
```

the wait time of a task in *waitQueue* is also reduced. In FRaft, the next task in *processQueue* will be handled immediately.

Algorithm 1 provides a high level description of the FRaft log replication approach. The procedure *FollowerProcess* depicts the tasks sent by leader processing of followers. A single thread's processing method used to append log sequence to disk and apply to memtable are shown in the procedure *Replay*. The leader in FRaft, like it in Raft, takes charge of sending log entries set to followers with batching and pipelining optimizations. This approach brings a new problem that log entries have not been applied to memtable but have been applied in the followers so that we may read an updated state which have not been applied in the leader from the follower. We solve this problem by a special apply strategy.

3.2 New Apply Strategy

In order to achieve higher speed of applying log entries, we adopt multiple threads(apply workers) for applying. When apply workers are running, log entries are applied to memtable in parallel so that the update states are chained to the update list. However, these updated states which should agree on sequentiality is invisible at this moment. We push the applied log index into a sorted *publishPool* to ensure that the update states are published in sequences. Consequently, these undetermined update states cannot be accessed by clients although we apply

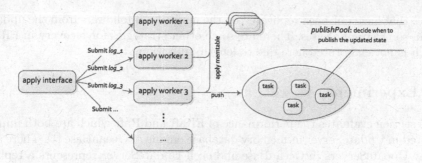

Fig. 4. Apply strategy in the followers.

them to local memtable. We leverage the *publishPool* to publish the update states. When the follower receives *commitIndex* from the leader, it will get relevant tasks which $task.logIndex <= commitIndex$ from *publishPool* to publish them. Figure 4 shows the processing in detail.

4   Recovery

The most common types of failure are network anomaly and server failure. In FRaft, the recovery strategy is a bit different from Raft as we actually apply the undetermined log entries to memtable ahead of time in followers.

Network anomaly. Network anomaly is mainly caused by the network jitter, delay or network packet that may lead to some abnormal phenomena for the leader that it cannot receive the responses from followers, and cannot apply the transaction; for the follower, network anomaly may lead receiving a discontinuous log sequence. The leader won't apply the transaction or retry to send *AppendEntries* RPCs when it doesn't receive the most majority of members' responses, because there is no way to know whether the followers receive the log entries or not. When the leader sends the next RPC combined with *commintIndex*, the followers check if the log entries is consecutive. If it is discontinuous, the followers will ask for the leader to obtain the missing log entries. The follower won't response to leader until the log is consecutive.

Server failure. Recovery strategy should ensure that the system is restored to the latest state before fault and there is no serious problem of data loss when a random server fails. If a server restarts as the leader, it will apply the log before *commitIndex* firstly, and then wait for log entries from *commitIndex* to *lastestIndex* are obtained by all followers before providing service. Otherwise, if a server restarts as a follower, it will obtain the log entries between its own *commitIndex* and the leader's *commitIndex* firstly.

Special case. In FRaft, there exists a bit difference from Raft as the log entries applied to memtable ahead of time in followers so that the update states are chained to the update list, although it is invisible. We need to revoke update states that have not been replicated on the majority of followers from the update list in some specific cases. There are some other strategies for recovery in FRaft which is not elaborated here due to length limitations.

5 Experiment

This section evaluates the performance of FRaft and Raft which are both implemented in UpdateServer in-memory database engine of OceanBase [1]. There are many UpdateServers in OceanBase and each UpdateServer represents a replica where transactions are dealt with by the leader UpdateServer, log entries are replicated on all follower UpdateServers. When the leader fails, new leader can be elected from the remaining servers to takeover the system that guarantees consistency and high availability.

The experiment is divided into two parts. Comparison of FRaft and Raft from several aspects is evaluated in part 1 by running on the public benchmark YCSB 0.7.0 [9]. In part 2, primary-copy log replication and FRaft implementation based on OceanBase are compared in case of banking business by using Sysbench test tools [3], which aims to illustrate that FRaft ensures strong data consistency without sacrificing too much performance in action production.

In part 1, there are 3 UpdateServers(3 replicas) in the system. The configuration of each server is shown in the Table 1. We conduct an experiment in part 2 to compare the performance of FRaft and primary-copy running the realistic banking production workload with 3 UpdateServers. There are two transactions in the realistic banking production: *contract* and *pay* where including one update statement, two query statements in *contract* and 5 query statements and 3 update statements in *pay*.

Table 1. Server setup

Type	Description
CPU	Intel(R) Xeon(R) E5-2640*2 2.3 GHZ 20cores/CPU
Memory	504 GB
Network	Intel Corporation I350 Gigabit Ethernet
Operating System	Red Hat Enterprise Linux Server release 6.2 (Santiago)

Throughput
The first set of experiment measures the amount of throughput that the system can support. Throughput here is the number of successfully committed operations per second. Clients send replace auto-commit transactions to the leader UpdateServer. Results are shown in Figs. 5 and 6, FRaft's peak throughput is 84172 operations per second(ops) for 150 connections and performance tends to close this value. Conversely, the peak throughput of Raft in this set of experiment is 64300 ops for 125 connections. Generally, performance of FRaft has an

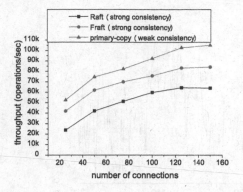

Fig. 5. Write transaction throughput.   **Fig. 6.** Transaction latency.

improvement of 1.3x than Raft, thanks to a change in a way FRaft replicated log on followers. On the other hand, we compared the throughput with primary-copy replication. In primary-copy replication, the leader does not have to wait for the majority of followers' responses before applying transactions. Although this approach benefits a lot because of low latency, it cannot provide high availability when there is a crash in the leader. Results show that FRaft sacrifices 20% performance to provide strong data consistency and high availability in such a case.

Figure 6 shows the latency of two log replication approaches with the increases of number of connections. Raft and FRaft's performance has an upward trend while Raft spends more than $4\,\mu s$ than FRaft in average. This is because the leader on Raft waits for the follower longer than that on FRaft, leading to increase the whole transaction latency. We will discuss the amount of time Raft and FRaft spend on the follower in the next set of experiment. There has been a sharp rise in latency when the number of connections reaches to 180 connections, this is because the system is congested.

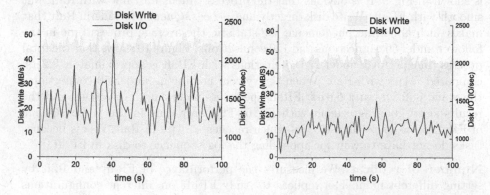

Fig. 7. The behavior of disk in Raft.   **Fig. 8.** The behavior of disk in FRaft.

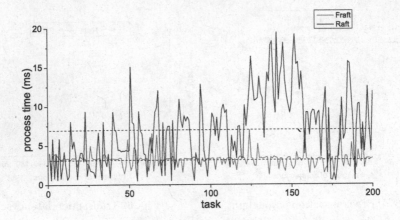

Fig. 9. Process time of follower.

Follower's criteria. FRaft and Raft have different log replication strategies on the followers, thus we evaluate the performance of the followers by observing the following criteria. (1) FRaft appends log entries to buffer firstly and then writes a batch to disks, instead of appending log to disk directly in Raft. Thus it is important to observe the disk I/O when running high workloads. We show the relevant criteria for disk of Raft and FRaft in Figs. 7 and 8. The duration of this experiment last 100 s. We monitor the run state of disk by nmon, which is a computer performance system monitor tool for the AIX and Linux operating systems developed by IBM. Results show that the amount of data written to disk reach to 50MB per seconds which are similar in Raft and FRaft. However, Raft has a disk IOPS close to 1500 and FRaft has a IOPS just below 1200. This is because FRaft writing more log entries to disk once, compared to Raft that writes less log entries once, has a less disk IOPS. (2) Another important criteria as evaluation standard of Raft and FRaft is the average follower's process time per task. We gathered the processing time of the task for a period of time, result is shown in Fig. 9. It is obvious that the process time is unstable with Raft, but smooth with FRaft. Write disk directly and *processQueue* blocking in Raft that makes such abnormal phenomena. We statistic the average process time in the follower under 50 connections and 120 connections. Figure 10 shows that the total process time per task under 120 connections with FRaft is approximately 3.2 μs, compared 7.4 μs with Raft. When the system is running under 50 connections, the value is 2.2 μs and 6.6 μs. FRaft spent less than 4 μs on average, and the results show good agreement with Fig. 6. The wait time in the *processQueue* of FRaft is no more than 0.03 μs which reaches to 4 μs in Raft, this is because tasks do not have to wait for appending the log sequence to disk in FRaft.

Number of replicas. We measures the performance of FRaft and Raft by setting different number of replicas to study FRaft on different configurations in this set of experiments. Results are shown in Fig. 11. FRaft and Raft behave similarly and maintain a close throughput under several settings. This is because

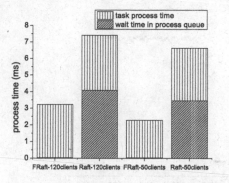

Fig. 10. Average process time of follower.

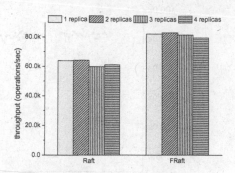

Fig. 11. The effect of increasing the number of replicas on the throughput.

log replication on replicas is in parallel, the number of replicas will not cause too much impact on the performance. However, there is still a slight downward trend of the operations per second with the number of replicas increases in both approaches. This is because the leader should wait for more followers' responses to apply transactions when more servers with different processing time participant in the system. It is obvious that the commit latency is capped by the slowest follower out of the majority followers responded to the leader.

Write to read ratio. As different applications have different read and write behavior, this set of experiment aims to evaluate whether FRaft is suitable for read- or write-intensive workloads, and the results of different write to read ratio are shown in Fig. 12. With the increase of the ratio of read transactions, throughput of FRaft is always higher than Raft.

Fault-tolerance. We observe the behavior of the system in the case of server failures in this set of experiment through throughput to judge whether the system

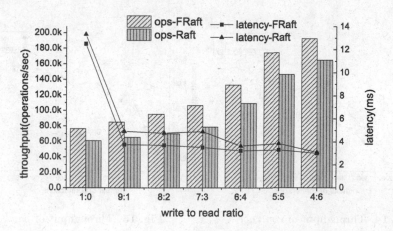

Fig. 12. The effect of the ratio of writes to reads on performance.

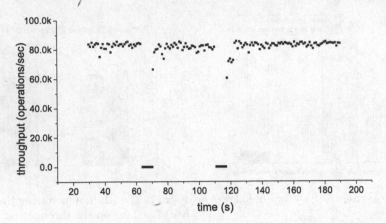

Fig. 13. The effect of server fault on the system.

can provide service. The experiments last 3 min. We kill UpdaterServer process of the leader with the *kill* command after 64 s passed by. Then 6 s later, new leader is elected and the system becomes alive. Results of this set of experiment are shown in Fig. 13. Another exception occurred in 111 s, the system spends 6 s to be alive again. The results prove that FRaft is able to detect anomalies and recover quickly under high workload that makes the system can serve normally.

Workloads. In this set of experiments, we evaluate the performance of FRaft under the realistic banking production workload, compared with primary-copy replication. Results are shown in Figs. 14 and 15. The peak transactions per second (tps) of *contract* in primary-copy is 30177, which reaches to 26701 tps, a 11.52% decrease, for FRaft. Consider the second transactions *pay*, the performance are 8447 tps in primary-copy, 8158 in FRaft with 650 connections, declined 3.42%. The results indicate that FRaft sacrifices not too much performance on high consistency and availability basis.

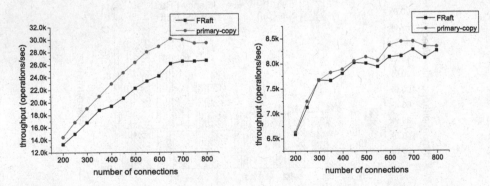

Fig. 14. Throughput of *contract*.   **Fig. 15.** Throughput of *pay*.

6   Related Works

The architecture of running transactions on top of replicated data store was proposed by Gifford [14] in 1981 which has been widely used in various systems like Google Spanner, by making use of Paxos for log replication. Besides Paxos replication, primary-copy replication [23] proposed by Michael Stonebraker in 1979 also has been frequently used in systems, i.e. MySQL. The leader does not have to wait for the followers' responses in primary-copy replication to commit transactions. Primary-copy replication has a low transaction latency but it cannot guarantee high availability and strong data consistency when there is a crash on the system. Paxos as a consensus protocol have been widely used in many systems [6,10,21] in which log replication is sponsored by leader node. The leader appends log entries to disk and applies them to memtable until it receives responses indicating log entries are persistent from the majority of followers in the first phase. In the next round, the leader send asynchronous commit message which contains last commit log index (LSN) to followers, followers apply the log before LSN to memtable.

Raft is a consensus algorithm designed as an alternative to Paxos, which was meant to be more understate than Paxos by means of separation logic while also guarantee data consistency and high availability. Comparing with Paxos, the leader will send a *AppendEntries* RPC which packs log sequence and last commit log index (*commitIndex*) to the followers to reduce a network transmission.

There are many works about improving performance of Paxos in engineering. SMARTER [6] was implemented by Microsoft based on SMART with batching and pipelining client requests where pipelining means that it allows multiple proposal run simultaneously, which almost likes slide window mechanism of TCP/IP; batching which was adopted by many systems [4,6,8] means that packaging several requests into one task, the disk and network overhead distributed on multiple requests. Nuno Santos et al. [22] made some efforts on tanning Paxos for high-throughput with batching and pipelining. There are many parameters like the network latency, bandwidth, the speed of the nodes and the properties of applications that affect the performance of these two optimizations in Paxos, thus they present an analytical model that can be used for gathering the parameters for tanning Paxos to achieve higher performance.

7   Conclusion

In this work, we present an efficient log replication approach FRaft to optimize the process of flushing log entries to disk and applying them to replicas. Instead of writing log to disk directly in the standard Raft, FRaft writes log entries to memory buffer firstly as much as possible. This enables more aggressive batch disk I/O and can handle the received tasks in a non-blocking manner. Moreover, FRaft avoids additional memory copy through applying log entries ahead of time. Experimental results show that the throughput of RFaft is 1.3 times that of the Raft and has a lower transaction latency.

Acknowledgements. This work is partially supported by National High-tech R&D Program (863 Program) under grant number 2015AA015307, National Science Foundation of China under grant numbers 61432006 and 61672232, and Guangxi Key Laboratory of Trusted Software (kx201602). The corresponding author is Peng Cai.

References

1. OceanBase. https://github.com/alibaba/oceanbase/
2. Raft. https://raft.github.io/
3. Sysbench. https://launchpad.net/sysbench/
4. Ananthanarayanan, R., Basker, V., Das, S., Photon, A.G., et al.: fault-tolerant and scalable joining of continuous data streams. In: Proceedings of the ACM SIGMOD, pp. 577–588 (2013)
5. Bartoli, A., Calabrese, C., Prica, M., Muro, E.A.D., Montresor, A.: Adaptive message packing for group communication systems. In: On The Move to Meaningful Internet Systems, pp. 912–925 (2003)
6. Bolosky, W.J., Bradshaw, D., Haagens, R.B., Kusters, N.P., Li, P.: Paxos replicated state machines as the basis of a high-performance data store. In: Proceedings of the 8th USENIX Symposium on Networked Systems Design and Implementation, NSDI (2011)
7. Carmeli, B., Gershinsky, G., Harpaz, A., Naaman, N., Nelken, H., Satran, J., Vortman, P.: High throughput reliable message dissemination. In: Proceedings of the 2004 ACM Symposium on Applied Computing (SAC), pp. 322–327 (2004)
8. Chandra, T.D., Griesemer, R., Redstone, J.: Paxos made live: an engineering perspective. In: Proceedings of the Twenty-Sixth Annual ACM Symposium on Principles of Distributed Computing, pp. 398–407 (2007)
9. Cooper, B.F., Silberstein, A., Tam, E., Ramakrishnan, R., Sears, R.: Benchmarking cloud serving systems with YCSB. In: Proceedings of the 1st ACM Symposium on Cloud Computing, SoCC 2010, pp. 143–154 (2010)
10. Corbett, J.C., Dean, J., Epstein, M., Spanner, A.F., et al.: Google's globally distributed database. ACM Trans. Comput. Syst. **31**, 8:1–8:22 (2013)
11. Dwork, C., Lynch, N.A., Stockmeyer, L.J.: Consensus in the presence of partial synchrony. J. ACM **35**, 288–323 (1988)
12. Friedman, R., Hadad, E.: Adaptive batching for replicated servers. In: 25th IEEE Symposium on Reliable Distributed Systems, pp. 311–320 (2006)
13. Friedman, R., van Renesse, R.: Packing messages as a tool for boosting the performance of total ordering protocols. In: Proceedings of the 6th International Symposium on High Performance Distributed Computing, pp. 233–242 (1997)
14. Gifford, D.K.: Information storage in a decentralized computer system. Univ. Microfilms (1982)
15. Junqueira, F.P., Reed, B.C., Serafini, M.: Zab: high-performance broadcast for primary-backup systems. In: Proceedings of the 2011 IEEE/IFIP 41st International Conference on Dependable Systems and Networks, pp. 245–256. IEEE Computer Society (2011)
16. Lamport, L.: The part-time parliament. ACM Trans. Comput. Syst. **16**(2), 133–169 (1998)
17. Mao, Y., Junqueira, F.P., Marzullo, K.: Mencius: building efficient replicated state machines for wans. In: Proceedings of the 8th USENIX Conference on Operating Systems Design and Implementation, pp. 369–384. USENIX Association (2008)

18. Moraru, I., Andersen, D.G., Kaminsky, M.: Paxos quorum leases: fast reads without sacrificing writes. In: Proceedings of the ACM Symposium on Cloud Computing, pp. 22:1–22:13. ACM (2014)

19. Oki, B.M., Liskov, B.H.: Viewstamped replication: a new primary copy method to support highly-available distributed systems. In: Proceedings of the Seventh Annual ACM Symposium on Principles of Distributed Computing, PODC 1988, pp. 8–17. ACM (1988)

20. Ongaro, D., Ousterhout, J.K.: In search of an understandable consensus algorithm. In: 2014 USENIX Annual Technical ConferenceATC, pp. 305–319 (2014)

21. Rao, J., Shekita, E.J., Tata, S.: Using paxos to build a scalable, consistent, and highly available datastore. PVLDB, pp. 243–254 (2011)

22. Santos, N., Schiper, A.: Tuning paxos for high-throughput with batching and pipelining. In: 13th International Conference Distributed Computing and Networking, pp. 153–167 (2012)

23. Stonebraker, M.: Concurrency control and consistency of multiple copies of data in distributed ingres. IEEE Trans. Softw. Eng. **3**, 188–194 (1979)

New Word Detection
in Ancient Chinese Literature

Tao Xie[✉], Bin Wu, and Bai Wang

Beijing Key Laboratory of Intelligent Telecommunication Software and Multimedia,
School of Computer Science, Beijing University of Posts and Telecommunications,
Beijing 100876, China
{xietao0222,wubin,wangbai}@bupt.edu.cn

Abstract. Mining Ancient Chinese corpus is not as convenient as modern Chinese, because there is no complete dictionary of ancient Chinese words which leads to the bad performance of tokenizers. So finding new words in ancient Chinese texts is significant. In this paper, the Apriori algorithm is improved and used to produce candidate character sequences. And a long short-term memory (LSTM) neural network is used to identify the boundaries of the word. Furthermore, we design word confidence feature to measure the confidence score of new words. The experimental results demonstrate that the improved Apriori-like algorithm can greatly improve the recall rate of valid candidate character sequences, and the average accuracy of our method on new word detection raise to 89.7%.

Keywords: New word detection · Ancient chinese literature · Apriori-like · Neural network · Word confidence

1 Introduction

Detecting new words in corpus has great significance in natural language processing (NLP), and it is indispensable to word segmentation, named entity recognition and other tasks. According to study of Ma and Chen (2003) [4], more than 62% of word segmentation errors derive from word out of dictionary.

Nowadays, Chinese new word detection mostly focuses on modern Chinese corpus. The research in ancient Chinese corpus is very limited. However, ancient Chinese is quite different from modern Chinese in several ways, such as words, phrases and syntactic structure. Generally, technique used in modern Chinese word detection may not be suitable for ancient Chinese. Moreover, many words used today, cannot be treated as a word in ancient Chinese. For example, the word 可以 in modern Chinese means 'can'. Meanwhile, in ancient Chinese, it has to be treated as two separate words, when these two words occur together, they mean 'can rely on'.

Deng et al. (2016) [7] has proposed an unsupervised method for simultaneously discovering and segmenting words and phrases from domain-specific Chinese texts. Although it was useful for segmenting ancient Chinese texts, the

© Springer International Publishing AG 2017
L. Chen et al. (Eds.): APWeb-WAIM 2017, Part II, LNCS 10367, pp. 260–275, 2017.
DOI: 10.1007/978-3-319-63564-4_21

granularity of word segmentation was uneven. There were a lot of segmentation ambiguities, and the effect of mining low frequency words was limited. However, a lot of new words in the ancient Chinese literature belong to low frequency words and there are few segmentation ambiguities in the ancient Chinese words. For this reason, a novel model was proposed to detect new words of ancient Chinese literature in this paper.

This paper puts forward a novel model using improved Apriori-like algorithm and LSTM neural network to mine new words in ancient Chinese texts. Apriori-like algorithm was used to generate candidate character sequences. Traditional Apriori algorithm can hardly find low frequency new words and may produce a large amount of noise words. We have improved the original Apriori algorithm and proposed the rule of word formation for low frequency words, which increased the recall rate of valid candidate character sequences greatly.

Recently, neural network models have increasingly been used for NLP tasks for their ability to minimize the effort in feature engineering. Long short-term memory (LSTM) neural network was used to identify the boundaries of candidate character sequences in this paper. We use LSTM network to acquire segmentation probability between two characters. Then, candidate character sequences were classified to new words and noise words based on their segmentation probability sequences.

Our major contributions include:

1. We propose a novel model called AP-LSTM, which combining Apriori-like algorithm and LSTM neural network together. This model improves the accuracy of detecting new words of ancient Chinese literature greatly.
2. Improved Apriori-like algorithm breaks the bottleneck of identifying low frequency new words, and greatly reduces the redundant items of candidate character sequences.
3. We propose word confidence feature to measure the probability score of new words, which indicates how likely the new word would be a valid word.

The rest of the paper is organized as follows. Section 2 discusses related research. Section 3 gives an overview of AP-LSTM model. Details of improved Apriori-like algorithm and LSTM neural network model are discussed in Sects. 4 and 5. Section 6 shows the filtering rule of new words and word confidence. Experiments and results are described in Sects. 7, and 8 summarizes the conclusion and future works of our method.

2 Relevant Work

Generally speaking, Chinese new word detection interweaves with Chinese word segmentation, particularly in Chinese NLP. In these works, new word detection is considered as an integral part of segmentation, where new words are identified as the most probable segments inferred by the probabilistic models. Typically models include conditional random fields proposed by Peng et al. (2004) [12], and a combined model trained with adaptive online gradient descent based on feature frequency information (Sun et al. 2012 [13]).

Another line is regarding new word detection as a separate task. The first genre of such studies is to employ complex linguistic rules or knowledge. For example, Justeson and Katz (1995) [11] extracted technical terminologies from documents using a regular expression. Chen and Ma (2002) [5] have combined morphological and statistical rules to detect Chinese new word. These methods require engineering of linguistic features and their scalability is poor. The second genre of the studies is to use statistical methods and regarded new word detection as multi-word expression extraction. The first model for quantifying multi-word association is Pointwise Mutual Information (PMI) (Chruch and Hanks 1990) [6]. Zhang et al. (2009) [16] has proposed Enhanced Mutual Information (EMI) which measures the cohesion of n-gram using the frequency of its own and the frequency of each sub-word. Bu et al. (2010) [3] has proposed a new feature named multi-word expression distance (MED). However, the capacity of these statistical methods to detect low frequency words is limited. And multi-features fusion was used frequently to improve the precision and reliability of the recognition.

In addition, user behavior data has recently been explored to find new word. Zheng et al. (2009) [18] has utilized user typing custom in Sogou Chinese Pinyin input method to detect new words. Zhang et al. (2010) [17] has used dynamic time warping to detect new words from query logs. These works performed well, however, they were restrained by the unavailability of expensive commercial resources.

Zhang et al. (2014) [15] has proposed a pragmatic quantitative model to analyze and estimate the performance of new word detection. Huang et al. (2014) [10] has proposed statistical measures to quantify the utility of a lexical pattern and detect new sentiment words. Their works heavily focused on evaluation of new word detection.

All of these works above mainly focused on modern Chinese. In ancient Chinese, the performance of their methods is limited to the in-depth linguistic knowledge and widely distribution of low frequency words.

3   Problem Statement

Before discussing about the model proposed in this paper, we need to figure out which words belong to new words in ancient Chinese and identify our problem. As for "new word", it is a word with specific meaning firstly. Second, they occur very rarely in modern Chinese corpus. Finally, they are characterized with the ancient era and various regions, which involved a classical Chinese words, poetry vocabulary, terminology and so on. So the description of new words in ancient Chinese is as follow:

New words in ancient Chinese are sequences of characters which contain a clear semantic interpretation and the historical characteristics. Meanwhile they are not included in the standard dictionary.

Based on the above definition, we can know the meaning of new words in ancient Chinese. And we designed a standard modern Chinese lexicon (used in rest paper) to filter out the noise words, including modern Chinese and stop words. Furtherly, we can present the new word detection problem formally as follows:

Definition 1. *For an given ancient Chinese corpus C and a standard dictionary D. New word detection problem aims at finding all new words W which are out of D. It consists of generating candidate new words A and the filtering of noise new words T. In other words, new word set $(A - T)$ is the final result.*

4 An Overview of AP-LSTM Model

The AP-LSTM model consists of two steps: generation step and selection step. Generation step is used to produce candidate new words in the original corpus. Noise new words are filtered out during selection step. We utilized improved Apriori-like algorithm to generate candidate character sequences and used LSTM neural network to recognize the boundaries of words. Finally, we pruned the candidate new words based on the filtering rule. In addition, word confidence could be used to measure the score of a candidate new word being a valid word. New words can be further mined based on this feature. The procedure of AP-LSTM is as follows (Fig. 1):

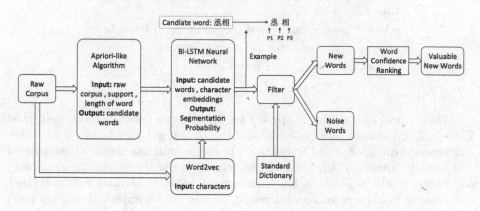

Fig. 1. Flow of AP-LSTM

For example, the word 丞相 (prime minister) generated by Apriori-like algorithm will acquire three segmentation probabilities computed by LSTM. Then, we classified it to new word or noise word based on filtering rule and standard dictionary. Finally, word confidence score is measured by its segmentation probabilities. And we can mine more valuable new words based on word confidence score ranking.

5   Improved Apriori-Like Algorithm

Traditional Apriori algorithm was applied in association rule and frequent item-sets mining. Since it was proposed by Agrawal et al. [1] in 1994. It not only influenced the association rule mining community, but also affected other data mining fields. In recent years, some researchers have applied Apriori algorithm to NLP tasks and used it to generate domain-specific words. Besides, Wang et al. (2006) [14] has used Apriori algorithm to generate frequent items, then filter frequent items by confidence and acquire specific words. However, this method will generate a lot of noise words and redundant items.

In this paper, we improved Apriori algorithm from two aspects and used it to generate candidate new words $\mathcal{A}$:

1. Candidate generation.
2. Finding low frequency new words.

5.1   Candidate Generation

Original Apriori algorithm contains the join step and the pruning step. Join opera-tion is used to generate candidate items, as shown in Table 1.

Table 1. Candidate generation in Apriori

Join L_{k-1} with L_{k-1}
select $P\,(p_1 p_2 \cdots p_{k-1})$ and $Q\,(q_1 q_2 \cdots q_{k-1})$ from L_{k-1}
if $p_1 p_2 \cdots p_{k-2} = q_1 q_2 \cdots q_{k-2},\ p_{k-1} \neq q_{k-1}$
insert $p_1 p_2 \cdots p_{k-1} q_{k-1}$ into C_k

When Apriori algorithm is applied to Chinese new word detection. In the Table 1, L_{k-1} denotes candidate words of length k-1, p_1, q_1, ..., p_{k-1}, q_{k-1} are characters in candidate word P and Q. C_k denotes candidate character sequences of length k. Although this method can mining frequency candidate character sequences, it will generate a large amount of noise words and will not take the order of characters into consideration. Therefore, we improved original join operation, as shown in Table 2. And a specific example of join operation between original Apriori algorithm and our Apriori-like algorithm is shown in Fig. 2.

In the Apriori-like algorithm, we presumed that the character sequences is ordered. It means every non-empty subsequence of frequent itemsets is ordered. The improved Apriori algorithm do not need pruning step, According to the two properties of Apriori:

1. All the non-empty subset of frequent itemsets are frequent itemsets.
2. All the superset of non-frequent itemsets must be non-frequent itemsets.

Table 2. Candidate generation in new Apriori

Join L_{k-1} **with** L_{k-1}
select $P(p_1 p_2 \cdots p_{k-1})$ **and** $Q(q_1 q_2 \cdots q_{k-1})$ **from** L_{k-1}
if $p_2 p_3 \cdots p_{k-1} = q_1 q_2 \cdots q_{k-2}$
insert $p_1 p_2 \cdots p_{k-1} q_{k-1}$ **into** C_k

Original Apriori algorithm:
P: 昭化軍節 度
Q: 昭化軍節 使 $\longrightarrow$ 昭化軍節 度使

Improved Apriori algorithm:
P: 化軍節度 使
Q: 昭 化軍節度 $\longrightarrow$ 昭 化軍節度 使

Fig. 2. Example of join operation

Obviously, candidate frequent itemsets generated by new join operation can satisfy the two properties mentioned above. For example, k-frequent item p only has two non-empty subsequences ("$p_1 p_2 \cdots p_{k-1}$" and "$p_2 p_3 \cdots p_k$"), which are both coming from (k-1)-frequent itemsets. So Apriori-like algorithm can greatly reduce the noise words and take the order of characters into account. Figure 3 shows the comparison results of two Apriori algorithm.

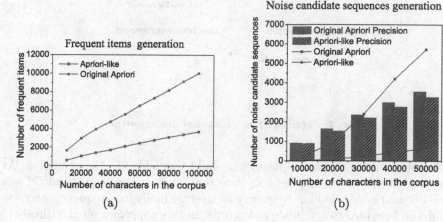

Fig. 3. Histogram denotes the number of valid words. (The experiment corpus used here is Song Poetry.)

Figure 3(a) shows that the number of frequent items generated by Apriori-like algorithm has been greatly reduced compared with the original Apriori algorithm. And experiment demonstrated that the Apriori-like algorithm can eliminate a lot of noise new words, almost without any impact on accuracy. Result is shown in Fig. 3(b). In addition, this method is more accordant with word formation rule of characters.

5.2 Finding Low Frequency New Word

Although new Apriori-like algorithm can find a lot of valid candidate new words (Here valid new words denotes word with specific semantics), there are still plenty of new words can not been mined by algorithm. Then we found that a big part of words in ancient Chinese Literature only occur once or twice in the whole corpus. Frequent itemsets generation is based on support, thus new Apriori-like algorithm cannot find low frequency new words of ancient Chinese literature. For this reason, we defined word formation rule for low frequency words.

Rule 1: *Low frequency character sequence made of frequent itemsets is more likely to be a new word.*

Based on this rule, we further improved Apriori algorithm. The algorithm flow is as follows.

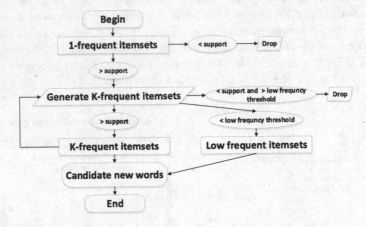

Fig. 4. Algorithm flow of Apriori-like algorithm

As shown in Fig. 4, low frequency threshold was added to the algorithm. At every step of the iteration, k-frequent item that support less than threshold was found out and added into low frequency itemsets. Finally low frequency itemsets would also be added to candidate new words. In this paper, we set the threshold to 2. Later our experiment demonstrated that the Apriori-like algorithm can increase recall rate of valid candidate character sequence in ancient Chinese literature greatly.

6 Long Short-Term Memory Neural Network Model

Traditional new word detection usually artificially constructs word features to classify candidate character sequences. However, these features often cannot comprehensively acquire information of word. But methods of neural network and Embedding can address the bottleneck of feature engineering. LSTM neural network (Graves et al. [9]) is an extension of the recurrent neural network (RNN). Since LSTM neural can keep the previous import information in memory cell and avoid the limitation of window size of local context, it is widely applied to NLP tasks. In this paper, LSTM neural network was used to identify the boundary of candidate new words.

6.1   Character Embeddings

The first step of processing symbolic data using neural network is representing them into distributed vectors, namely embeddings (Bengio et al. [2]). Formally, in new word detection task, we have a character dictionary C of size $|C|$. Each character $c \in C$ is represented as a real-valued vector (character embeddings) $v_c \in R^d$ where d is the dimensionality of the vector space. The character embeddings are then stacked into an embedding matrix $M \in R^{d \times |C|}$. For a character $c \in C$, the corresponding character embedding $v_c \in R^d$ is retrieved by the lookup table layer.

6.2   Segmentation Probability Model

We regard acquiring probability between two characters as character-based sequence labeling problem. Each character context is labeled as one of 1, 0 to indicate the segmentation. 1 represents segmented and 0 represents non-segmented, where size of context is even. So segmentation probability denotes the probability of being cut in the middle of the character context.

LSTM achieved great success in many sequence labeling tasks. LSTM are the same as RNNs, except that the hidden layer updates are replaced by purpose-built memory cells. As a result, they may be better at finding and exploiting long range dependencies in the data. To get better effect, we employed bi-directional LSTM model (Graves et al. [8]) to get information of both sides of a word. The bi-LSTM architecture is shown in Fig. 5.

For every candidate character sequence generated by new Apriori-like algorithm, firstly we get its character context in the corpus. For example, we assume that the window size is 4. As for character sequence $C_t C_{t+1}$, we find its per-position in the source text and acquire its two adjacent characters on the left and on the right respectively. So we can get characters window $C_{t-2} C_{t-1} C_t C_{t+1} C_{t+2} C_{t+3}$. If there is no adjacent character, we replace it with symbol 'Padding'. Then, we can acquire three input context $C_{t-2} C_{t-1} C_t C_{t+1}$, $C_{t-1} C_t C_{t+1} C_{t+2}$, $C_t C_{t+1} C_{t+2} C_{t+3}$ respectively. Finally we can get the segmentation probabilities of these three contexts by LSTM. Thus we can get three segmentation probabilities of candidate character sequence $C_t C_{t+1}$ in per-position of the source corpus. The segmentation probability denotes the internal segmentation probability or boundary segmentation probability of a candidate word.

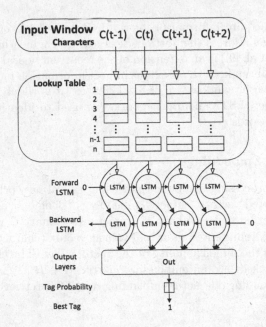

Fig. 5. The architecture of bi-LSTM model for segmentation probability

6.3 Training

We implement the neural network using the Keras framework. Character embeddings are trained by off-the-shelf tools word2vec. Training and inference are done based on segmented experiment corpus. The initial states of the LSTM are zero vectors.

Given a training set D, the regularized objective function is the loss function $J(\theta)$ including a ℓ_2-norm term:

$$\ell_i(\theta) = \begin{cases} 0 & (y_i = Y_i) \\ 1 & (y_i \neq Y_i) \end{cases} \tag{1}$$

$$J(\theta) = \frac{1}{|D|} \sum_{(x_i, y_i) \in D} \ell_i(\theta) + \frac{\gamma}{2} \parallel \theta \parallel_2^2 \tag{2}$$

Where x_i, y_i are input and output of train set, Y_i denotes tag computed by the network, θ is the parameter set of our model, γ is a regularization parameter. We train our network to minimize the loss function using a generalization of gradient descent called subgradient method.

7 New Word Detection

In this section, we proposed a filtering rule and a statistical feature to filter out the noise new words T of candidate new words $\mathcal{A}$.

7.1   Filtering Rule

For every candidate character sequence generated by Apriori-like algorithm, Firstly we filter it based on a standard dictionary, if the standard dictionary contains it, we filtered out it and regarded it as a known word. Then, we would acquire candidate new words.

For every candidate new word, it may occur one or more times in the testing set. And there will be an input context in per-position. For every input context, it will have a segmentation probability sequence computed by LSTM network.

In the ancient Chinese literature, there are a lot of characters can be a single word. For this reason, we filter candidate character sequence based on below rule:

Rule 2: *For per-input context of character sequence, if there is one of context that both of its left and right adjacency segmentation probability are more than 0.5, means that it is segmented, we classify it to new word.*

7.2   Word Confidence

In order to analyze filtered new words better. We propose a word confidence (WC) feature, a probability score to measure how likely the new word is a valid word. It consists of two parts: Branching Probability (BP) and Cohesiveness Probability (CP).

$$BP(s) = \sqrt[2]{\sqrt[n]{\prod_{i=1}^{n} P_l i} * \sqrt[n]{\prod_{j=1}^{n} P_r j}} \tag{3}$$

$$CP(s) = \sqrt[n+l]{\prod_{i=1}^{n} \prod_{c=1}^{l-1} P_c} \tag{4}$$

$$WC(s) = \mu * BP(s) + (1 - \mu) * (1 - CP(s)) \tag{5}$$

Where n is the number of context of new word s, for each context, s has a segmentation probability sequence, P_l is its left adjacency segmentation probability and P_r is its right adjacency segmentation probability, P_c denotes internal segmentation probability. l is the length of new word s.

We add μ to the calculation of score because we find BP score is more important than CP score when defining whether new word s is a valid word.

8   Experiments

8.1   Datasets

In this paper, we mainly used two datasets, Song Poetry and History of the Song Dynasty, the most representative ancient Chinese literature. We crawled 19387 Song Poetry, consisting of around 2 million Chinese characters, from http://www.gushici.org, the largest ancient poetry website in China. And we acquired the History of the Song Dynasty, abbreviated as HSD, from experiment of Deng et al. [7]. HSD contains 496 chapters and about 5.3 million Chinese characters. Then, we asked four annotators to segment 30,000 random selected sentences in Song Poetry and 32,000 random selected sentences in HSD. If there was a disagreement, discussions were required to make the final decision.

In addition, we regarded the dictionary of jieba (a popular open source project in Github.com, contains 584429 known words) as a standard lexicon $\mathcal{D}$ to filter segmentation result. Then we manually selected words with specific semantics and acquired 14891 new words in Song Poetry and 12132 new words in HSD.

8.2   Candidate New Words Generated by Apriori-Like Algorithm

First, new Apriori-like algorithm was used to generate candidate character sequences in Song Poetry corpus and HSD corpus. In this experiment, the support were all set to 5. We set length of frequent itemsets to 5 for Song Poetry and set length of frequent itemsets to 10 for HSD, and the thresholds of low frequency were all set to 2. 99782 and 72255 candidate character sequences were generated in Song Poetry and HSD respectively. Then we analyzed these character sequences. Result is shown in Fig. 6.

Number of words generated by Apriosr-like denotes how many character sequences belong to words of corpus. From Fig. 6, we can find that improved Apriori-like algorithm can mine low frequency new words efficiently. And the recall rate of valid candidate character sequences has improved greatly.

8.3   Segmentation Probability by LSTM

We splitted 80% of tagged sequences as training set, left 20% as test set. Meanwhile, we set input context size to 4 and empirically set dropout rate to 0.26.

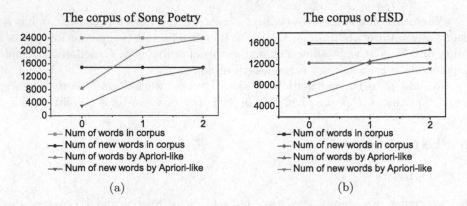

Fig. 6. X-axis denotes the threshold of low frequency. Y-axis is the number of words generated by new Apriori-like algorithm.

Dropout is used to avoid over-fitting problem. We tested performance based on different character embedding dimensions and sizes of LSTM units in Song Poetry corpus and HSD corpus. Result is shown in Table 3.

Table 3. Performance of LSTM network

Song poetry corpus					HSD corpus				
Hiddens	Emb	P	R	F	Hiddens	Emb	P	R	F
150	50	90.72	90.61	90.66	150	50	91.80	91.61	91.66
128	100	90.96	90.86	**90.91**	128	100	91.67	91.50	91.55
150	100	90.71	90.66	90.68	150	100	92.12	92.01	92.05
200	100	90.83	90.76	90.79	200	100	91.90	91.73	91.77
150	150	90.30	90.22	90.06	150	150	92.06	91.91	91.96
150	200	90.68	90.52	90.60	150	200	92.19	92.09	**92.12**

From Table 3, we found that LSTM performed the best when character embeddings dimension is 200 and size of LSTM units is 150 in the corpus of HSD. In the corpus of Song Poetry, LSTM had the best performance when character embedding dimension is 128 and size of LSTM units is 100. Thus, we used this two sets of parameters to conduct the later experiments.

8.4 Detect All New Words in Corpus

We classified candidate character sequences generated by Apriori-like algorithm to training set and test set corresponding to LSTM network. In the test set, we ran improved Apriori-like algorithm and used the parameters of the previous experiments. Then we got 13905 candidate character sequences in Song Poetry and 12265 candidate character sequences in HSD. Then, the dictionary of jieba was used to filter candidate character sequences, we got 10173 candidate new words in Song Poetry and 4017 candidate new words in HSD.

We utilized LSTM network to compute segmentation probability of per-context of candidate new word. Then we filtered the result generated by LSTM network based on filtering rule. Result is shown in Table 4.

As we can learn from the Table 4, our recall rate is relatively low. It mainly caused by the accuracy of LSTM network. Wrong tags will misled the filtering result. In addition, we analyzed the invalid new words and found that some new words did not occur in the test set but occur in the training set and un-annotated corpus, which further illustrates our method is effect and precise.

8.5 Word Confidence Analysis

We further analyzed the results of new word detection in two corpus. Word confidence score of each candidate new word was computed and ranked in Fig. 7.

Table 4. Result of new word detection

Corpus	Song Poetry	HSD
New word in corpus	2349	1778
Valid new word by our method	2107	1559
Detecting new word by our method	2852	2193
Precision	89.70	87.68
Recall	73.88	71.74
F1 value	81.02	78.92

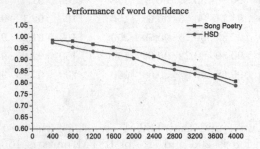

Fig. 7. Result of word confidence score ranking. X-axis is the number of words returned (K), and Y-axis is precision of valid new word.

It can be found that our word confidence feature is effect in word recognition. When we obtained top 2000 words of ranking candidate new words in the two corpus, the accuracy could be up to 90%. And we listed the top 20 new words ranked by word confidence score. It is shown in Table 5.

As is shown in Table 5, the quality of ranking result was very high. Compound words (e.g., 槛菊 it means 'Chrysanthemum outside the railing') and low frequency words (e.g., 亓氏 it is a name) could be discovered well.

Table 5. Top 20 new words in corpus

Song Poetry	HSD
1 流霞 5 沙汀 9 嫩香 13 孤帏 17 万斛	1 亓氏 5 是歲 9 交州 13 升黜 17 岢嵐
2 寒灯 6 宝辇 10 苹花 14 岁华 18 井梧	2 孫覺 6 畿內 10 暴骸 14 臺諫 18 徽樂
3 洞户 7 残蝉 11 红茵 15 画檐 19 幽闺	3 浮漏 7 龜茲 11 賢妃 15 熟戶 19 伏誅
4 槛菊 8 万灵 12 九仪 16 庭宇 20 星闱	4 銀帛 8 冢宰 12 郊宮 16 夏兵 20 熒惑

8.6   AP-LSTM vs. Other Technique

We compared AP-LSTM with the current state-of-the-art open source Chinese segmentation tools (Ansj, ICTCLAS, and Stanford Chinese-word-segmenter) in

Song Poetry corpus. And we also compared our result with TopWords model (Deng et al. [7], they showed that the accuracy of their model has reached 90%) in HSD corpus.

We used segmentation tools to segment corpus and filtered segmentation result with the dictionary of jieba. Then we counted the number of new words found by segmenter. In fairness to all tools, we transform the ancient Chinese to simplified Chinese by Opencc[1] in the first group of experiments. Comparison results is shown in Table 6.

Table 6. Comparative results of AP-LSTM with other Technique on new word detection

Corpus	Song Poetry				HSD	
Category	AP-LSTM	Ansj	ICTCLAS	Chinese-word-segmenter	AP-LSTM	TopWords
Noise words	242	2219	2219	1551	219	372
Valid new words	**2107**	130	130	798	**1559**	1406
Low frequency words	**243**	50	45	77	**207**	112
Precision	**89.70%**	5.53%	5.53%	33.97%	**87.68%**	79.08%

As shown in the Table 6, the performance of AP-LSTM is much better than other open source Chinese segmentation tools in Song Poetry corpus, which illustrated there were great difference between modern Chinese and ancient Chinese. The tools and models of modern Chinese may not be suitable for ancient Chinese. In the corpus of HSD, AP-LSTM performs better than TopWords, we compared the result of discovering words and found that our method could find more low frequency words and reduce segmentation ambiguity. In the result of TopWords, there was segmentation ambiguity in many words, which results in the various segmentation forms of a same word and the severe dependence of the accuracy of TopWords on sample data.

8.7 Experiment on Tokenizer

As we have detected new words in the corpus, we added all these new found words in the dictionary of tokenizer, and compared its performance with that of original tokenizer. The experiment is operated on the segmented subset of corpus and its result is as follow (Table 7):

From the table above, the adding of unknown words greatly improves the performance of tokenizer.

---

[1] https://github.com/BYVoid/OpenCC.

Table 7. Segmentation evaluation

Segmentation evaluation	Evaluation		
	Precision	*Recall*	*F*1
Original tokenizer	0.6544	0.6149	0.6340
Add new words	0.8213	0.8025	0.8117

9   Conclusion

In this paper, we proposed a new word detection model based on improved Apriori-like algorithm and LSTM network in ancient Chinese literature. The AP-LSTM model can find low frequency new words efficiently. And word confidence feature can further mine new words which makes result more accurate. Finally, we detected new words in two of the most representative ancient Chinese literature, Song Poetry and History of the Song Dynasty. Experiments show the effectiveness of AP-LSTM model.

In the future work, we will add more character features to character embeddings, which can improve the performance and accuracy of neural network.

Acknowledgement. This work is supported in part by the National Basic Research (973) Program of China (No. 2013CB329606). The authors would like to thank Xinyu Wu, Chunzi Wu, Chang Liu and Zhao Tang for their help in tagging, and Bin Wu for his advice to this paper.

References

1. Agrawal, R., Srikant, R., et al.: Fast algorithms for mining association rules. In: Proceedings of the 20th International Conference on Very Large Data Bases, VLDB. vol. 1215, pp. 487–499 (1994)
2. Bengio, Y., Ducharme, R., Vincent, P., Jauvin, C.: A neural probabilistic language model. J. Mach. Learn. Res. **3**(Feb), 1137–1155 (2003)
3. Bu, F., Zhu, X., Li, M.: Measuring the non-compositionality of multiword expressions. In: Proceedings of the 23rd International Conference on Computational Linguistics, pp. 116–124. Association for Computational Linguistics (2010)
4. Chen, A.: Chinese word segmentation using minimal linguistic knowledge. In: Proceedings of the Second SIGHAN Workshop on Chinese Kanguage Processing, vol. 17, pp. 148–151. Association for Computational Linguistics (2003)
5. Chen, K.J., Ma, W.Y.: Unknown word extraction for Chinese documents. In: Proceedings of the 19th international conference on Computational Linguistics, vol. 1, pp. 1–7. Association for Computational Linguistics (2002)
6. Church, K.W., Hanks, P.: Word association norms, mutual information, and lexicography. Comput. Linguist. **16**(1), 22–29 (1990)
7. Deng, K., Bol, P.K., Li, K.J., Liu, J.S.: On the unsupervised analysis of domain-specific Chinese texts. In: Proceedings of the National Academy of Sciences, p. 201516510 (2016)

8. Graves, A., Fernández, S., Schmidhuber, J.: Bidirectional LSTM networks for improved phoneme classification and recognition. In: Duch, W., Kacprzyk, J., Oja, E., Zadrożny, S. (eds.) ICANN 2005. LNCS, vol. 3697, pp. 799–804. Springer, Heidelberg (2005). doi:10.1007/11550907_126
9. Graves, A., Schmidhuber, J.: Framewise phoneme classification with bidirectional lstm and other neural network architectures. Neural Netw. 18(5), 602–610 (2005)
10. Huang, M., Ye, B., Wang, Y., Chen, H., Cheng, J., Zhu, X.: New word detection for sentiment analysis. In: ACL (1), pp. 531–541 (2014)
11. Justeson, J.S., Katz, S.M.: Technical terminology: some linguistic properties and an algorithm for identification in text. Natural Lang. Eng. 1(01), 9–27 (1995)
12. Peng, F., Feng, F., McCallum, A.: Chinese segmentation and new word detection using conditional random fields. In: Proceedings of the 20th International Conference on Computational Linguistics, p. 562. Association for Computational Linguistics (2004)
13. Sun, X., Wang, H., Li, W.: Fast online training with frequency-adaptive learning rates for Chinese word segmentation and new word detection. In: Proceedings of the 50th Annual Meeting of the Association for Computational Linguistics: Long Papers, vol. 1, pp. 253–262. Association for Computational Linguistics (2012)
14. Wang, L.X., Wang, J.D., Wang, J.: Approach for lexicon updating based on data mining. Appl. Res. Comput. 12, 062 (2006)
15. Zhang, H., Shi, S.: Which performs better for new word detection, character based or Chinese word segmentation based? In: 2014 International Conference on Asian Language Processing (IALP), pp. 10–14. IEEE (2014)
16. Zhang, W., Yoshida, T., Tang, X., Ho, T.B.: Improving effectiveness of mutual information for substantival multiword expression extraction. Expert Syst. Appl. 36(8), 10919–10930 (2009)
17. Zhang, Y., Sun, M., Zhang, Y.: Chinese new word detection from query logs. In: Cao, L., Zhong, J., Feng, Y. (eds.) ADMA 2010. LNCS, vol. 6441, pp. 233–243. Springer, Heidelberg (2010). doi:10.1007/978-3-642-17313-4_24
18. Zheng, Y., Liu, Z., Sun, M., Ru, L., Zhang, Y.: Incorporating user behaviors in new word detection. In: IJCAI. vol. 9, pp. 2101–2106. Citeseer (2009)

Identifying Evolutionary Topic Temporal Patterns Based on Bursty Phrase Clustering

Yixuan Liu[✉], Zihao Gao[✉], and Mizuho Iwaihara[✉]

Graduate School of Information, Production and Systems, Waseda University,
Fukuoka 808-0135, Japan
liuyixuan@ruri.waseda.jp, gao_jihao@suou.waseda.jp,
iwaihara@waseda.jp

Abstract. We discuss a temporal text mining task on finding evolutionary patterns of topics from a collection of article revisions. To reveal the evolution of topics, we propose a novel method for finding key phrases that are bursty and significant in terms of revision histories. Then we show a time series clustering method to group phrases that have similar burst histories, where additions and deletions are separately considered, and time series is abstracted by burst detection. In clustering, we use dynamic time warping to measure the distance between time sequences of phrase frequencies. Experimental results show that our method clusters phrases into groups that actually share similar bursts which can be explained by real-world events.

Keywords: Topic evolution · Temporal pattern · Burst detection · DTW · Clustering

1 Introduction

Over the past decade, numerous online collaboration systems have appeared and thrived on the Internet. Prime examples include Wikipedia, Yahoo! Answers, Mechanical Turkbased systems [3]. They enlist a large number of people to supply information and solve problems. Users become the main strength of contributors. Wikipedia is an open, multilingual Internet encyclopedia written collaboratively by volunteers around the world [11], and end users can also edit articles. Each edit revision is saved and all the revisions are available to the public. We can utilize the revision history to discover trends of topics.

In this paper, we particularly focus on how representative phrases change their frequencies in revisions, to find whether bursty edits occur phrases. We further present a method for clustering phrases by similarity on bursty time series, where we expect that topics in one cluster share similar temporal patterns in edit history. One cluster may contain multiple real-world events which are related each other. Such fine-grained temporal relationship is difficult to be found on topic models over a static corpus or a temporal series of corpora.

Unlike traditional text clustering works, we solve the problem by focusing on the changes of phrase frequencies in revisions. We also discuss extraction of candidate

© Springer International Publishing AG 2017
L. Chen et al. (Eds.): APWeb-WAIM 2017, Part II, LNCS 10367, pp. 276–284, 2017.
DOI: 10.1007/978-3-319-63564-4_22

phrases that have significant temporal features, where additions and deletions of a phrase are separately evaluated.

In Wikipedia edit history, articles of events or persons are edited over years, and edits of articles can be bursty or gentle, and the peak time can be shifted. Therefore we need to adopt a flexible function to measure temporal similarities between phrases. Major contributions of this paper are summarized below:

- We define a number of scoring functions to find candidate phrases that are having significant temporal features, and define the cumulative phrase frequency, which is effective to obtain time series of edit activities of a phrase in articles.
- We apply burst detection on the time series of each phrase to reduce minor details and simply the time series. Then we apply k-means to cluster phrases, where similarity is defined by burst patterns, and dynamic time warping (DTW) [2] is utilized.

2 Related Work

Kalogeratos et al. [5] proposed an algorithm to improving text clustering algorithm by term coburstiness. The algorithm first constructs a bursty term correlation graph, then applying graph partitioning technique to find clusters based on bursts and inter-burst relationships. The fundamental problem in [5] is that it is irrespective of bursts of edits. Kleinberg [6] proposed a burst model where a burst is defined as a rapid increase of a term's frequency of occurrences. If the term frequency is encountered at an unusual high rate, then the term is labeled as 'bursty'. The work is used to identify bursts in text stream and produce state labels of bursts. Tran et al. [10] engaged in temporal text mining domain and proposed to represent an event by entities, which is instructive for event representation during burst event detection in Wikipedia. However, though a number of research activities on burst, it is difficult to compare methods because of a lack of common procedures today. Subašic [9] build up an evaluation system for temporal text mining methods, which makes it possible to measure the effectiveness of temporal text mining for news.

3 Selecting Historically Significant Phrases

To find evolution of topics in a Wikipedia category, we need to monitor a relatively long period of edit history. We utilize key phrases to represent candidate topics. We discuss how to detect phrases that have bursty surge of edits in an article and a collection of articles. Wikipedia articles are edited repeatedly to reflect a chain of new events. Phrases occurring in such burst edits shall be detected as bursty phrases. We discuss selecting phrases that are semantically representative and having significant edit activities in their history.

3.1 POS Tagging and Filtering

In order to cluster shifty topics, we first detect key phrases that could represent topics in one article. In this step, POS (part of speech) Tagger and Chunker provided by Apache openNLP [4] is used to apply POS tagging and chunking function upon the revisions of Wikipedia articles, to produce various POS labels to each chunk.

3.2 Decay Phrase Frequency

Next we introduce a number of temporal features to refine candidate phrases. Revisions of one article are created over time, where various phrases are added or deleted, and phrase frequencies, namely the times one phrase occur in one revision, dynamically change over time. Traditional TF-IDF does not capture such temporal factors of document revisions.

In order to give weights to phrases through the whole edit history, one way is to compute the term frequency with a decay factor, like Aji [1]. However, their model ignores revision quality, and decay factor is applied based on revision counts, instead of time interval. We should give a higher weight to phrases appearing in long-lived revisions because that long-lived revisions are accepted by editors as trustworthy and high-quality. Let $t(r)$ denote the time point of revision r and $f(p, r_i)$ denote the frequency of phrase p in i^{th} revision. We propose our timespan-weighted phrase frequency over history as:

$$PFH = \frac{f(p, r_i)}{i^\rho} \cdot \frac{t(r_{i+1}) - t(r_i)}{t(r_n) - t(r_1)} \tag{1}$$

Here, $\rho > 0$ is a decay factor, and r_i is the ith revision of the article, $i = 1,..,n$. We evaluate how widely a term occurs in the considered article set as below:

$$rf(p, D) = \frac{|r \in D : p \in r| + 1}{N} \tag{2}$$

Here, N is the total number of articles in the corpus D, $|r \in D : p \in r|$ is the number of revisions in which phrase p occurs. In case p is not in the corpus, we add one to the numerator to avoid division by zero.

3.3 Survival Rate

We define the survival rate of a phrase p in an article a as:

$$SR(p) = e^{\frac{scale(p)}{|R|}} \cdot \frac{contain(p)}{scale(p)} \tag{3}$$

Here, $|R|$ is the number of revisions of article a, $scale(p)$ is the number of revisions of article a in the period between p appears first and p appears last, and $contain(p)$ is the number of revisions of article a containing p. The first part measures how many revisions the phrase survived over the history, and the second part measures how long the phrase appears without interruptions during its lifespan. The survival rate is independent from text length. Phrases having long and non-interrupting lifespans are highly scored by the survival rate.

Combining the features we proposed so far, we define the following weight function for selecting historically-significant phrases:

$$W_{p_i} = PFH(p_i, R_x) \cdot SR(p_i) \cdot rf(p_i, R_x) \tag{4}$$

4 Abstracting Time Series of Bursty Phrases

4.1 Time Series Modeling

Let $S = [s_1, \ldots, s_n]$ be a corpus, which is a sequence of target articles $s_1, \ldots, s_n$. The corpus can be chosen based on certain topics, such as from a Wikipedia category.

The *revision frequency* of a phrase is the times the phrase occurs in one revision. As new revisions are created over time, certain phrases increase their revision frequency, while others keep stable, or decrease, and sometimes fluctuate. To detect edit activities, we should focus only on changes of revision frequency, which is caused by additions and deletions of a phrase. Let r_i and r_{i-1} be two adjacent revisions of an identical article. The frequency difference of phrase p between revision r_i and r_{i-1} is given by:

$$\delta(p, r_i) = pf(p, r_i) - pf(p, r_{i-1}) \tag{5}$$

Here pf is the revision frequency of phrase p in the article. When $\delta(p, r_i) > 0$ holds, additions of p are more than deletions of p in creating revision r_i from r_{i-1}.

Usually, the length of one articles grows over time, and the frequency of a phrase increases as well. On the other hand, deleting a phrase p can happen when p is overwritten by new contents, due to obsoleteness, error corrections, new facts, new concepts, etc. Also, edit fights between editors can cause deletions of a phrase. Furthermore, a real-world event can trigger a rush of edits, causing fluctuations of phrase frequencies. Thus additions and deletions of a phrase indicate distinct phenomena, so that we should introduce a burst detection which treats additions and deletions separately. We define the adding effect af of phrase p from revision ri to revision rj as:

$$af(p, r_i, r_j) = \begin{cases} \delta(p, r_i) \cdot e^{-\lambda_1 (j-i)}, & \delta(p, r_i) > 0 \; and \; j > i \\ 0, & \delta(p, r_i) \leq 0 \; or \; j \leq i \end{cases} \tag{6}$$

We also define the *deleting effect df* of phrase p from revision r_i to revision r_j as:

$$df(p, r_i, r_j) = \begin{cases} \delta(p, r_i) \cdot e^{-\lambda_2 (i-j)}, & \delta(p, r_i) < 0 \; and \; j < i \\ 0, & \delta(p, r_i) \geq 0 \; or \; j \geq i \end{cases} \tag{7}$$

Here, we assume that additions and deletions of preceding revisions i are affecting revision j with an exponential decay, based on revision number difference $j-i$, with the decay parameters $\lambda_1 > 0$ and $\lambda_2 > 0$. By having these separate decay parameters, we can control relative weighting between adding and deleting revisions over a phrase.

Finally, we define *cumulative phrase frequency* $Pf(p, r_j)$ below, which sums up adding and deleting effects of revisions before j.

$$PF(p, S) = [Pf(p, r_{11}), \ldots, Pf(p, r_{nm})]$$
$$Pf(p, r_j) = \delta(p, r_{11}) + \sum_{i=1}^{n} [af(p, r_i, r_j) + df(p, r_i, r_j)] \tag{8}$$

Here, r_{mn} is the nth revision of the mth article in corpus S. We take the sum of cumulative phrase frequencies for every article containing p and for every revision in a time unit of one week, into one element $Pf(p, r_{ij})$, and construct the time series $PF(p, S)$ for temporal clustering.

4.2   Burst Detection and Time Series Abstraction

Cumulative phrase frequency captures trends of edit activities, but its time series $PF(p, S)$ still contains noises and spikes, which makes difficult to cluster similar edit histories. To overcome this problem, we apply Kleinberg's burst detection algorithm [6], to convert the cumulative phrase frequencies into burst levels. The burst detection algorithm finds a state sequence where each symbol b_i corresponds to a burst level of non-negative integers. After burst detection, we obtain a time series $TS_p = [(s_1, x_1), \ldots, (s_n, x_n)]$ on phase p, where s_i is the burst level at time point x_i. After this time series abstraction, we can easily detect bursts co-occurring in phrases.

4.3   Temporal Clustering by Dynamic Time Warping Measure

In order to uncover the temporal dynamics of key phrases in Wikipedia, we apply the temporal clustering method dynamic time warping (DTW) [1] to the abstracted time series of edit histories. Since the time sequences on phrases have sparse and random bursts, and their durations are diverse, time shift on peaks of frequencies is necessary. The classical Euclidean distance metric which compares at exact time points is not suitable in this situation.

For a pair of phrases p and q, let TS_1 and TS_2 be their time sequences of cumulative phrase frequencies. We define the distance $d(i, j)$ between the ith component (s_{1i}, x_{1i}) of TS_1 and the jth component (s_{2j}, x_{2j}) of TS_2 as $d(i, j) = \sqrt{((s_{1i} - s_{2j})^2 + (x_{1i} - x_{2j})^2)}$. Based on the local cost matrix by $d(i, j)$, dynamic time warping paths are calculated by dynamic programming, which yield DTW distances. We carry out k-means clustering [12], to cluster phrases having similar burst patterns. Since burst positions and durations are varying, in k-means clustering we adopt the DTW metric, instead of the Euclidean metric.

5 Experiments

To confirm the effect of cumulative phrase frequency, we present in Fig. 1 the result on phrase "airstrikes," which is in the candidate phrases selected by the weighting function in Sect. 3, from article "American-led intervention in Syria."

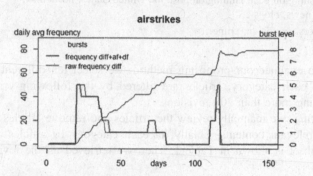

Fig. 1. Detected bursts of "airstrikes" with both effects in "American-led intervention in Syria" (Color figure online)

The red curve in Fig. 1 shows burst levels detected on cumulative phrase frequencies, where additions and deletions are evaluated separately with decay parameters, while the blue curve shows burst levels detected on changes of raw phrase frequencies. We can find that the red curve has more detailed burst levels and burst intervals, responding to sudden changes of phrase frequencies. On the other hand, although the blue curve has reduced noises, a number of bursts are not detected. Note that we can control the burst sensitivity by changing the decay parameters of additions and deletions.

Table 1. Sample of dataset articles in experiments

Title	Time span
Barack Obama	2004/9/30-2016/2/13
Democratic Party (United States)	2001/9/21-2014/12/4
Donald Trump presidential campaign, 2016	2014/8/4-2016/3/22
Donald Trump	2011/8/28-2016/3/3
Hillary Clinton	2007/1/24-2016/5/6
United States Presidential election, 2016	2014/9/28-2016/3/12
United States Presidential election, 2012	2010/6/15-2014/7/19
United States Presidential election, 2008	2004/12/15-2011/6/7
Republican Party (United States)	2010/6/14-2015/10/7
Republican Party presidential primaries, 2016	2010/6/14-2015/10/7

Table 2. Clustering results on phrases by cummlative phrase frequencies

Cluster	Phrases	Number of phrases
1	Trump, Rand Paul, Ted Cruz, his campaign, reuters	5
2	Illegal immigrant, ISIS, Syria, Mac Cain...	40
3	Washington post, Huffington post, the United States, news max, fox news, blogs...	9
4	Actors, campaign, crime...	46

In order to evaluate our proposing method, we select ten different categories in Wikipedia. In each category, articles are filtered by the following criteria: English language, having more than 200 revisions.

After filtering, we manually review the articles and remove articles with repeated contents or irrelevant contents. Finally, in each category 10 articles are filtered as datasets. A subset is shown in Table 1. Then we retrieved all the revisions of each article.

We selected historically significant phrases by the method described in Sect. 3, and produced time sequences by their cumulative phrase frequencies. After simplifying the time sequences by burst detection, temporal clustering was carried out.

Figure 2 and Table 2 show a part of phrases in each cluster of the temporal clustering, where phrases having similar burst patters are grouped into the same cluster.

We choose a week as a unit of time series, which gives us reasonable revision counts per unit time. Temporal clustering is applied over revisions created in the 167 weeks from Aug. 1st, 2012 to Oct. 6th, 2015. We tried several times and four clusters were observed. The burst detection step realizes removal of most of spikes and noises, so that only notable bursts are observed. In cluster 1, the patterns can be linked to real-world events. For example, in November 2012, Ted Cruz won general election. He is the first Hispanic American to serve as a U.S. Senator representing Texas. Then in 2015, he declared to join the election again so the phrase bursts at 2015 summer. Cluster 2 only has three peaks, which are related to phrases about isolated events, including illegal immigrants. In Sept. 2015, a new policy for immigration was in effect, and the relationship between this event and the burst is obvious. Cluster 3 is similar to cluster 1, but in cluster 3 two main bursts share the same level, which can be linked to political news in 2012 and 2015 on presidential elections. In cluster 4, there is no obvious burst, and most phrases in this cluster are common nouns which are quite frequent in articles.

We compared temporal clustering results without burst detection, and clustering by Euclidian distance. Due to space limitation, we show overviews of the results. When the burst detection step was omitted, the time sequences contained sharp peaks and noises. Also, when the Euclidean metric was used instead of DTW, time shifting ability was lost. In both cases, cluster qualities were inferior than Table 2, and resulting clusters contain phrases that are difficult to link to real-world events. We therefore conclude that both burst detection and DTW steps are necessary in our scheme.

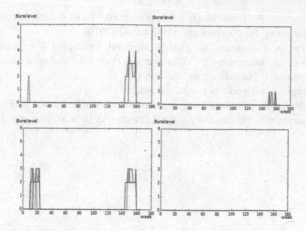

Fig. 2. Temporal patterns of evolutionary topics

6 Conclusion and Future Work

In this paper, we proposed a novel approach for capturing topical trends, by analyzing changes of phrase frequencies in edit revisions of articles. We combine burst detection and DTW for temporal clustering of phrases, where phrases that were edited around the same time periods were grouped. Bust detection of phrases are utilized to simplify phrase time series. Our experimental results show that the proposed method produces meaningful clusters of phrases that share similar burst patterns. In future, we will focus on improving clustering qualities and efficiency.

References

1. Aji, A., Wang, Y., Agichtein, E., et al.: Using the past to score the present: extending term weighting models through revision history analysis. In: Proceedings 19th ACM International Conference on Information and Knowledge Management, pp. 629–638 (2010)
2. Adwan, S., Arof, H.: On improving dynamic time warping for pattern matching. Measurement **45**(6), 1609–1620 (2012)
3. Doan, A., Ramakrishnan, R., Halevy, A.Y.: Crowdsourcing systems on the world-wide web. Commun. ACM **54**(4), 86–96 (2011)
4. http://opennlp.apache.org/
5. Kalogeratos, A., Zagorisios, P., Likas, A.: Improving text stream clustering using term burstiness and co-burstiness. In: Proceedings of the 9th Hellenic Conference Artificial Intelligence, p. 16. ACM (2016)
6. Kleinberg, J.: Bursty and hierarchical structure in streams. Data Mining Knowl. Discov. **7**(4), 373–397 (2003)
7. Liu, Y., Gao, Z., Iwaihara, M.: Identifying evolutionary topic temporal patterns based on bursty phrase clustering. DEIM Forum C5-1, March 2017
8. A Press: Wikipedia and Artificial Intelligence: An Evolving Synergy

9. Subašic, I., Berendt, B.: From bursty patterns to bursty facts: the effectiveness of temporal text mining for news. In: Proceedings of the ECAI (2010)

10. Tran, T., Ceroni, A., Georgescu, M., et al: Wikipevent: leveraging Wikipedia edit history for event detection. In: International Conference on Web Information Systems Engineering. Springer International Publishing, pp. 90–108 (2014)

11. Wikipedia: http://en.wikipedia.org/wiki/Wikipedia

12. Yang, J., Leskovec, J.: Patterns of temporal variation in online media. In: Proceedings of the 4th ACM International Conference on Web Search and Data Mining, pp. 177–186 (2011)

Personalized Citation Recommendation via Convolutional Neural Networks

Jun Yin and Xiaoming Li[✉]

School of Electronic Engineering and Computer Science, Peking University,
Beijing 100871, People's Republic of China
{jun.yin,lxm}@pku.edu.cn

Abstract. Automatic citation recommendation based on citation context, together with consideration of users' preference and writing patterns is an emerging research topic. In this paper, we propose a novel personalized convolutional neural networks (p-CNN) discriminatively trained by maximizing the conditional likelihood of the cited documents given a citation context. The proposed model not only nicely represents the hierarchical structures of sentences with their layer-by-layer composition and pooling, but also includes authorship information. It includes each paper's author into our neural network's input layer and thus can generate semantic content features and representative author features simultaneously. The results show that the proposed model can effectively captures salient representations and hence significantly outperforms several baseline methods in citation recommendation task in terms of recall and Mean Average Precision rates.

Keywords: Convolutional neural networks · Citation recommendation · Personalization

1 Introduction

With the impressive expanding speed of research papers producing annually, finding out which papers to cite is not a trivial task. For newcomers in a research area, this is especially a challenging task. Under such circumstances a system automatically recommending candidate citations is in great need. Figure 1 shows an example of citation recommendation. Given a piece of citation context and a set of papers in the left part of Fig. 1, the right part is the recommended result that we expect our proposed method outputs. Some researchers have been already aware of the necessity of citation recommendation and they have figured out various algorithms [2,7,9,14,16].

Unfortunately, simply solving out the citation recommendation task via lexical matching encounters various problems or difficulties with observation that the same concept is often expressed using different expressions and word usages in different papers. Latent semantic models such as latent semantic analysis (LSA) are able to represent documents. Based on this, probabilistic topic models such as probabilistic LSA (PLSA), Latent Dirichlet Allocation(LDA), and

© Springer International Publishing AG 2017
L. Chen et al. (Eds.): APWeb-WAIM 2017, Part II, LNCS 10367, pp. 285–293, 2017.
DOI: 10.1007/978-3-319-63564-4_23

Bi-Lingual Topic Model(BLTM) are developed and applied in semantic matching. Recent days deep learning models have shown great potential on learning effective representations and achieved state-of-the-art performance in computer vision [17] and natural language processing applications [1]. In spite of deep model's more appealing capability in automatically learning feature representation, they are inferior to shallow models such as in collaborative filtering from considering implicit authorship and similarity relationship between items. This calls for integrating deep learning with context aware recommendation by performing deep learning collaboratively.

Our research contributions can be summarized as follows:

(1) To the best of our knowledge, this work is a successful attempt in applying the CNN-like method to citation recommendation task. This method avoids a lot of time-consuming feature engineering to represent semantic content and author preference.
(2) We propose a novel p-CNN which includes paper's author into input layer and thus can generate semantic content features and representative author features simultaneously. Personalization citation recommendation is achieved via this approach.

2   Related Work

2.1   Citation Recommendation

There have been different approaches dealing with citation recommendation task, which can be grouped into three categories. The first category uses a graphical framework. Each research paper is represented as a node with citation relationship regarded as links between them. The recommendation task is cast as link prediction [7,9,13,16]. The second category usually utilizes various kinds

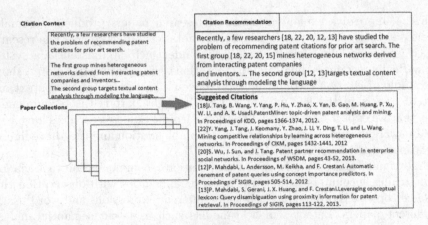

Fig. 1. One example of citation recommendation

of content based semantiç analysis techniques. [14] applys Topic Model. [8] uses translation model. [2] formalizes this problem under the retrieval framework. From IR view of point, citation context is regarded as query while the target paper is retrieved document corpus. The third category concentrates on personalized aspect of this task, in which collaborative filtering is a widely used method. Diverse kinds of features are designed [18] to enhance recommendation performance yet have the disadvantage of energy-consuming.

2.2 Deep Learning

Recently, various kinds of deep learning methods have achieved great success in speech, image and natural language processing [1]. By exploiting deep architectures, deep learning techniques are capable to discover the hidden structures from training data and features at different levels of abstractions useful for the target tasks. Convolutional neural networks (CNN) are designed utilize layers with convolving filters that are applied to local features [6]. CNN models have subsequently been utilized and demonstrated to achieve excellent results in NLP tasks, such as question answering [19], search query retrieval [12], sentence modeling and matching [4,5], and several other traditional NLP tasks [1].

3   Problem Formulation

We formalize this task as context-aware recommendation problem. The input query is short sentences (**citation context**). The task is to return a ranked list of **cited papers** as candidates to users. Deliberately avoiding from heuristic feature designing, we turn to deep convolutional network architecture to learn sentence and author representation under supervised learning framework.

4   Our Proposed Approaches

The model architecture is illustrated in Fig. 2. By using convolution and max-pooling architecture, local contextual information at the word n-gram level is modeled first. Then, salient local features in a word sequence and author information are combined to form a global low-dimensional feature vector. Finally, a multi-layer perceptron (MLP) is introduced here as a nonlinear similarity function computing the matching degree by comparing the global feature vector [5].

Input Layer. The input layer takes words in the form of embedding vectors. In our work, we set the maximum length of sentences to 50 words. For sentences shorter than that, we put zero padding.

For citation context, we get input layer denoted as l_t by concatenating each word embedding of citation context and its author. For cited documents the input layer is generated by concatenating its content and author. More specifically,

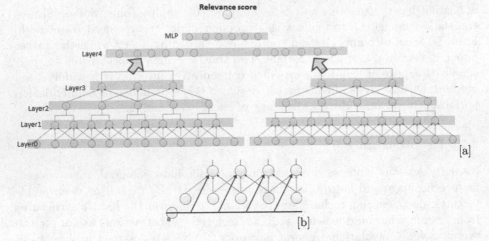

Fig. 2. (a)We use CNN mapping word sequence to a low-dimensional vector in semantic space; (b)The convolution units of multi- modal convolution layer of p-CNN

for each citation context cc, the input layer with its window indexed by t to upcoming convolution layer is given by the concatenated vector

$$l_t = \left[\sum_i \{a_{cc}\}, w_t^T, w_{t+1}^T, w_{t+2}^T \right] \qquad (1)$$

Here we simply average the author vectors as an expansion in the input layer. The author vector is initialized as the average of title word vectors from his or her ever published papers. The author vectors are model parameters and updated during the training phrase with respect to the model cost. Our experiments show that learning task-specific author vectors through fine-tuning offers performance gains compared with static author vectors.

Convolution. The convolution operation can be viewed as extracting local features from a sliding window of width k_1. Contextual feature vector h_t in this layer is computed by:

$$h_i^{(l)} = \sigma(W^{(l)} h_i^{(l-1)} + b^{(l)}) \qquad (2)$$

where: $h_i^{(l)}$ is the output vector of feature maps for location i in Layer l, $\sigma(\cdot)$ is the activation function, W^l is the parameters on Layer l, $h_i^{(l-1)}$ is the segment of Layer-$(l-1)$ for the convolution layer at location i. For the first convolution layer: $h_i^{(0)} = x_{i:i+k_1-1} = [x_i^T, x_{i+1}^T, ..., x_{i+k_1-1}^T]$ concatenates the word vectors for k_1 width of sliding window from input sentence x.

Max-Pooling. To retain only the most useful local features produced by the convolutional layers, we take the max-pooling in every non-overlapping two-unit windows following each convolution. It computes as follows:

$$h_i^{(l)} = max\{h_t^{(l-1)}(i), h_t^{(l-1)}(i+1)\}, \qquad (3)$$

MLP. These are fully connected layers. The input of which is the concatenation of fixed length vector of convolution-max pooling output from both citation context and cited paper. MLP can be seen as a nonlinear score function computing the probability of citation context references cited papers [11,12].

4.1 Training

We employ a discriminative training strategy with a large margin objective. The model parameters are learned to minimize the pair-wise loss function written as follows:

$$Loss(\theta) = max\{0, 1 + s(cc, D_j^-) - s(cc, D^+)\} \tag{4}$$

where $s(cc, D)$ is the predicted relevance score for pair (cc, D) computed as the output of MLP. The negative sampling process chooses the documents written by totally different authors as negative samples, which gets better performance than random negative sampling.

The dimension of word vector and author vector is 100. The network architecture of all CNN-related methods is configured as two for convolution, two for pooling, and two for MLP. Dropout proved to be such a good regularizer shown in our experiment, which is also consistent with [5].

5 Experiments

5.1 Dataset

We use the same dataset as [18]. Following [15], one citation context consists of three sentences around the a citation placeholder. We combine the title and abstract as the content of cited papers. The dataset includes 73236 citation relationships which send out by the papers from the 10 seed venues[1]. We filtered 35362 distinct words and 2,191 distinct authors who appear in our data set more than 5 times. Rare words and authors appearing less than 5 times are replaced by "UNK_W" and "UNK_A".

5.2 Baseline Methods

We compare our proposed methods with several popular methods as follows. All the baseline models are trained on the same training set as our proposed model.

Language model (**LM**): is one-gram language model.

Translation Model (**TM**): [8] proposed a translation model to overcome the language gap between citation contexts and cited papers.

word2vec: word2vec[2] [10] is an unsupervised model. We calculate the matching degree of citation context and cited documents with cosine similarity followed by sum of words as input.

---

[1] ACL, CIKM, EMNLP, ICDE, ICDM, KDD, SIGIR, VLDB, WSDM, WWW.
[2] https://code.google.com/p/word2vec/.

DSSM: [3] compute the (query, document) pair score by cosine similarity between their semantic vectors via conventional CNN.

CNN: [1] uses principled CNN to get sentence representation without consideration of author information.

5.3   Experimental Results

The evaluation metrics aims to assess the positions of the true citations in the ranking list for each given context. We report standard metrics of recall and MAP on the test set as the experimental results. The main results are summarized in Table 1. We can see that our p-CNN is the best performer, beating other methods in terms of MAP and Recall@10. Besides, supervised learning on citation relationship is essential for obtaining superior personalized citation recommendation performance, i.e., p-CNN and CNN all outperforms word2vec. By comparing p-CNN with CNN, we also see the effectiveness of learned author feature vectors.

We then further investigated the performance of p-CNN using recall at different positions with experimental results shown in Fig. 3. We can see p-CNN

Table 1. Performance of citation recommendation

	MAP	Recall@10
rand	0.007	0.000
LM	0.299	0.376
TM	0.501	0.594
word2vec	0.657	0.779
DSSM	0.710	0.801
CNN	0.723	0.832
p-CNN	**0.762**	**0.857**

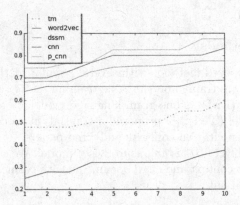

Fig. 3. Recall value of every position at one to ten.

is stable on recall and achieve the best performance at all positions. Specifically, we observe that even at position one, the p-CNN still can have effective recommendation.

5.4 Parameter Analysis

We investigate the parameter sensitivity to test the stability of our model with results shown in Fig. 4. We can see that CNN and p-CNN are not sensitive to dimension n when d is greater than 100. Actually, $d = 100$ is enough to make our models achieve their approximate optimal performance, while increasing d results in no higher performance but training time consuming.

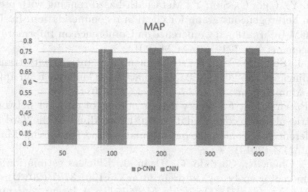

Fig. 4. Parameter sensitivity w.r.t. dimension.

6 Conclusions

Our work is a successful attempt in applying the CNN-like method to citation recommendation task. Our proposed p-CNNs approach is able to automatically learn the representative features for each author and integrate them into computing relevance score between citation context and candidate cited documents. Experimental results show that the proposed convolutional neural network can significantly improve the recommendation performance.

Acknowledgments. This work has been partially supported by the 973 Program under Grant No. 2014CB340405 and National Natural Science Foundation of China under Grant No. U1536201.

References

1. Collobert, R., Weston, J., Bottou, L., Karlen, M., Kavukcuoglu, K., Kuksa, P.: Natural language processing (almost) from scratch. J. Mach. Learn. Res. **12**, 2493–2537 (2011)

2. He, Q., Pei, J., Kifer, D., Mitra, P., Giles, L.: Context-aware citation recommendation. In: Proceedings of the 19th International Conference on World Wide Web, pp. 421–430. ACM (2010)
3. Huang, P.S., He, X., Gao, J., Deng, L., Acero, A., Heck, L.: Learning deep structured semantic models for web search using clickthrough data. In: Proceedings of the 22nd ACM International Conference on Information & Knowledge Management, pp. 2333–2338. ACM (2013)
4. Kalchbrenner, N., Grefenstette, E., Blunsom, P.: A convolutional neural network for modelling sentences. arXiv preprint arXiv:1404.2188 (2014)
5. Kim, Y.: Convolutional neural networks for sentence classification. arXiv preprint arXiv:1408.5882 (2014)
6. LeCun, Y., Bottou, L., Bengio, Y., Haffner, P.: Gradient-based learning applied to document recognition. Proc. IEEE **86**(11), 2278–2324 (1998)
7. Liu, X., Yu, Y., Guo, C., Sun, Y.: Meta-path-based ranking with pseudo relevance feedback on heterogeneous graph for citation recommendation. In: Proceedings of the 23rd ACM International Conference on Conference on Information and Knowledge Management, pp. 121–130. ACM (2014)
8. Lu, Y., He, J., Shan, D., Yan, H.: Recommending citations with translation model. In: Proceedings of the 20th ACM International Conference on Information and Knowledge Management, pp. 2017–2020. ACM (2011)
9. Mahdabi, P., Crestani, F.: Query-driven mining of citation networks for patent citation retrieval and recommendation. In: Proceedings of the 23rd ACM International Conference on Conference on Information and Knowledge Management, pp. 1659–1668. ACM (2014)
10. Mikolov, T., Chen, K., Corrado, G., Dean, J.: Efficient estimation of word representations in vector space. arXiv preprint arXiv:1301.3781 (2013)
11. Shen, Y., He, X., Gao, J., Deng, L., Mesnil, G.: A latent semantic model with convolutional-pooling structure for information retrieval. In: Proceedings of the 23rd ACM International Conference on Conference on Information and Knowledge Management, pp. 101–110. ACM (2014)
12. Shen, Y., He, X., Gao, J., Deng, L., Mesnil, G.: Learning semantic representations using convolutional neural networks for web search. In: Proceedings of the 23rd International Conference on World Wide Web. pp. 373–374. ACM (2014)
13. Strohman, T., Croft, W.B., Jensen, D.: Recommending citations for academic papers. In: Proceedings of the 30th Annual International ACM SIGIR Conference on Research and Development in Information Retrieval, pp. 705–706. ACM (2007)
14. Tang, J., Zhang, J.: A discriminative approach to topic-based citation recommendation. In: Theeramunkong, T., Kijsirikul, B., Cercone, N., Ho, T.-B. (eds.) PAKDD 2009. LNCS, vol. 5476, pp. 572–579. Springer, Heidelberg (2009). doi:10.1007/978-3-642-01307-2_55
15. Tang, X., Wan, X., Zhang, X.: Cross-language context-aware citation recommendation in scientific articles. In: Proceedings of the 37th International ACM SIGIR Conference on Research & Development in Information Retrieval, pp. 817–826. ACM (2014)
16. Wang, S., Lei, Z., Lee, W.C.: Exploring legal patent citations for patent valuation. In: Proceedings of the 23rd ACM International Conference on Conference on Information and Knowledge Management, pp. 1379–1388. ACM (2014)
17. Wang, X., Wang, Y.: Improving content-based and hybrid music recommendation using deep learning. In: Proceedings of the 22nd ACM International Conference on Multimedia, pp. 627–636. ACM (2014)

18. Liu, Y., Yan, R., Yan, H.: Guess what you will cite: personalized citation recommendation based on users' preference. In: Banchs, R.E., Silvestri, F., Liu, T.-Y., Zhang, M., Gao, S., Lang, J. (eds.) AIRS 2013. LNCS, vol. 8281, pp. 428–439. Springer, Heidelberg (2013). doi:10.1007/978-3-642-45068-6_37
19. Yin, J., Jiang, X., Lu, Z., Shang, L., Li, H., Li, X.: Neural generative question answering. arXiv preprint arxiv:1512.01337 (2015)

A Streaming Data Prediction Method Based on Evolving Bayesian Network

Yongheng Wang[✉], Guidan Chen, and Zengwang Wang

College of Information Science and Electronic Engineering,
Hunan University, Changsha 410082, China
{wyh, chengd2015, wangzw}@hnu.edu.cn

Abstract. In the Big Data era, large volumes of data are continuously and rapidly generated from sensor networks, social network, the Internet, etc. Learning knowledge from streaming Big Data is an important task since it can support online decision making. Prediction is one of the useful learning task but a fixed model usually does not work well because of the data distribution change over time. In this paper, we propose a streaming data prediction method based on evolving Bayesian network. The Bayesian network model is inferred based on Gaussian mixture model and EM algorithm. To support evolving model structure and parameters based on streaming data, an evolving hill-climbing algorithm is proposed which is based on incremental calculation of score metric when new data is arrived. The experimental evaluations show that this method is effective and it outperforms other popular methods for streaming data prediction.

Keywords: Streaming data · Prediction · Evolving Bayesian network

1 Introduction

In the Big Data era, large volumes of data are continuously and rapidly generated from sensor networks, social network, the Internet, etc. Recently streaming big data processing attracts more attention. The less time we use to make decision, the more value we can get from the whole processing. Therefore, learning knowledge from Big Data and making decision on line is a very important task. Predictive analytics is an important learning method since it can identify events before they have happened, so that some unwanted events can be eliminated or their effects mitigated. Many models and algorithms have been used in predictive analytics. In this paper, we use Bayesian networks model since it has rich mathematical basis and it can explain to the user how the system came to its conclusions.

Currently predictive analytics technology for streaming data has some challenges. First, in streaming data prediction the system can only process data on single-pass and cannot control over the order of samples that arrive over time. The second challenge is call "concept drift" which reflects a situation that the input and/or output concepts do not follow a fixed and predictable data distribution. A fixed model learned from historical data may not handle the concept drift issue well. The last challenge is that, event streams often have high incoming rate. Streaming data prediction is usually used to

L. Chen et al. (Eds.): APWeb-WAIM 2017, Part II, LNCS 10367, pp. 294–302, 2017.
DOI: 10.1007/978-3-319-63564-4_24

support online decision support system which need high performance even for large scale distributed event streams.

To address these challenges, in this paper we propose a Streaming Data Processing method based on Evolving Bayesian Networks (SDP-EBN), which supports single-pass processing of streaming data and uses evolving hill-climbing algorithm to support model structure evolving.

2 Related Work

Predictive analytics has been studied for many years and a series of models and algorithms are proposed. Among these models, deep neural networks and Bayesian network are more popular for streaming data. Huang et al. proposed a predictive analytics method based on deep belief networks [1]. Predictive analytics with deep learning model usually can achieve good accuracy, but the training of the model is complex and it prone to the problem of over fitting.

As a valid model for uncertain knowledge representation and inference, Bayesian Network is widely used in predictive analytics. Zhu et al. proposed a Bayesian network model for traffic prediction using prior link flows [2]. Evolving systems are developed to address the concept drifts in data stream. Angelov et al. proposed an evolving intelligent system which supports learning and adjusting model from data stream [3]. Recently many incremental machine learning methods [4] are proposed which are suitable for learning from data stream with concept drifts. Another related work is evolving neural network which uses evolutionary algorithms to optimize the structure and parameters of neural networks [5].

Learning Bayesian network structure from data is an NP-hard problem as the number of possible structures grows super-exponentially by the number of nodes. Approximate structure learning algorithms are usually categorized into two groups of constraint-based and search-and-score (S&S) approaches. The commonly used score metrics include Bayesian Information Criterion (BIC) [6] and Bayesian Dirichlet equivalent (BDe) [7], etc.

To address the concept drifts in streaming data, evolving Bayesian network is needed. For abrupt drifts, a common idea for evolving Bayesian network is to learn different models and switch among models when data distribution is changed [8]. Incremental drifts are more difficult to be handled. Some researchers use sampling-based method for evolving Bayesian network [9]. However, since we only need best network structures in streaming data prediction, heuristic search-based methods [10, 11] will typically find them more quickly. Compared to the related work, our work considers how to calculate the score metric incrementally so that the model does not need to be trained from the scratch. Furthermore, the current hill-climbing algorithms that starts from only the highest local peak is prone to miss best result near other local peaks. Our algorithm searches top-k peaks parallel which can get better accuracy without losing much performance.

3   The SDP-EBN Method

3.1   Bayesian Network Model for Streaming Data Prediction

The BN model for streaming data prediction is shown in Fig. 1 which has two dimensions: event type and time. The event type is related to some attribute of objects, e.g., the position of objects, or some state of the system, e.g., the traffic flow of roads. Assume the event type number is N, in a prediction task the state of node (i,t) is related to a set of states before time t (parent nodes). Let $x_{i,t}$ represent the state variable of node (i,t) and $pa(i,t)$ represent the set of parent nodes of (i,t). N_P denotes the number of nodes in $pa(i,t)$. $N_T\Delta_t$ is the time window which means we only consider nodes from time $t-N_T\Delta_t$ to $t-\Delta_t$ as the parents of nodes in time t. The set of state variables for $pa(i,t)$ is $X_{pa(i,t)} = \{x_{j,s} \mid (j,s) \in pa(i,t)\}$. According to the BN theory, the joint distribution of all nodes in the network can be represented as:

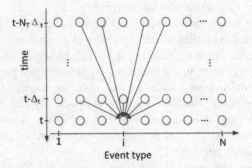

Fig. 1. The streaming data and Bayesian network structure

$$p(X) = \prod_{i,t} p(xi, t \mid Xpa(i,t))$$ (1)

The conditional probability $p(xi, t \mid Xpa(i,t))$ can be calculates as:

$$p(xi, t \mid Xpa(i,t)) = p(xi, t, Xpa(i,t))/Xpa(i,t)$$ (2)

Since it is not easy to calculate the joint distribution $p(x_{i,t}, X_{pa(i,t)})$, we use Gaussian Mixture Model (GMM) [12] to approximate it like the following:

$$p(x_{i,t}, X_{pa(i,t)}) = \sum_{m=1}^{M} \pi m gm(x_{i,t}, X_{pa(i,t)} \mid \mu m, \Sigma m)$$ (3)

where M is the number of Gaussian models and $g_m(\cdot \mid \mu_m, \Sigma_m)$ is the m-th Gaussian distribution with $(N_P + 1) \times 1$ vector of mean values μ_m and $(N_P + 1) \times (N_P + 1)$ covariance matrix Σ_m. Using EM algorithm we can infer the parameters

$\{\pi_m, \mu_m, \Sigma_m\}_{m=1}^M$ from historical data [12]. Then the conditional distribution $p(x_{i,t} \mid X_{pa(i,t)})$ can be derived from $p(x_{i,t}, X_{pa(i,t)})$ and the prediction $\hat{x}i, t$ can be calculated from $X_{pa(i,t)}$ with minimum mean square error (MMSE) method. In SDP-EBN, the GMM parameters can be updated based on EM algorithm. Therefore, we mainly consider how to evolve the structure of the Bayesian networks.

3.2 Bayesian Network Structure Learning

Since search-based methods are more suitable for evolving than constraint-based methods, we use a search-and-score method with BIC metric [6] score function. The BIC metric is defined as following:

$$g_{BIC}(G:D) = \sum_{i=1}^{n} \sum_{j=1}^{q_i} \sum_{k=1}^{r_i} m_{ijk} \log \theta_{ijk} - \sum_{i=1}^{n} \frac{q_i(r_i-1)}{2} \log m \quad (4)$$

where D is the samples, $m = |D|$, $q_i = |X_{pa(i)}|$, r_i is the number of discrete states of x_i, m_{ijk} represents the times in D that x_i is k when the parent of x_i is j, $m_{ij} = \sum_{k=1}^{r_i} m_{ijk}$, $\theta_{ijk} = m_{ijk}/m_{ij}$, satisfying constraints $0 \le \theta_{ijk} \le 1$, $\sum_{k=1}^{r_i} \theta_{ijk} = 1$, n is the number of variables.

BN structure learning is the task of selecting a BN structure G that explains a given data set D. It has been shown that searching the huge graph space for the optimal structure is a NP-hard problem. Therefore, greedy local search method is used in this paper. To calculate the BIC metric incrementally, an algorithm is proposed as shown in Algorithm 1.

Algorithm 1: Incremental calculation of BIC metric with new data (BN structure is not changed)

Input: BN structure G, old data set D, new data get ΔD
Output: new metric BIC_new

```
1    Δedge ←{e | edge e is affected by ΔD}
2    if |Δedge|/EdgeNumber(G) > θ) then
3        calculate BIC_new directly from skeleton
4        Return BIC_new
5    Δp ← {i,j,k | m_ijk is changed by ΔD}
6    Δli ← 0 //li means the likelihood part of BIC
7    for each pair (i,j) in Δp
8    |   m_ij_new ← 0
9    |   for each k in (i,j,k) triple of Δp
10   |       m_ij_new ←m_ij_new + m_ijk
11   |   for each k in (i,j,k) triple of Δp
12   |       Δli ← m_ijk · log(m_ijk/m_ij_new) - m_ijk_old · log(m_ijk_old/m_ij_old)
13   |   update m_ij with m_ij_new
14   end for
15   li_new ← li_old + Δli
16   BIC_new ← li_new – penalty_old
17   return BIC_new
```

The hill-climbing search algorithm (HCS) works in a step-by-step fashion to generate a model. At each step, it makes the maximum possible improvement in an objective quality function. In our work the objective quality function is the BIC metric. Based on the property of the BN model in Fig. 1, we define a simple but effective one edge neighborhood (OEN). An OEN of a given model *M* is the set of all the alternative models that can be built from M through adding or removing one edge, $\text{OEN}(M) = \{M'|M' = \text{op}(M) \land \text{op} \in \{\text{add an edge, remove an edge}\}\}$.

In Algorithm 1 we assume the BN structure is static. In order to calculate the BIC metric incrementally with one edge change when data is not changed, we use the following equation:

$$BIC_{new} = BIC_{old} + \sum_{k=1}^{r_v} m_{uvk} log\theta_{uvk} - \frac{q_v(r_v - 1)}{2} logm \tag{5a}$$

$$BIC_{new} = BIC_{old} - \sum_{k=1}^{r_v} m_{uvk}\_old log\theta_{uvk}\_old + \frac{q_v\_old(r_v\_old - 1)}{2} logm \tag{5b}$$

where (5a) is for adding an edge and (5b) is for removing an edge.

To generate BN structure from data D without considering new incoming data, we use a max-min hill-climbing (MMHC) algorithm.

3.3 Evolve Bayesian Network Structure

We use an evolving MMHC algorithm (EMMHC) to evolve BN structure as shown in Algorithm 2. The current BN structure G_{cur} has been learnt by the MMHC algorithm. Existing methods usually merge the new data into existing data which makes the system has bad scalability. The BIC score of the BN structure is updated using Eqs. 5a, 5b, and the skeleton is also updated. The EMMHC algorithm tries to find the optimized BN structure give an incremental data set ΔD. If the change of score when applying ΔD is smaller than a threshold value δ, then the current BN structure need not be changed (lines 3–4). The BIC(G_{new}, ΔD) is calculated using Algorithm 1. The EMMHC algorithm creates k subtasks to incrementally search for optimized BN structure starting from the structures with top-k score in the localMinList (lines 5–6). The subtasks are executed parallel and the result with lowest score is selected.

Algorithm 2: EMMHC

input: current BN structure G_{cur}, statistic information in skeleton I_{stat}, incremental data set ΔD, localMinList
output: new BN structure G_{new}
1   currentScore ←BIC(G_{cur})
2   G_{new} ← G_{cur}
3   **if** (|BIC(G_{new}, ΔD) - currentScore |/currentScore < δ) **then**
4   return G_{new}
5   **for each** G in top-k of localMinList
6   create a new subtask by calling LocalSearch with parameters G, I_{stat}, ΔD, currentScore
7   run all sub tasks parallel and wait the result from all sub tasks
8   G_{min} ← the BN structure with the lowest score in the sub tasks
9   **if** (BIC(G_{min}) < BIC(G_{new})) **then**
10   G_{new} ← G_{min}
11   **return** G_{new}

The subtask LocalSearch is shown in Algorithm 3. The idea behind this algorithm is that, since only part of the graph is affected by the new data, we only need to add or remove edges inside the affected part to minimum the score. This algorithm has two stages: expansion stage and contraction stage. During the expansion stage, an edge is repeatedly added that can minimum the score function until the score cannot be decreased. During the contraction stage, an edge is repeatedly removed if the score function is not increased. In line 2, the getCandidateEdgesToAdd function gets the node pairs that in affectedNodes but have no edge between them (edge can be added according to the structure of Fig. 2). In line 3, the getCandidateEdgesToRemove function gets the edges that are in current BN structure and affected by incremental data set ΔD.

Algorithm 3: LocalSearch

input: current BN structure G_{cur}, statistic information in skeleton I_{stat}, incremental data set ΔD, current score value currentScore
output: new BN structure G_{new}

```
1      affectedNodes ← getAffectedNodes(ΔD)
2      candidateEdgesToAdd ← getCandidateEdgesToAdd(affectedNodes, G_cur)
3      candidateEdgesToRemove ← getCandidateEdgesToRemove (affectedNodes, G_cur)
4      expansionStage, contractionStage ← true
5      while(expansionStage)
6      |   minScore ← the min score of the BN by adding an edge in
       |       candidateEdgesToAdd
7      |   if (minScore < currentScore)
8      |       G_new ← the BN structure that get main score
9      |   else
10     |       expansionStage ← false
11     end while
12     while(contractionStage)
13     |   minScore ← the min score of the BN by removinging an edge in
14     |       candidateEdgesToRemove
15     |   if  minScore ≯ currentScore then
16     |       G_new ← the BN structure that get min score
17     |   else
18     |       contractionStage ← false
19     end while
20     return G_new
```

---

4 Experimental Evaluations

We evaluated the EBN method separately since it is the key part of this paper. We draw samples from a known BN, apply structure learning on the synthetic data, and compare the learned structure with the original one. Our original BN structure contains 783 event types (nodes) and 1752 edges. The time window N_T is set to 15. In order to evaluate the accuracy of BN structure learning and updating, we use the Structural Hamming Distance (SHD) [13]. Servers with 2 Xeon E3 processors and 16 GB memory is used in the experiments for EBN.

We first evaluate the parameters δ and θ. The sample number for phase p_0 is 50,000 and the data window for each dynamic phase contains 2,000 samples. The phase number n is set to 30. The average SHD values and total running time values for different δ and θ are shown in Table 1. We can see when δ increases, the accuracy is decreased but the running time is also decreased. From Table 1 we can also find the accuracy is almost not affected by θ.

In the next experiment, we compare the accuracy and performance of our EBN method with a typical sampling-based method nsDBN [9] and a recent search-based method iHCMC [11]. We evaluated the accuracy and performance of the three methods for different window size and total data size. The accuracy evaluation result of the 3 methods with different window size and data size is shown in Fig. 2. We also evaluated the performance of the 3 methods with different window size and the result is shown in Fig. 3.

Table 1. Average SHD values and total running time values for different δ and θ. The values inside the parenthesis are total running time values (seconds)

δ θ	0.01	0.02	0.03	0.04	0.05	0.06	0.07	0.08
0.1	157 (83.4)	161 (79.3)	167 (77.6)	178 (75.4)	195 (72.6)	213 (68.7)	226 (64.4)	245 (52.5)
0.2	159 (80.6)	157 (77.5)	174 (76.3)	187 (74.4)	192 (70.2)	205 (66.6)	230 (63.1)	239 (49.7)
0.3	162 (78.4)	154 (74.4)	178 (73.3)	188 (72.4)	188 (69.1)	214 (63.9)	232 (61.2)	238 (47.2)
0.4	148 (77.7)	169 (72.3)	162 (71.9)	172 (70.0)	205 (67.7)	218 (62.8)	219 (61.8)	244 (46.3)
0.5	151 (77.5)	155 (75.7)	168 (74.5)	171 (72.2)	191 (68.6)	203 (64.4)	218 (63.9)	251 (47.8)
0.6	161 (81.8)	170 (79.9)	158 (77.7)	168 (72.8)	206 (69.9)	208 (67.3)	223 (66.1)	248 (52.7)
0.7	149 (86.2)	165 (84.9)	155 (82.4)	182 (78.4)	191 (73.2)	213 (69.7)	225 (67.4)	239 (54.9)
0.8	153 (89.2)	158 (87.6)	163 (84.3)	175 (80.3)	187 (75.3)	215 (73.3)	233 (69.8)	245 (58.9)

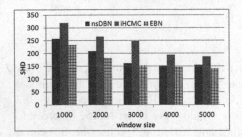

Fig. 2. The accuracy of 3 methods with different window size and data size. Dynamic phase number n is fixed at 30.

Fig. 3. The performance of 3 methods with different window size and data size

We evaluated the complete SDP-EBN method using both real and simulated data. The real application data comes from the PEMS traffic monitoring network[1]. We also developed a traffic simulation system based on the road traffic simulation package

---

[1] PeMS project, https://pems.eecs.berkeley.edu/.

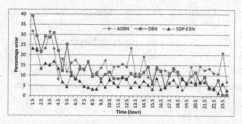

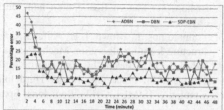

Fig. 4. Percentage error of the three methods on PEMS data

Fig. 5. Percentage error of the three methods on SUMO data

SUMO [14]. Our method is compared with Adaptive Dynamic Bayesian Network (ADBN) [8] and Deep Belief Network (DBN) [1]. The accuracy evaluation result is shown in Figs. 4 and 5. From all experiments, we can see SDP-EBN outperforms other popular methods when processing data with distribution drift.

Acknowledgments. This work is supported by the National Natural Science Foundation of China (No.61371116).

References

1. Huang, W., Song, G., Hong, H., et al.: Deep architecture for traffic flow prediction - deep belief networks with multitask learning. IEEE Trans. Intell. Transp. Syst. **15**(5), 2191–2201 (2014)
2. Zhu, S., Cheng, L., Chu, Z.: A Bayesian network model for traffic flow estimation using prior link flows. J. Southeast Univ. (English edn.) **29**(3), 322–327 (2013)
3. Angelov, P.: Autonomous Learning Systems: From Data Streams to Knowledge in Real-time. Wiley Press (2013)
4. Li, T.: PICKT: a solution for big data analysis. In: Proceeding of the 10th International Conference on Rough Sets and Knowledge Technology, Tianjin, China, pp. 15–25 (2015)
5. Yevgeniy, B., Olena, V., Iryna, P., et al.: Fast learning algorithm for deep evolving GMDH-SVM neural network in data stream mining tasks. In: Proceeding of the IEEE First International Conference on Data Stream Mining & Processing, Lviv, Ukraine, pp. 257–262 (2016)
6. Shtarkov, Y.M.: Universal sequential coding of single messages. Probl. Inf. Transm. **23**(3), 3–17 (1987)
7. Heckerman, D., Geiger, D., Chickering, D.M.: Learning bayesian networks: the combination of knowledge and statistical data. Mach. Learn. **20**(3), 197–243 (1995)
8. Pascale, A., Nicoli, M.: Adaptive Bayesian network for traffic flow prediction. In: Proceeding of the Statistical Signal Processing Workshop (SSP), pp. 177–180. IEEE (2011)
9. Robinson, J.W., Hartemink, A.J.: Learning non-stationary dynamic Bayesian networks. J. Mach. Learn. Res. **11**, 3647–3680 (2010)
10. Yue, K., Fang, Q., Wang, X., et al.: A parallel and incremental approach for data-intensive learning of Bayesian networks. IEEE Trans. Cybern. **45**(12), 2890–2904 (2015)
11. Acharya, S., Lee, B.: Causal network construction over event streams. Inf. Sci. **261**, 32–51 (2014)

12. Bishop, C.: Pattern Recognition and Machine Learning, 2nd edn. Springer, New York (2010)
13. Tsamardinos, I., Brown, L.E., Aliferis, C.F.: The max-min hill-climbing Bayesian network structure learning algorithm. Mach. Learn. **65**(1), 31–78 (2006)
14. Behrisch, M., Bieker, L., Erdmann, J., et al.: Sumo - simulation of urban mobility: an overview. In: Proceeding of the Third International Conference on Advances in System Simulation (SIMUL 2011), Barcelona, Spain, 23 October 23, pp. 63–68 (2011)

A Learning Approach to Hierarchical Search Result Diversification

Hai-Tao Zheng[(✉)], Zhuren Wang, and Xi Xiao

Tsinghua-Southampton Web Science Laboratory, Graduate School at Shenzhen,
Tsinghua University, Shenzhen, China
{zheng.haitao,xiaox}@sz.tsinghua.edu.cn,
wang-zr14@mails.tsinghua.edu.cn

Abstract. The queries in search engine that issued by users are often
ambiguous. By returning diverse ranking results we can satisfy different
information needs as far as possible. Recently, a hierarchical structure
are proposed to represent user intents instead of a flat list of subtopics.
Although the hierarchical diversification model performs better than previous models, it utilizes a predefined function to calculate the diversity
score, which may not reach the optimal result. The model's parameters need to be tuned manually and repeatedly without intention, which
cause a time-consuming problem. In this paper, we introduce a learning
based hierarchical diversification model. Benefit from the learning model,
the parameter values are determined automatically and more optimal.
Experiments show that our approach outperform several existing diversification models significantly.

1 Introduction

Search result diversification [1,2,10,11] is an effective way to solve the query
ambiguation problem. The diversification model regards the problem as a combination of relevance score and diversity score. Diversity score are often calculated
based on the user intents. Previous models regard the user intents as a flat list
of subtopics. However, the flat list of subtopics cannot match the actual intents
in evaluation tasks [4,8] good enough. The hierarchical structure of subtopics
are proposed to solve this problem.

The hierarchical diversification methods perform better than previous diversification methods, but it just utilizes a predefined function to calculate the
diversity score. There exists some parameters to tune. Usually researchers have
to set a series of repeatedly experiments to find the suitable parameter. It is
hard to reach optimal value by manual tuning, and it causes a time-consuming
problem.

In this paper, we introduce a learning approach to hierarchical search result
diversification called L-HSRD. Firstly, we redefine the loss function as the generation probability of sequential selection for a ground truth list. Then Stochastic
gradient descent are employed to optimize the value of weight. Finally we derive
our ranking function to generate the diverse list sequentially.

© Springer International Publishing AG 2017
L. Chen et al. (Eds.): APWeb-WAIM 2017, Part II, LNCS 10367, pp. 303–310, 2017.
DOI: 10.1007/978-3-319-63564-4_25

We demonstrate L-HSRD is more excellent than other diversification models in terms of official evaluation metrics including α-NDCG [6], ERR-IA [3] and NRBP [7]. Additionally, we conduct a series of experiments to illustrate the robustness of our method, which get a outstanding performance.

The main contributions of our work are listed as follows:

1. L-HSRD is the first method introducing the learning mechanism for the hierarchical search result diversification. We conduct inference for the loss function based on its sequential selection model, which solves the parameter tuning problem at the same time.
2. We put forward a series of instructive different features based on the hierarchical structure of subtopics.
3. We conduct extensive experiments to verify that L-HSRD achieves excellent performance comparing with the existing diversification models.

2 Related Work

Search result diversification model can be categorized as implicit approaches and explicit approaches. The implicit approaches includes MMR [2]. MMR selecs the document iteratively, and meanwhile, both content-based relevance and diversity should be considered. It is considered as a low effective model [11,14].

Explicit approaches extract the aspects explicitly and make use of them to calculate the diversity score. The algorithms such as IA-select [1], xQuAD [14] and RxQuAD [15] are proposed to reduce redundancy. These methods select the document that covering more novel aspects. The PM-1 and PM-2 [9] models mainly consider the proportionality of aspects and produce the diverse result by virtue of the proportionality of aspects.

In addition, as for learning model, Zhu et al. [17] proposed a learning model without considering the aspects underlying the query. Yue et al. [16] proposed Structural SVMs to model the diversity but discarded the relevance.

3 Learning Approach to Hierarchical Search Result Diversification

3.1 Definition of Ranking Function

At first, We generate the subtopics like Hu [12] and get the relevance score $P(q_i|d)$ between the subtopic q_i and each document d. In our model, the ranking function we select the "local-best" document in the round i is given as follows:

$$f_i(d, D \backslash S_{i-1}) = \lambda_r P(d|q) + \lambda_d^T F(S_{i-1} \cup d), \tag{1}$$

where d means current document to be considered in the sequential selection process, D denotes the candidate document set, S_{i-1} denotes the documents already selected in previous $i-1$ round, q denotes the query, $F(S_{i-1} \cup d)$ stands for the feature function, represented by the feature vector of $(f_1, f_2, \ldots, f_T)$, T stands for the number of features.

Feature Definition. We define our features on the document set $S_{i-1} \cup d$. In our work, we provide several representative features for the learning process, which are shown as follow:

- **Features defined between the query and first-level subtopics.** Average, minimum, maximum, standard deviation of relevance score $P(q_i|d)$ for a document d to a subtopic q_i from level 1.
- **Features defined between the first-level subtopics and second-level subtopics.** Average, minimum, maximum, standard deviation of relevance score $P(q_i|d)$ for a document d to a subtopic q_i from level 2.
- **Features defined only on the first-level subtopics.** Firstly we define the set of documents which possess a highest score of relevance for first-level subtopics as St_1. (all the documents possess a relevance score for all the subtopics in level 1 and level 2). The features are defined as the entropy of all the documents d in St_1: $P_{entropy}(d) \stackrel{def}{=} -\sum_{w \in d} P(w|d) \log P(w|d)$, where w is a term and $p(w|d)$ is the probability that w appears in d (given by the language model).
- **Features defined only on the second-level subtopics.** The features are defined as the entropy of all the documents d in St_2 like above.

3.2   Definition of Loss Function

The ranking process is a sequential selection, we define the loss function as the likelihood loss of the generation probability:

$$L(f(X,C),Y) = -\log P(Y|D) = -\log[P(y(1)|D)P(y(2)|D\backslash S_1)\ldots P(y(n)|D\backslash S_{n-1})], \quad (2)$$

where X stands for the feature, C represents the weight, the Y is the final result, $y(1),\ldots,y(n)$ is the ground truth, n represents the top n result, the index i denotes its ranking position, D is the initial retrieved documents D_{init}, S_{i-1} denotes the result set we had iteratively selected in the last $i-1$ round, the probability $P(y(i)|D\backslash S_{i-1})$ represents the probability that selecting the document $y(i)$ under the condition of $D\backslash S_{i-1}$.

The above sequential definition approach can be well captured by the Plackett-Luce Model [13]. We can derive every step in our generation process, which is shown as:

$$P(Y|D) = \prod_{i=1}^{n} P(y(i)|D\backslash S_{i-1}) = \prod_{i=1}^{n} \frac{exp(f_i(y(i), D\backslash S_{i-1}))}{\sum_{k=i}^{n} exp(f_i(y(k), D\backslash S_{i-1}))}, \quad (3)$$

Given the training data $\{(X,C,Y)^{(Tr)}\}$ (Tr denotes for the number of training samples), the total loss function is formulized as follows:

$$L(f(X,C),Y) = -\sum_{i=1}^{T_r}\sum_{j=1}^{n} \log(\frac{exp(\lambda_r P(y(j)|q) + \lambda_d^T F(S_{j-1} \cup y(j)))}{\sum_{k=j}^{n} exp(\lambda_r P(y(k)|q) + \lambda_d^T F(S_{j-1} \cup y(k)))}) \quad (4)$$

3.3 Learning and Prediction

As for the training, we generate the training data and optimize the loss function. Then, we use the trained ranking function to re-rank and predict the result.

To generate the ground truth training data, we construct a list $y(i)$ which maximize the ERR-IA metrics. In the algorithm, at the i-th step in loop structure, we select the document d from $D\backslash S_{i-1}$ to maximize ERR-IA score and update the $D\backslash S_{i-1}$ by adding the document d. We get the final training data by recording the best document int every step.

Nextly, stochastic gradient descent method are applied to optimize the loss function. At every step, we calculate the gradient and update the value. The gradient for step i in loop structure at training set D_{init} is computed as follows:

$$\Delta\lambda_r^{(i)} = \sum_{j=1}^{n} \left(\frac{\sum_{k=j}^{n} P(y(k)|q)exp(\lambda_r P(y(k)|q) + \lambda_d^T F(S_{i-1} \cup y(k)))}{\sum_{k=j}^{n} exp(\lambda_r P(y(k)|q) + \lambda_d^T F(S_{j-1} \cup y(k)))} \right.$$
$$\left. - \frac{P(y(j)|q)exp(\lambda_r P(y(j)|q) + \lambda_d^T F(S_{j-1} \cup y(j)))}{exp(\lambda_r P(y(j)|q) + \lambda_d^T F(S_{j-1} \cup y(j))} \right) \quad (5)$$

$$\Delta\lambda_d^{(i)} = \sum_{j=1}^{n} \left(\frac{\sum_{k=j}^{n} F(S_{j-1} \cup y(k))exp(\lambda_r P(y(k)|q) + \lambda_d^T F(S_{j-1} \cup y(k)))}{\sum_{k=j}^{n} exp(\lambda_r P(y(k)|q) + \lambda_d^T F(S_{j-1} \cup y(k)))} \right.$$
$$\left. - \frac{F(S_{j-1} \cup y(j))exp(\lambda_r P(y(j)|q) + \lambda_d^T F(S_{j-1} \cup y(j)))}{exp(\lambda_r P(y(j)|q) + \lambda_d^T F(S_{j-1} \cup y(j)))} \right) \quad (6)$$

Finally, the sequential prediction method is used to predict the result. In algorithm, at the i-th step in loop structure, we select the best document d from $D\backslash S_{i-1}$ to maximize our ranking function and update the candidate set $D\backslash S_{i-1}$ by adding document d. Then we predict the final diverse ranking list by recording the best document in every step.

4 Experiments

4.1 Experimental Setup

Dataset. We use TREC web track (WT2009 for short), WT2010, WT2011 as our dataset. We do our evaluation on the ClueWeb09 Category B retrieval collection[1]. Our query set contains of 150 queries, from TREC web track 2009 (WT2009) [5], TREC web track 2010 (WT2010), TREC web track 2011 (WT2011) (50 for each).

Evaluation Metrics. Three mainly evaluation metrics are used to evaluate the performance of our method: α-NDCG, ERR-IA, NRBP (computed at cutoff 50). To measure the robustness, we use Win/Loss ratio metrics. The evaluation metrics reported at different cutoffs. We use $\{5, 10, 20\}$ as our cutoffs to set up our experiments. The α are set to 0.5 in our experiments.

---

[1] http://www.lemurproject.org/clueweb09.php/.

Baseline Methods. We use the Indri[2] to conduct our retrieval run with its default parameter configuration. The krovetz stemmer and stopword removal are applied both in the index and retrieval time. All of the search result diversification methods are applied based on the top-50 retrieved documents.

We compare L-HSRD with some baseline models as follows:

- **QL.** The Query-likelihood language model is used for indri search engine as an initial retrieval method. We use it to provide the initial top 1000 documents for our diversification method.
- **MMR.** A classical implicit diversification model.
- **xQuAD.** xQuAD is a popular explicit diversification model which focus on the redundancy of aspects.
- **PM2.** PM2 is a popular explicit diversification model. PM2 generates the result set according to the aspects proportionality.
- **HxQuAD.** HxQuAD is a hierarchical diversification model based on xQuAD [12].
- **SVMDIV.** SVMDIV is a learning model for search result diversification [16]. We get the source code from the svmdiv homepage[3] provided by the author.

There exists a single parameter λ to tune in baselines 2–5 (corresponding to MMR, xQuAD, PM2, HxQuAD), we divide the data into 5 parts randomly and perform a 5-fold cross validation to train λ through optimizing ERR-IA. In our model, we also perform a 5-cross validation with a ratio of 3:1:1 for training, validation and prediction for the test query on each year. The final result are calculated over all the folds.

4.2 Experimental Results

Diversification Analysis. Table 1 shows the result of the diversification evaluation in terms of α-nDCG, ERR-IA, and NRBP. The best result per baseline is highlighted in bold.

It is noted that L-HSRD always performs best in α-nDCG and ERR-IA. It improves the initial retrieval ranking method with gains up to 34.34%, 48.43%, 20.66%, in terms of α-nDCG on WT2009, WT2010, WT2011 respectively. The improvement of L-HSRD over the MMR in terms of α-nDCG is up to 35.51%, 48.12%, 23.91% on WT2009, WT2010, WT2011 respectively. The improvement of L-HSRD over the xQuAD in terms of α-nDCG is up to 40.06%, 32.16%, 20.44% on WT2009, WT2010, WT2011 respectively, and the improvement of L-HSRD over the PM2 is up to 24.47%, 39.31%, 20.17% on WT2009, WT2010, WT2011 respectively. It indicates that our learning approach tackles the diversity measurement problem more effectively with the consideration of integrate different level of diversity features based on the hierarchical subtopics. Besides the non-learning model, the improvement of L-HSRD over the SVMDIV in terms of α-nDCG is up to 20.15%, 29.79%, 16.45% on WT2009, WT2010, WT2011

---

[2] http://www.lemurproject.org/indri.php.
[3] http://projects.yisongyue.com/svmdiv/.

Table 1. Diversification performance using the official evaluation metrics for WT2009, WT2010, WT2011

Year	Experiment	ERR-IA@20	α-nDCG@20	NRBP
2009	QL	0.1376	0.2548	0.1008
	MMR	0.1405	0.2526	0.1070
	xQuAD	0.1411	0.2444	0.1113
	PM2	0.1482	0.2750	0.1101
	SVMDIV	0.1531	0.2849	0.1219
	HxQuAD	0.1653	0.3025	0.1372
	L-HSRD	**0.2084**	**0.3423**	**0.1894**
2010	QL	0.1484	0.2445	0.1092
	MMR	0.1494	0.2450	0.1129
	xQuAD	0.1732	0.2746	0.1326
	PM2	0.1599	0.2605	0.1175
	SVMDIV	0.1698	0.2796	0.1158
	HxQuAD	0.1807	0.2924	0.1303
	L-HSRD	**0.2248**	**0.3629**	**0.2011**
2011	QL	0.3288	0.4454	0.2802
	MMR	0.3253	0.4337	0.2834
	xQuAD	0.3235	0.4462	0.2812
	PM2	0.3316	0.4472	0.2831
	SVMDIV	0.3429	0.4615	0.2923
	HxQuAD	0.3606	0.4860	0.3107
	L-HSRD	**0.4216**	**0.5374**	**0.3501**

respectively. It shows that considering relevance and different type of features in diversity measurement is helpful in diversification. As for the traditional hierarchical diversification method, the improvement of L-HSRD over the HxQuAD in terms of α-nDCG is up to 13.16%, 24.11%, 10.58% on WT2009, WT2010, WT2011, respectively. HxQuAD only use a predefined function to measure the diversity score, and the parameter may not be optimal because it needs to be tuned manually. Our learning model tackles the parameter tuning problem in an automatic fasion and reach optimal result.

Robustness Analysis. An effective search result diversification method should not only outperforms other models in terms of diversity metrics, but also maintains a high level of robustness. We set up series of experiments on robustness research to study the Win/Loss behaviour.

The Win/Loss ratio are proposed by Yue et al. [16] and Dang et al. [9] for entirety robustness measurement. It denotes whether the model improve or hurt the result when comparing with the basic relevant baseline QL [9,16] in terms of evaluation metrics. In our experiment, ERR-IA is used to calculate the Win/Loss ratio.

Table 2. Win/Loss ratio

experiment	WT2009	WT2010	WT2011	Total
MMR	20/18	24/17	23/16	67/51
xQuAD	23/18	23/16	24/14	70/48
PM2	22/20	26/14	25/14	73/48
HxQuAD	27/15	30/10	31/10	88/35
L-HSRD	**28/14**	**30/9**	**32/9**	**90/32**

It can be inferred that L-HSRD model performs best with its ratio of 2.65 from the Table 2. It reflects the remarkable robustness of L-HSRD model comparing with other diversification models. This confirms the overall performance of our model is not restricted to a small subset, it still work in the whole dataset for three years data.

5   Conclusion and Future Work

In this paper, we propose a learning based approach to hierarchical search result diversification. We pay our attention to the hierarchical diversification models and introduce the learning approach to address this as a learning problem. We have demonstrated the effectiveness of L-HSRD comparing with other diversification models. We find our model achieve considerable results in terms of official diversity metrics on three years in TREC web track dataset. To prove its robustness, we set the experiment about Win/Loss ratio. We believe L-HSRD will play an important role to improve the search result diversification method.

There exists a number of directions to be explored in the future. We are looking forward to take some considerable steps to make L-HSRD achieve convergency as quick as possible. Meanwhile, we are looking forward to make use of the deep learning technology to improve the hierarchical search result diversification.

Ackonwloedgements. This research is supported by National Natural Science Foundation of China (Grant No. 61375054), Natural Science Foundation of Guangdong Province Grant No. 2014A030313745, Basic Scientific Research Program of Shenzhen City (Grant No. JCYJ20160331184440545), and Cross fund of Graduate School at Shenzhen, Tsinghua University (Grant No. JC20140001).

References

1. Agrawal, R., Gollapudi, S., Halverson, A., Ieong, S.: Diversifying search results. In: Proceedings of the Second ACM International Conference on Web Search and Data Mining, pp. 5–14. ACM (2009)
2. Carbonell, J., Goldstein, J.: The use of MMR, diversity-based reranking for reordering documents and producing summaries. In: Proceedings of the 21st Annual International ACM SIGIR Conference on Research and Development in Information Retrieval, pp. 335–336. ACM (1998)

3. Chapelle, O., Metlzer, D., Zhang, Y., Grinspan, P.: Expected reciprocal rank for graded relevance. In: Proceedings of the 18th ACM Conference on Information and Knowledge Management, pp. 621–630. ACM (2009)
4. Clarke, C.L., Craswell, N., Soboroff, I.: Overview of the TREC 2009 web track. Technical report, DTIC Document (2009)
5. Clarke, C.L., Craswell, N., Soboroff, I.: Preliminary report on the TREC 2009 web track. In: Proceeding of TREC (2009)
6. Clarke, C.L., Kolla, M., Cormack, G.V., Vechtomova, O., Ashkan, A., Büttcher, S., MacKinnon, I.: Novelty and diversity in information retrieval evaluation. In: Proceedings of the 31st Annual International ACM SIGIR Conference on Research and Development in Information Retrieval, pp. 659–666. ACM (2008)
7. Clarke, C.L.A., Kolla, M., Vechtomova, O.: An effectiveness measure for ambiguous and underspecified queries. In: Azzopardi, L., Kazai, G., Robertson, S., Rüger, S., Shokouhi, M., Song, D., Yilmaz, E. (eds.) ICTIR 2009. LNCS, vol. 5766, pp. 188–199. Springer, Heidelberg (2009). doi:10.1007/978-3-642-04417-5_17
8. Collins-Thompson, K., Macdonald, C., Bennett, P., Diaz, F., Voorhees, E.M.: TREC 2014 web track overview. Technical report, DTIC Document (2015)
9. Dang, V., Croft, W.B.: Diversity by proportionality: an election-based approach to search result diversification. In: Proceedings of the 35th International ACM SIGIR Conference on Research and Development in Information Retrieval, pp. 65–74. ACM (2012)
10. Dou, Z., Hu, S., Chen, K., Song, R., Wen, J.R.: Multi-dimensional search result diversification. In: Proceedings of the Fourth ACM International Conference on Web Search and Data Mining, pp. 475–484. ACM (2011)
11. Drosou, M., Pitoura, E.: Search result diversification. ACM SIGMOD Rec. **39**(1), 41–47 (2010)
12. Hu, S., Dou, Z., Wang, X., Sakai, T., Wen, J.R.: Search result diversification based on hierarchical intents. In: Proceedings of the 24th ACM International on Conference on Information and Knowledge Management, pp. 63–72. ACM (2015)
13. Marden, J.I.: Analyzing and Modeling Rank Data. CRC Press, London (1996)
14. Santos, R.L., Macdonald, C., Ounis, I.: Exploiting query reformulations for web search result diversification. In: Proceedings of the 19th International Conference on World Wide Web, pp. 881–890. ACM (2010)
15. Vargas, S., Castells, P., Vallet, D.: Explicit relevance models in intent-oriented information retrieval diversification. In: Proceedings of the 35th International ACM SIGIR Conference on Research and Development in Information Retrieval, pp. 75–84. ACM (2012)
16. Yue, Y., Joachims, T.: Predicting diverse subsets using structural SVMs. In: Proceedings of the 25th International Conference on Machine Learning, pp. 1224–1231. ACM (2008)
17. Zhu, Y., Lan, Y., Guo, J., Cheng, X., Niu, S.: Learning for search result diversification. In: Proceedings of the 37th International ACM SIGIR Conference on Research & Development in Information Retrieval, pp. 293–302. ACM (2014)

Demo Papers

TeslaML: Steering Machine Learning Automatically in Tencent

Jiawei Jiang[1,2]($\boxtimes$), Ming Huang[2], Jie Jiang[2], and Bin Cui[1]

[1] Key Lab of High Confidence Software Technologies (MOE), School of EECS,
Peking University, Beijing, China
{blue.jwjiang,bin.cui}@pku.edu.cn
[2] Tencent Inc., Shenzhen, China
{jeremyjiang,andymhuang,zeus}@tencent.com

Abstract. In this demonstration, we showcase TeslaML, the machine learning (ML) platform in Tencent Inc. TeslaML offers an interactive and visual workspace for users to create an ML pipeline via dragging, placing, and connecting the implemented modules. For the non-experts, TeslaML provides many ready-to-use ML modules to build an ML pipeline without any programming. Besides, TeslaML abstracts many existing ML systems as system modules. The integration of various systems enables the experienced users to use their preferred systems, to test new algorithms, and to obtain the most efficient execution. Furthermore, TeslaML provides many schedulers to meet different scheduling requirements.

1 Introduction

Machine learning (ML) has shown its spectacular capability in the past decade. As the size of many industrial-scale datasets have overwhelmed the processing capability of a single machine, it is inevitable to deploy ML tasks over decentralized clusters. From the systematic perspective, we find that there are two fundamental issues for distributed ML systems—*how to render distributed ML easy-to-use for the practitioners?* and *how to efficiently execute ML tasks in a distributed environment?*

Motivation 1: Ease of Use. A crucial aspect of a distributed ML system is to provide easy-to-use programming APIs or tools for the users. The first commonly used approach, adopted by systems like Hadoop and Spark, is to provide low-level dataflow abstractions such as map-reduce [1]. Alternatively, the second avenue, chosen by MLlib, GraphX and Tensorflow, offers high-level programming abstractions on the top of low-level APIs [7]. To lower the barrier of entry for the majority of the users, some platforms like Azure and BigML establish a visual and interactive approach [2]. The users can construct an ML task by choosing the provided ML-related modules. Nevertheless, this programming-free solution only empowers black-boxed services so that the users cannot debug the performance or test new algorithms. Furthermore, since many ML tasks evolve over time or hyper-parameters, it is a common phenomenon to execute

© Springer International Publishing AG 2017
L. Chen et al. (Eds.): APWeb-WAIM 2017, Part II, LNCS 10367, pp. 313–318, 2017.
DOI: 10.1007/978-3-319-63564-4_26

an ML pipeline many times. Unfortunately, Azure and BigML do not provide an automatic infrastructure to address this iterative nature. At a higher level, an ideal ML platform should both promote the ease of use for the nonspecialist users and furnish the experienced designers with flexible programming interfaces and various schedulers.

Motivation 2: Efficient Execution. Existing ML systems cannot tackle diverse ML workloads efficiently in a holistic way. First, an ML pipeline generally prefers different systems for different phases. For the phase of data ingress, parallel dataflow engines like Hadoop and Spark are prevalent due to their highly efficient dataflow parallelisms. For the phase of model training, however, some ML-oriented systems reveal more merits since their parallel architectures facilitate the iterative execution of distributed ML tasks. For instance, DistBelief prevents the single-point bottleneck of Spark during iterative model aggregations by using the parameter server architecture [3,4] to partition the model over several machines. For the phase of model evaluation, systems like PySpark can visually assess the trained model. Second, each ML system has strengths and weaknesses, depending on the ML algorithm which it runs. For example, MLlib is ill-suited for the iterative computation in the presence of high-dimensional model; Dist-Belief cannot support general graph computation efficiently; GraphX would not exploit computing efficiencies available in the structured graphs typically found in deep networks; and Tensorflow reveals inefficiency in a distributed setting.

Demonstration Features: In this paper, we showcase TeslaML, a platform that concurrently addresses the aforementioned challenges. The key features of our demonstration can be summarized below:

- **Visual and Interactive Deployment of ML Pipelines.** In TeslaML, the front-end subsystem can be accessed through a web browser. A visual interface with a pallet of modules enables the users to construct an end-to-end ML pipeline by dragging, placing, and connecting these preprogrammed modules.
- **Ready-to-use ML Modules.** TeslaML provides a wide variety of ready-to-use ML algorithms as modules, along with other modules that help with data input, output, and visualization. TeslaML makes it easy for the non-experts to conduct data analytics with ML techniques in a programming-free manner.
- **Systems Modules.** Many popular ML systems are integrated into TeslaML as system modules. The experienced users are allowed to run ML tasks with their preferred systems by uploading their own programs.
- **Pipeline Schedulers.** To facilitate the iterative executions of many ML tasks, TeslaML offers several schedulers—manual, automatical, periodical, rerunning, and tuning schedulers.

2   System Overview

2.1   Front-End Subsystem

Figure 1(a) shows the front-end subsystem of TeslaML. The visual interface that can be accessed through a web browser consists of four areas — the module panel, the canvas, the configuration panel, and the toolbar.

Module Panel: There are four tabs inside the module panel, each of which refers to one category of modules.

Input modules. The input modules offer visual approaches to handling the input data. TeslaML can read the input data from a dataset on the file system or a table in the database.

System modules. A plenty of prevailing systems are abstracted as system modules in TeslaML. To name a few, TeslaML unifies Spark, Angel [5], Tensorflow, Mariana [8], Caffe, Torch, PySpark, and XGBoost.

Algorithm modules. Most of, if not all, common ML algorithms are pre-programmed as ready-to-use algorithm modules. We implement these modules with Spark (for feature extraction, feature transformation, classification, regression, clustering, and graph algorithms) and Tensorflow (for deep learning algorithms).

Output modules. There are two categories of output modules. The visualization modules help the users see the distribution of input data. The evaluation modules evaluate the quality of the trained model. As illustrated in Fig. 1(a), a ROC curve appears when the mouse hovers over the ROC module.

Canvas: The users can drag modules from the module panel and place them onto the canvas. Then, the users can connect two modules with an arrow. Through this drag-place-and-connect operation, a `job-flow` is created as a DAG inside which each node represents a `job` in TeslaML.

Configuration Panel: The configuration panel will emerge at the right side of the canvas by clicking any module on the canvas. In this panel, the related

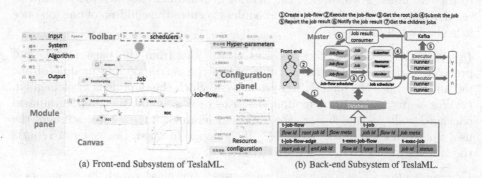

(a) Front-end Subsystem of TeslaML.   (b) Back-end Subsystem of TeslaML.

Fig. 1. System architecture of TeslaML.

configurations of the module can be preset, such as the input data path, the data format, the output data path, the training hyper-parameters, and the resource configurations (e.g., the number of physical machines and the maximal memory budget). For the system module, the users can upload their own packaged codes through the configuration panel.

Toolbar: At the top of the canvas is the toolbar with which the users can schedule the job-flow with five schedulers. (1) Manual scheduler empowers the users to start and kill each job manually. (2) Automatically parallel/serial scheduler executes the job-flow in parallel or in serial. The parallel mode concurrently starts the sibling jobs once their parent jobs are finished, whereas the serial mode runs the sibling jobs successively. (3) Periodical scheduler periodically executes the job-flow in five granularities—minute, hour, day, week, and month, helping the ML models evolving over time. (4) Rerunning scheduler runs the previous instances of one job-flow within a time range. (5) Tuning scheduler repeatedly runs the job-flow with a series of hyper-parameters which are set in the configuration panel.

2.2   Back-End Subsystem

The back-end subsystem of TeslaML is implemented with the classical master-slave architecture in which there are a master node and several executor nodes.

Database Tables: As Fig. 1(b) shows, the information of job-flows and jobs is stored in MySQL. **t-job-flow** stores the static metadata of each job-flow created by the user. **t-job** stores the static metadata of each job. **t-job-flow-edge** stores the edges between connected jobs in a job-flow. We say the ending job of one edge is the child job of the starting job. **t-exec-job-flow** records the scheduler type and status of each running job-flow instance. **t-exec-job** records the status of each running job instance.

Master: The master node obtains the metadata of in-flight job-flows from the database. Afterwards, the job-flow scheduler analyzes the structure of each job-flow. Based on the chosen scheduler, the job scheduler submits the active jobs to the executors. For instance, the periodical scheduler executes the job-flow at regular intervals, and the parallel scheduler executes the children of one job once it is successfully finished.

Executor: The executor node receives active jobs from the master node and proceeds as follows: (1) The executor parses the corresponding module of one job and identifies the underlying system with which the module is implemented. (2) For a preprogrammed module, the executor submits the preprogrammed packaged code to the cluster. For a system module, the executor submits the user uploaded code to the cluster. (3) If the job is finished, the executor reports the running result of the job to the master.

Communication: The communication between the master and the executors is governed by Kafka, a distributed streaming system which is extensively used to reliably transfer data between systems. On the producer side, the executor sends the running result of finished jobs to Kafka. On the consumer side, the master receives the messages from Kafka, extracts the running result of the job, and notifies the job-flow scheduler. If the job succeeds, the job-flow scheduler will schedule the children jobs of the finished job. However, if the job fails somehow, the scheduler will cease the job-flow and report the failure information to the front-end subsystem where the users can see the status of the job-flow.

3   Demonstration Features

The attendees will participate in a live demonstration of four phases of our platform: ❶ the interactive front-end, ❷ the creation of an ML pipeline with the ready-to-use modules, ❸ the creation of an ML pipeline with the system modules, and ❹ the execution of an ML pipeline with different schedulers.

In phase ❶, we showcase the front-end of TeslaML. The attendees can access an interactive and visual workspace through a web browser. The attendees will experience the front-end, including the canvas, module panel, configuration panel, and toolbar. We will show how to drag-place-and-connect the provided modules.

In phase ❷, we showcase how to easily build an ML pipeline from scratch. The attendees will be invited to choose ready-to-use modules for different stages of an ML pipeline, including data input, model training, and model evaluation, to create a job-flow on the canvas. On the configuration panel, the attendees can set the related configurations of each module. For example, the attendees can create an ML pipeline to train a convolutional neural network by—(1) dragging and placing a dataset module onto the canvas, and setting the data path of the mnist dataset on the configuration panel, (2) placing a sampling module and a normalizing module to transform the input data, (3) putting a module of LeNet [6] on the canvas, (4) choosing an evaluation module to calculate the precision, and (5) connecting these modules to construct a complete pipeline.

In phase ❸, we showcase how to customize an ML pipeline with system modules. With the same manner described in phase ❷, the attendees can create an ML pipeline with the system modules that offer interfaces to run the existing popular systems. We will show that the attendees can upload their own codes and set the hardware configurations on the configuration panel. For example, the attendees can use Spark and Angel, the parameter server system of Tencent Inc., to train a logistic regression model, respectively. The trained dataset is a 400 GB real-world dataset from Tencent Video which consists of 100 million instances. Each instance contains 60 million features. On the configuration panel, the attendees can set the number of physical machine to 50 and set the maximal memory of each machine to 20 GB. According to the empirical results, Spark needs 40 min to run an epoch, while Angel only needs 1.6 min, bringing a 25× speedup. The performance improvement verifies that the parameter server based system is more suitable for high-dimensional datasets by addressing the single-point bottleneck during the model aggregation.

In phase ❹, we showcase how the implemented schedulers execute the user-defined job-flows. The attendees can run the job-flow created in phase ❷ and ❸ manually or automatically. After each execution, the attendees can see the running result of each job by right clicking the corresponding module. In addition to these basic schedulers that are applied per execution, the attendees are invited to experience the other three schedulers that are employed to iteratively run a job-flow.

Acknowledgements. This research is supported by the National Natural Science Foundation of China under Grant No. 61572039, 973 program under No. 2014CB340405, Shenzhen Gov Research Project JCYJ20151014093505032, and Tecent Research Grant (PKU).

References

1. Dean, J., Ghemawat, S.: MapReduce: simplified data processing on large clusters. CACM **51**(1), 107–113 (2008)
2. Halperin, D., Teixeira de Almeida, V., Choo, L.L., et al.: Demonstration of the Myria big data management service. In: SIGMOD, pp. 881–884 (2014)
3. Jiang, J., Cui, B., Zhang, C., Yu, L.: Heterogeneity-aware distributed parameter servers. In: SIGMOD (2017)
4. Jiang, J., Jiang, J., Cui, B., Zhang, C.: TencentBoost: a gradient boosting tree system with parameter server. In: ICDE (2017)
5. Jiang, J., Yu, L., Jiang, J., Liu, Y., Cui, B.: Angel: A new largescale machine learning system. NSR (2017)
6. LeCun, Y., Bottou, L., Bengio, Y., Haffner, P.: Gradient-based learning applied to document recognition. Proc. IEEE **86**(11), 2278–2324 (1998)
7. Meng, X., Bradley, J., Yuvaz, B., et al.: MLlib: machine learning in Apache Spark. JMLR **17**(34), 1–7 (2016)
8. Zou, Y., Jin, X., et al.: Mariana: tencent deep learning platform and its applications. PVLDB **7**(13), 1772–1777 (2014)

DPHSim: A Flexible Simulator
for DRAM/PCM-Based Hybrid Memory

Dezhi Zhang, Peiquan Jin(✉), Xiaoliang Wang, Chengcheng Yang,
and Lihua Yue

University of Science and Technology of China, Hefei, China
jpq@ustc.edu.cn

Abstract. In this paper, we demonstrate a flexible simulator for DRAM and
PCM based hybrid memory systems. PCM (Phase Change Memory) as a new
kind of non-volatile memories has received much attention from both academia
and industry. While PCM has many new features, such as fast read speed,
non-volatility, and low power consumption, how to use PCM at memory hierarchy still remains a problem. In addition, at present PCM chips and storage
devices are not available. This makes it hard to evaluate PCM-related algorithms. Thus, we design a flexible simulator named *DPHSim* to provide a test
bed for PCM-related studies. The unique features of *DPHSim* are manifold.
(1) It supports various kinds of memory hierarchy, including DRAM-only,
PCM-only, PCM as the cache of DRAM, and hybrid memory (PCM and DRAM
are both used as main memory). (2) It provides flexible configuration options for
workloads generation and system setting. (3) It offers user-friendly memory
allocation APIs for users to simulate memory hierarchy and evaluate performance on *DPHSim*. After an overview on the features and architecture of
DPHSim, we present a case study of *DPHSim* to demonstrate its feasibility and
flexibility.

Keywords: PCM · Hybrid memory · Simulator · Performance evaluation

1 Introduction

Phase change memory (PCM) has emerged as a new kind of memory with attractive
features, such as fast read speed, high density, low power consumption, and
non-volatility [1]. These properties make PCM be a candidate of future memories that
aim to replace existing memories like DRAM. However, PCM has a low write-speed
and limited write-endurance, which makes it unpractical to completely replace DRAM
with PCM at this moment [2]. This also opens a research problem, i.e., how to use
PCM at memory hierarchy. However, so far PCM-based storage devices are not
available. This makes it hard to carry out researches upon real PCM devices as well as
on PCM-related memory systems, which calls for PCM-related test beds. Further, even
if PCM devices will be available in the future, we still need a simulation tool to provide
different configuration options of memory hierarchy, so that we can devise algorithms
towards specific memory hierarchy. Such simulators can enable researchers to explore
the impact of memory algorithms on different types of memory systems.

L. Chen et al. (Eds.): APWeb-WAIM 2017, Part II, LNCS 10367, pp. 319–323, 2017.
DOI: 10.1007/978-3-319-63564-4_27

In this paper, we propose a simulator called *DPHSim* for evaluating DRAM/PCM-based hybrid memory systems. It provides high configurability, fine-grained timing modules, and particularly a friendly user interface for hybrid memory allocation. *DPHSim* has a few unique features. First, it supports various kinds of memory hierarchy, including DRAM-only, PCM-only, PCM as the cache of DRAM, and hybrid memory (PCM and DRAM are both used as main memory). Second, it provides flexible configuration options for workloads generation and system setting. Finally, it offers user-friendly memory allocation APIs for users to simulate memory hierarchy and evaluate performance on *DPHSim*.

There are some PCM-oriented simulators, such as PCRAMsim [3] and NVMain [4]. However, PCRAMsim onlys support PCM-only simulation. NVMain supports simulation on a hybrid system (both DRAM and PCM as main memory), but it cannot construct a hierarchical system (DRAM as the cache of PCM). In addition, all existing simulators cannot expose hybrid memory spaces to programmers. On the contrary, The *DPHSim* proposed in this paper offers hybrid memory allocation APIs and users can completely manage the memory spaces of all the involved memory.

2   Architecture of *DPHSim*

Figure 1 shows the architecture of *DPHSim*. It simulates the hierarchy consisting of CPU, cache, and hybrid memory composed of DRAM and PCM. The *CPU Module* receives instructions of a workload gathered by *Pin*, and transfers each instruction into a memory read and some memory writes (if needed). All requests go into the *Cache Module*, which uses a queue to implement cache hits or miss operations. The *Memory Module* receives the requests from the previous level and emulates the execution of them under a specific memory system. The hybrid memory part implemented by a *Memory Module* enables users to simulate different kinds of memory systems, including DRAM only, PCM only, and DRAM/PCM-based hybrid memory. The *Memory Module* offers a bus-like data structure with multiple entries, where the shift-in requests are placed without any assumed ordering. All requests, labeled with ids, statuses, timestamps, addresses, and access types, usually come from applications or other superiors. Once a request is executed and completed by the simulator, the corresponding entry will be free and available. Particularly, on account of the numbered entries, when all of them have been filled, the coming requests are stalled until an entry has become available.

(1) **Memory Controller.** The memory controller module begins with the selection of a request from the bus into a request queue according to a request ordering policy. The request queue then maps the physical address of the request to the memory address under an address mapping scheme. After that, a sequence of PCM commands associated with each memory request are generated and sent to the specified PCM cells. The PCM commands in *DPHSim* include *Row Access* (RAC), *Column Read* (CRC), *Column Write* (CWC), and *Row Pre-charge* (RPC).

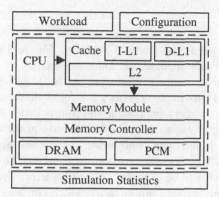

Fig. 1. Architecture of *DPHSim*.

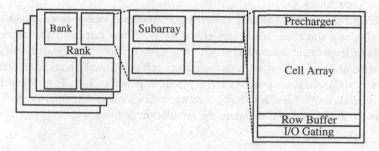

Fig. 2. Hierarchy of a memory array built in *DPHSim*

(2) **Memory Device Module.** As shown in Fig. 2, *DPHSim* simulates a layout of a memory array. In general, a memory array consists of several identical ranks that are the top-level structure modeled in *DPHSim*. Each rank is a fully-functional memory unit, and can be accessed independently. And multiple banks, which constitute a rank, operate simultaneously to fulfill a memory operation. Besides, each bank is composed of several subarrays which are elementary structures of memory.

(3) **Memory Allocation APIs.** The memory allocation APIs aim to expose memory spaces to programmers. Through the APIs, researchers can decide whether to satisfy a read or write request on a specific memory area, according to specific data placement algorithms. *DPHSim* provides a set of functions, including *Dmalloc()*, *Dfree()*, *Pmalloc()*, and *Pfree()* for memory allocaiton on DRAM or PCM.

We implement a user-controlled memory allocation based on splitting the virtual address range into two portions; one for DRAM and the other for PCM. For example, when we allocate memory space on DRAM by calling the function named *Dmalloc()*, the memory address translator enables that the generated virtual address is mapped to the DRAM address according to pre-defined virtual address range.

3 Demonstration

In this section, we present a case study of *DPHSim*. Figure 3 shows a screenshot of *DPHSim*. The demonstration includes the following steps.

(1) Using the hybrid memory allocation APIs. We use the proposed hybrid memory allocation APIs to implement user-managed data placement in programs. Although *DPHSim* allows users implement algorithms using the provided APIs, we will prepare a few programs that can be loaded directly in the demonstration.

(2) Generating workloads and configuring system settings. The *Open* button in the *Workload* panel is for choosing a workload, and the *Load* button is used for generating its trace which will be executed on the simulator. Then, users can configure the simulator through the *System Settings* panel, which contains detailed parameters of CPU, cache, and memory. *DPHSim* allows flexible modification on these parameters, such as memory architecture and device type. In addition, on the *Memory Parameters* panel, users can adjust memory parameters, including frequency, array organization, and timing parameters.

(3) Executing workloads and collecting statistics. Once the workload and the configuration are specified, users can either let *DPHSim* run until the workload is completed or implement a simulation within a limited time. The system will display detailed simulation results of each storage level (channel, rank, bank, and subarray). We provide a visualization panel (the right-bottom panel in Fig. 3) to show the histograms of reads and writes, row-buffer hits and misses on the channel-level. This can help users to understand and compare the simulation outputs.

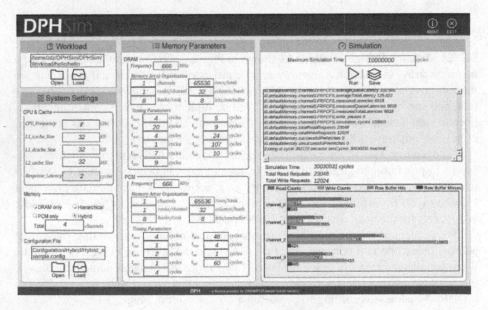

Fig. 3. Screenshot of *DPHSim*

Acknowledgements. This work is supported by the National Science Foundation of China (61472376 and 61672479). Peiquan Jin is the corresponding author.

References

1. Wu, Z., Jin, P., Yue, L.: Efficient space management and wear leveling for PCM-based storage systems. In: Wang, G., Zomaya, A., Perez, G.M., Li, K. (eds.) ICA3PP 2015. LNCS, vol. 9531, pp. 784–798. Springer, Cham (2015). doi:10.1007/978-3-319-27140-8_54
2. Chen, K., Jin, P., Yue, L.: Efficient buffer management for PCM-Enhanced hybrid memory architecture. In: Cheng, R., Cui, B., Zhang, Z., Cai, R., Xu, J. (eds.) APWeb 2015. LNCS, vol. 9313, pp. 29–40. Springer, Cham (2015). doi:10.1007/978-3-319-25255-1_3
3. Dong, X., Jouppi, N., Xie, Y.: PCRAMsim: system-level performance, energy, and area modeling for phase-change RAM. In: ICCAD, pp. 269–275. ACM (2009)
4. Poremba, M., Zhang, T., Xie, Y.: NVMain 2.0: a user-friendly memory simulator to model (Non-) volatile memory systems. IEEE Comput. Archit. Lett. **14**(2), 140–143 (2015)

CrowdIQ: A Declarative Crowdsourcing Platform for Improving the Quality of Web Tables

Yihai Xi, Ning Wang(✉), Xiaoyu Wu, Yuqing Bao, and Wutong Zhou

School of Computer and Information Technology, Beijing Jiaotong University,
Beijing, China
{xiyihai,nwang,16112087,16125158,16125212}@bjtu.edu.cn

Abstract. Web tables provide us with high-quality sources of structured data. However, we could not use those valuable tables directly owing to various problems such as conflict data and missing headers. We present CrowdIQ, a scalable platform that integrates crowdsourcing technology for improving the quality of web tables. We design CrowdIQL, which is a declarative language aiming at helping requesters operate tables more exactly and flexibly. Crowdsourcing task is also optimized in this platform by providing candidate items and minimizing useless data, which help requesters to get higher quality tables with less cost.

1 Introduction

In the era of big data, there are a huge number of web tables on the Internet. However, web tables are often with various problems such as conflict data and missing headers. At present, machine algorithms used to solve these problems are not yet able to provide satisfactory accuracy and recall, especially for semantics recovery. Crowdsourcing has been used to solve problems that are difficult for computers. Considering the complexity of improving table quality, some crowdsourcing platforms like AMT [1], which focuses on micro tasks, are not good at those tasks. Others are designed to solve special kinds of table problems, e.g., CrowdFill [2] is designed to collect structured data while CrowdSR [3] is used to recover table headers. Furthermore, CrowdDB [4] and Deco [5] both serve as finding missing data for tables. However, none of them could provide a universal solution for existing and emerging table problems. We propose to develop CrowdIQ, a crowdsourcing platform for improving the quality of web tables. It has unique features as follows: (1) CrowdIQ is able to offer a quality inspection to pending tables and report potential problems to requesters. (2) CrowdIQ provides a declarative language CrowdIQL to operate tables more exactly and flexibly. (3) CrowdIQ supports customized request for improving quality of tables and external algorithms for optimizing crowdsourcing tasks.

© Springer International Publishing AG 2017
L. Chen et al. (Eds.): APWeb-WAIM 2017, Part II, LNCS 10367, pp. 324–328, 2017.
DOI: 10.1007/978-3-319-63564-4_28

2 Platform Overview

The architecture of CrowdIQ is depicted in Fig. 1. CrowdIQ first inspects the pending table after a requester uploads it. Then, the requester chooses CrowdIQL or graphical operations to define request. According to results of the parser, CrowdIQ generates Human Intelligence Task(HIT) as well as its UI template. At last, CrowdIQ integrates crowdsourcing answers and returns high-quality tables. In the following, we will introduce three core modules in details.

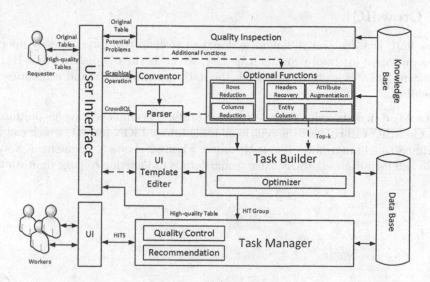

Fig. 1. CrowdIQ architecture

Optional Functions. This module consists of machine preprocessing algorithms and data minimization algorithms, which are designed to optimize crowdsourcing tasks with the help of Probase knowledge base. Requesters could invoke different functions to preprocess the pending tables by CrowdIQL. Top-k candidate items produced in preprocessing stage make crowdsourcing tasks easier by converting fill-in questions into multiple-choice questions. Data minimization algorithms based on the idea of clustering or sampling aim at prompting the crowd as less data as possible without sacrificing their answering accuracy. Our platform is scalable and could settle emerging table problems by providing plug-and-play service for new functions.

Task Builder. After receiving results from the Parser, Task Builder generates a corresponding group of HITs as well as UI templates according to candidate answers provided by machine algorithms. UI templates could be edited whenever the requester is not satisfied with them. The task builder also contains an optimizer which could implement cost control by arranging the order of HITs and prompting representative data to the crowd.

Task Manager. This module is designed to assign HITs to workers and return high-quality tables to requesters. It provides a quality control mechanism, which builds a cumulative contribution model to evaluate the performance of each worker and a decision-making model to deal with special tasks with uncertain number of answers. The mechanism is universal for different kinds of HITs, such as a mix of fill-in and multiple-choice questions. Task Manager also provides a recommendation mechanism based on similarity among HITs for improving the overall accuracy in CrowdIQ.

3   CrowdIQL

CrowdIQL is a declarative language designed to help requesters specify operations on tables for improving their quality, which is then translated into HITs. In the rest of this section, we sketch the data model, syntax and semantics of CrowdIQL.

Data Model. JavaScript Object Notation (JSON) is employed in this platform, i.e., CrowdIQ will convert the relational table into a JSON format, which can be modified to add attributes like table name. Figure 2 shows a fragment of a web table and its initialized JSON table (some data is hidden due to page limitation).

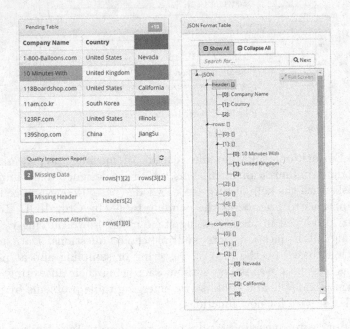

Fig. 2. Quality inspection for a web table

Syntax and Semantics. All elements in the relational table are treated as attributes. We will show how CrowdIQL executes basic operations with two running examples. The general syntax of CrowdIQL is as follows:

```
SELECT|INSERT|DELETE|UPDATE <table>.<attribute> [SHOWING <table>.<
attribute>] [USING [OUTER] <ALGORITHM(name)>ON<table>.<attribute>]
```

CrowdIQL introduces the keyword **SHOWING** to prompt the crowd with a few representative data instead of the whole table. Another keyword **USING** is designed to provide candidate items for crowdsourcing tasks by using machine algorithms. With the keyword **OUTER**, CrowdIQ is scalable and new functions could be put into service in a plug-and-play way.

Example 1: The command asks for recovering the third column label of the table in Fig. 2, while prompting the crowd with data only in the third column.

```
SELECT table.headers[2] SHOWING table.columns[2];
```

Example 2: The command asks for entity column (the column that contains entities) of the table while providing candidate items for crowdsourcing tasks by *entity_find* Function. Before that, attribute *entity_column* should be added into the JSON table by **INSERT**.

```
INSERT table.entity_column;
SELECT table.entity_column USING ALGORITHM(entity_find)
       ON table.columns;
```

4 Demonstration Overview

We plan to demonstrate CrowdIQ with some running examples, which were illustrated in Sect. 3. First, Quality Inspection module begins to use Probase to understand the pending table (e.g. "whether the table has headers?") and find problems about data format and structural integrity as in Fig. 2. Then the requester is allowed to write CrowdIQL to solve the problems for improving the quality of the web table. Optional functions in machine preprocessing and data minimization will be introduced by CrowdIQL to optimize tasks. The effectiveness of optimization in data size and question form would be illustrated by comparing the optimized task and the original one. UI templates of HITs could be edited to define the request more exactly. After finishing the tasks, the complete and high-quality table produced by CrowdIQ is available.

Acknowledgment. This work is supported by National Natural Science Foundation of China (Grant No. 61370060). We would also like to give our thanks for the support from Microsoft Research Asia.

References

1. Amazon Mechanical Turk. http://www.mturk.com
2. Park, H., Widom, J.: Collecting structured data from the crowd. In: SIGMOD (2014)
3. Liu, H., Wang, N., Ren, X.: CrowdSR: a crowd enabled system for semantic recovering of web tables. In: Dong, X.L., Yu, X., Li, J., Sun, Y. (eds.) WAIM 2015. LNCS, vol. 9098, pp. 581–583. Springer, Cham (2015). doi:10.1007/978-3-319-21042-1_67

4. Franklin, M.J., Kossmann, D., Kraska, T., et al.: CrowdDB: answering queries with crowdsourcing. In: SIGMOD (2011)
5. Parameswaran, A., Park, H., et al.: Deco: declarative crowdsouring. In: ACM International Conference on Information and Knowledge Management (2012)

OICPM: An Interactive System to Find Interesting Co-location Patterns Using Ontologies

Xuguang Bao, Lizhen Wang[✉], and Qing Xiao

Department of Computer Science and Engineering, School of Information
Science and Engineering, Yunnan University, Kunming 650091, China
lzhwang@ynu.edu.cn

Abstract. In spatial data mining, the usefulness of co-location patterns is strongly limited by the huge amount of delivered co-location patterns. Although many methods have been proposed to reduce the number of co-location patterns, most of them do not guarantee that the extracted co-location patterns are interesting for the user for being generally based on statistical information. This demonstration presents OICPM, an interactive system to discover interesting co-location patterns based on ontologies. With OICPM, the user can find his/her real interesting ones from a massive amount of co-location patterns efficiently within only a few rounds of selection, and the mined interesting co-location patterns are filtered in order for better decision.

Keywords: Spatial data mining · Co-location pattern mining · Interesting patterns · Interactive · Ontology

1 Introduction

Co-location pattern mining is a new branch studied in the spatial data mining recently. A spatial co-location pattern represents a subset of spatial features whose instances are frequently located in spatial neighborhoods. Spatial co-location patterns may yield important insights in many applications, such as Earth science, public health, biology, transportation, etc. [1].

Discovering interesting co-location patterns is an important task in spatial data mining. Most existing co-location pattern mining methods focus on efficiently computing co-location patterns which satisfy a pre-specified criterion. The massive amount of co-location patterns presented make the user confused. Thus, several methods were proposed in the literature in order to overcome this drawback such as co-location pattern concise representations [2, 3], redundancy reduction [4], etc. However, most of these methods do not guarantee that the extracted co-location patterns are interesting for the user, since pattern interestingness strongly depends on user knowledge and goals while most of these methods are generally based on statistical information.

In the Semantic Web field, ontology is considered as the most appropriate representation to express the complexity of the user knowledge. Thus, several works were proposed to perform co-location mining using ontologies. [5] described a system called

L. Chen et al. (Eds.): APWeb-WAIM 2017, Part II, LNCS 10367, pp. 329–332, 2017.
DOI: 10.1007/978-3-319-63564-4_29

OICRM to mine interesting co-location rules, and [6] gave a framework to mine interesting co-location patterns using ontologies.

OICPM (Ontology-based Interesting Co-location Pattern Miner) is a system to discover interesting co-location patterns based on [6]. Given a set of candidates (i.e., prevalent co-location patterns), our goal is to help a particular user interactively discover interesting co-location patterns according to his/her real interests. Instead of requiring the user to explicitly express his/her real interesting co-location patterns, we alleviate the user's burden by only asking him/her to choose a small set of sample co-location patterns according to his/her interest. OICPM designs an interactive process to find interesting co-location patterns and then has a filter to reduce the number of co-location patterns.

2 System Overview

Figure 1 shows the description of OICPM. OICPM takes a set of candidates as input. The candidates can be prevalent co-location patterns, closed co-location patterns [2], maximal co-location patterns [3] or other forms of co-location patterns. OICPM contains an interactive process (step 1–4) to get all the interesting co-location patterns and a filter to reduce the number of result set (step 5). Each step is described as below.

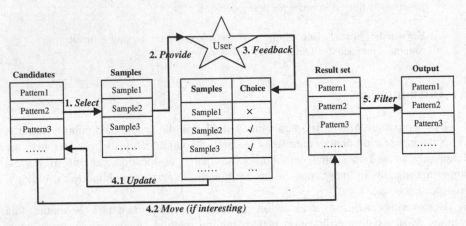

Fig. 1. Framework description

Step 1: A small collection (e.g., 10) of sample co-location patterns is selected from the candidates. The number of samples can be decided by the user.

Step 2: OICPM provides the user with the sample co-location patterns. The selection strategy is a greedy algorithm to find the most different co-location patterns [6].

Step 3: The user decides his/her interestingness on each sample co-location pattern and feedback his/her preference.

Step 4: This step contains 2 sub steps. The feedback information will be used to update the candidates (step 4.1), the update operation is based on the distance defined on two co-location patterns using ontologies [6]. During the updates, if one or more co-location patterns are determined interesting using ontologies, they will be moved from candidates to the result set (step 4.2). Note that the volume of candidates will be gradually reduced when updating the candidates. While finishing the operation of update, if the set of candidates becomes empty, which means that OICPM has found all the interesting co-location patterns, the interactive process will be finished and the filter process will start, otherwise, OICPM will go back to step 1 to continue the interaction (step1–step 4).

Step 5: A filter based on the down closure property of the prevalence of co-locations is designed to get the interesting co-location patterns from the result set.

3 Demonstration Scenarios

OICPM is well encapsulated with a friendly interface, what the user faces is only a simple user interface. In this demonstration, we use part of the data from points of interests (POI data) in Beijing to show the demonstration of OICPM.

Figure 2 shows the interface of OICPM in the interaction process. Figure 2(a) shows the main function of OICPM, an XML file with the description of ontology and a text file with candidates are required. The status bar shows the number of samples provided for the user which can be modified by the user. The *visualize* button shows the visualization of ontology as a tree structure shown in Fig. 2(b). Given the ontology description file and the candidates file, the interaction process can be performed once pressing the *run* button, if the user wants to interrupt the current process, the *stop* button can do it. Figure 2(c) shows the runtime information including the current operation, the time cost and the number of current candidates. Each round OICPM selects sample co-location patterns provided for the user. Figure 2(d) shows the interaction interface which requires the user to select his/her interesting sample co-location patterns.

Figure 3 shows the result interface after the interactive process. We can see that after the interactive process, the number of interesting co-location patterns is 642, while

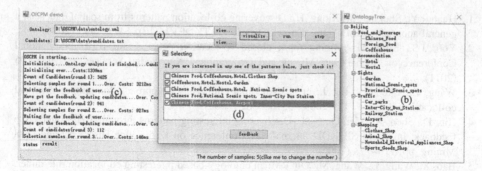

Fig. 2. Interface of OICPM in the process of interaction

a filter is performed on the result set, the final number of concise interesting co-location number is only 21, and the compression ratio is as high as 99.4%, which demonstrates

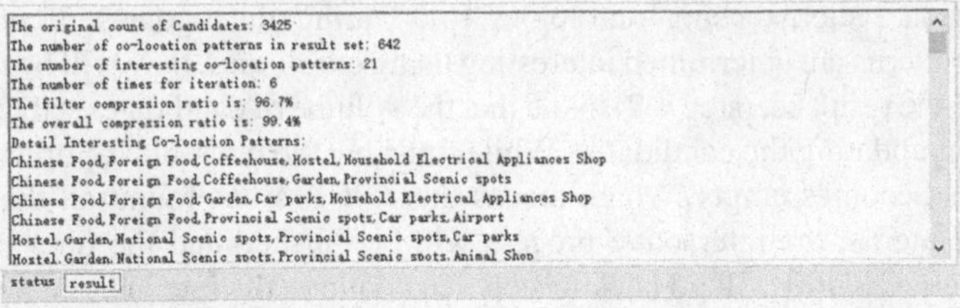

Fig. 3. Result interface of OICPM

the high efficiency of OICPM. With the number of features increases, the amount of prevalent co-locations is too massive for the user to make his/her decision, with OICPM, the user can efficiently get his/her real interesting co-location patterns in huge amount of co-location patterns.

4 Conclusion

In many co-location mining approaches, the algorithm or the measure is pre-designed and the user accepts the results passively. In this demonstration, we designed a system to find interesting co-location patterns interactively with the user. The demonstration scenarios showed the effectiveness and efficiency of our system.

Acknowledgements. This work was supported in part by grants (No. 61472346, No. 61662086) from the National Natural Science Foundation of China, by grants (No. 2016FA026, No .2015FB114) from the Science Foundation of Yunnan Province and by the Spectrum Sensing and borderlands Security Key Laboratory of Universities in Yunnan (C6165903).

References

1. Huang, Y., Shekhar, S., Xiong, H.: Discovering co-location patterns from spatial data sets: a general approach. IEEE Trans. Knowl. Data Eng. (TKDE) **16**(12), 1472–1485 (2004)
2. Yoo, J.S., Bow, M.: Mining top-k closed co-location patterns. In: IEEE International Conference on Spatial Data Mining and Geographical Knowledge Services, pp. 100–105 (2011)
3. Wang, L., Zhou, L., Lu, J., et al.: An order-clique-based approach for mining maximal co-locations. Inf. Sci. **179**(2009), 3370–3382 (2009)
4. Xin, D., Shen, X., Mei, Q., et al.: Discovering interesting patterns through user's interactive feedback. In: ACM SIGKDD International Conference on Knowledge Discovery and Data Mining, pp. 773–778 (2006)
5. Bao, X., Wang, L., Wang, M.: OICRM: an ontology-based interesting co-location rule miner. In: Asian Pacific Web Conference, pp. 570–573 (2016)
6. Bao, X., Wang, L., Chen, H.: Ontology-based interactive post-mining of interesting co-location patterns. In: Asian Pacific Web Conference, pp. 406–409 (2016)

BioPW: An Interactive Tool for Biological Pathway Visualization on Linked Data

Yuan Liu[1,3], Xin Wang[1,2,3(✉)], and Qiang Xu[1,3]

[1] School of Computer Science and Technology,
Tianjin University, Tianjin, China
{liuyuan0416,wangx,xuqiang3}@tju.edu.cn
[2] State Key Laboratory for Novel Software Technology,
Nanjing University, Nanjing, China
[3] Tianjin Key Laboratory of Cognitive Computing and Application,
Tianjin, China

Abstract. With the development of Linked Data, large amounts of biological pathway RDF data have been published on the semantic Web. To make these various datasets available to life scientists, we demonstrate an interactive tool, called BioPW, for the biological pathway Linked Data visualization. In contrast to showing the biological pathway data merely from one dataset, our tool, with the Open PHACTS Linked Data API and users' exploratory interaction, could clearly illustrate the biological pathways with their associated information from multiple perspectives of linked datasets.

Keywords: Linked data · Biological pathways · Visualization · Open PHACTS

1 Introduction

Linked Data is an approach to publishing structured data on the semantic Web [3]. This technology enables that data from different heterogenous sources can globally be well-integrated and better consumed.

Since Linked Data initiated, large volumes of biological pathway data have been published on the Linked Open Data (LOD) platform [1]. However, there are currently few tools for life scientists to retrieve and visualize the biological pathways in the Linked Data scenario, which basically means that the linked biological pathway data have not been fully consumed. A significant problem in the process of utilizing the biological Linked Data is about the integration of data from various heterogenous sources. As an essentially important part of biological Linked Data, biological pathway data consist of proteins, genes, metabolites, etc. Most of them are stored separately in different datasets and disconnected. It is an inconvenient and time-consuming task to search the different datasets manually. For example, Fig. 1 depicts the biological pathway *amino acid conjugation of benzoic acid*. To get all the information of the substances involved in

© Springer International Publishing AG 2017
L. Chen et al. (Eds.): APWeb-WAIM 2017, Part II, LNCS 10367, pp. 333–336, 2017.
DOI: 10.1007/978-3-319-63564-4_30

this biological pathway, life scientists have to spend much time on learning how to use tens of the datasets and collecting the information from them. In order to get the information of *AMP (adenosine monophosphate)*, for instance, which is a nucleotide that is used as a monomer in RNA, life scientists have to access WikiPathways, ChemSpider, DrugBank, HMDB, etc., to fetch the information.

Open PHACTS [2] is a European project that contributes to creating an integrated open pharmacological space (OPS) to promote the drug discovery research. Open PHACTS brings together the life science Linked Data sources, such as WikiPathways, ChEMBL, and so on. With Open PHACTS, life scientists have access to all of these Linked Data in a single and integrated platform.

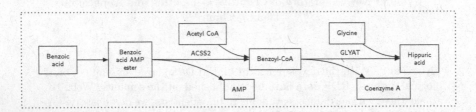

Fig. 1. The biological pathway *amino acid conjugation of benzoic acid*

However, the Open PHACTS platform only offers the Linked Data API to search for the biological pathway RDF data, which is still quite inconvenient to the life scientists. To this end, we have developed an interactive query and visualization tool, called BioPW, which can query and show the biological pathway Linked Data that is from multiple datasets step by step with users' exploratory interaction from multiple perspectives. To the best of our knowledge, BioPW is the first visualization tool that can interactively show the biological pathway Linked Data based on the Open PHACTS Linked Data API.

2   Demonstration

In this section, we present the demonstration scenarios by the following use cases.

Use case 1: Showing the biological pathway Linked Data query results. Input the biological pathway keywords and then click the search button. The visualization of the biological pathway results is shown in Fig. 2(a), which depicts the query result and the interactive visualization forms of a biological pathway. There are three parts on our user interface: (1) Part (i) is a summary for the biological pathway, including the name and description; (2) Part (ii) is a pie chart with a table that depicts the distribution of the substance types in this biological pathway and all the substances involved in it; (3) Part (iii) is the biological pathway graph representation. The nodes in Part (iii) represent the substances and the directed edges represent the interactions between them. To classify the different type substances in the biological pathway, we use different node colors to represent different type substances, for example, the light blue and orange nodes representing proteins and gene products, respectively.

Use case 2: Quickly locating the substance node in the biological pathway graph. By clicking the substance bar in the table, the mouse can locate to the node that represents the substance, as shown in Fig. 2(b). This function can be quite necessary especially when the graph is large.

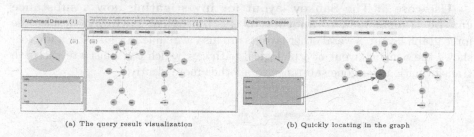

(a) The query result visualization   (b) Quickly locating in the graph

Fig. 2. Querying a biological pathway on Linked Data by using BioPW (Color figure online)

Use case 3: The visualization graph adjustment. The functionalities that our tool offers to adjust the visualization graph include: (1) users can adjust the distance and the charge force between nodes as they specify; (2) and the nodes can be dragged into any position, where they can remain fixed. Double clicking a node can unstick it. (3) The zoom function is also offered to zoom in and zoom out on the biological pathway graph for exploring the whole picture or parts of it. Figure 3 shows the process of adjusting the biological pathway graph.

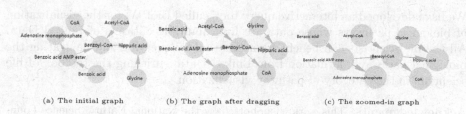

(a) The initial graph   (b) The graph after dragging   (c) The zoomed-in graph

Fig. 3. The visualization graph adjustment

Use case 4: Browsing the Linked Data from multiple datasets associated to the biological pathways. As shown in Fig. 4(a), the node can be highlighted and the tip box about the substance can be exhibited when the mouse is hovered over the node. And in the tip box, we can click the "More information" link to dive into the detailed information about the substance that the node represents. Figure 4(b) shows the detailed information about the substances that are involved in the biological pathway. As shown in Fig. 4(b), several properties

336 Y. Liu et al.

of *AMP* are listed that includes the structure image of the compound, the *Molecular formular*, *SMILES*, *Mol Weight*, etc. Furthermore, the dataset logos are used to indicate the data provenance of the properties, which means that the structure image is from WikiPathways, the *Molecular formular* is from ChemSpider, HMDB, DrugBank, and ChEMBL, and the *Mol Weight* information is from ChemSpider. By clicking the dataset logos, the Web pages in different data sources can be visited.

Use case 5: The Sankey layout for investigating how a substance can be transformed to another in the biological pathway. Besides the force graph layout, several other layouts are also available in BioPW. Figure 4(c) shows the Sankey layout of a biological pathway, which can illustrate the path transformation from one substance to another more clearly.

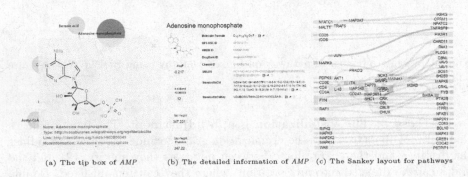

(a) The tip box of *AMP*   (b) The detailed information of *AMP*   (c) The Sankey layout for pathways

Fig. 4. Advanced features of BioPW

3 Conclusion

We have developed an interactive query tool, called BioPW, for the visualization of biological pathways whose data are from multiple datasets of Linked Data. All the life scientists can benefit from this tool by querying and navigating the biological pathway information from Linked Data, enriching the toolbox of life scientists in the category of pathway visualization.

Acknowledgments. This work is supported by the National Natural Science Foundation of China (61572353), the Natural Science Foundation of Tianjin (17JCY-BJC15400), and the Open Fund Project of State Key Lab. for Novel Software Technology (Nanjing University) (KFKT2015B20).

References

1. Andrejs Abele, J.P.M.: Linking open data cloud diagram. LOD Community (2017). http://lod-cloud.net/, 12
2. Gray, A.J.G.: Groth: Applying linked data approaches to pharmacology: Architectural decisions and implementation. Semant. Web **5**, 101–113 (2012)
3. WIKIPEDIA: Linked data. https://en.wikipedia.org/wiki/Linked_data

ChargeMap: An Electric Vehicle Charging Station Planning System

Longlong Xu[1], Wutao Lin[1], Xiaorong Wang[2], Zhenhui Xu[1], Wei Chen[1(✉)],
and Tengjiao Wang[1]

[1] Key Lab of High Confidence Software Technologies (MOE),
School of EECS, Peking University, Beijing 100871, China
{xllsniper,wtlin,pekingchenwei,tjwang}@pku.edu.cn,
xuzhpku@163.com
[2] Technology and Strategy Research Center,
China Electric Power Research Institute, Beijing 100192, China
xrwang@epri.sgcc.com.cn

Abstract. The deployment optimization of charging stations is meaningful for the promotion of electric vehicles. The traditional approaches
of charging station planning are mostly lack of comprehensive consideration. Some factors cannot be fully considered and effectively quantified in
those approaches, such as the load of power grid, charging demand, transportation cost, construction cost, etc. This demo presents ChargeMap,
a novel system of electric vehicle charging station planning which based
on a multi-factor optimization model. ChargeMap could attain a balance
among the factors that have great influence on charging station planning.
What's more, it delivers an apropos approach to quantify these factors.
In this demo, we bring forth the application of ChargeMap on the real
data sets of power grid, population, transportation and real estate of
Beijing, which delivers an effective solution to the optimization of charging station planning.

Keywords: Electric vehicle · Optimization · Charging station planning

1 Introduction

In the backdrop of alleviating pressure from oil resource shortage and environmental pollution across the globe, the electric vehicles (EVs) are garnering
increasing amount of attention. One of the factors that significantly impact the
growth of the EV market is access to public charging stations [2]. However, how
to strategically deploy the charging stations is a difficult task for urban planners
and electric utility companies [4].

Charging station planning is an intricate optimization issue which needs to
jointly consider the multiple factors such as the power grid, urban population,
economic benefits, charging demand, transportation, etc. Many approaches [1,3]
have been proposed to counter this issue, but most of them are lack of comprehensive consideration and could not perform efficiently for the quantification of
the factors above.

© Springer International Publishing AG 2017
L. Chen et al. (Eds.): APWeb-WAIM 2017, Part II, LNCS 10367, pp. 337–340, 2017.
DOI: 10.1007/978-3-319-63564-4_31

In this demo, we throw light on ChargeMap, a novel system which based on a multi-factor model for charging facility planning (MF-CFP). The goal of it is to minimize the overall cost of charging station placement in order to find an optimization planning scheme. ChargeMap attains a balance among the factors that have great influence on charging station planning, such as the load of power grid, traffic hot spots, frequent trajectories, charging demand, transportation cost, construction cost, etc. Furthermore, it delivers an apropos approach to quantify these factors. In this demo, we bring forth the application of ChargeMap on the large-scale data sets of power grid, urban population, transportation, Points-of-Interest (POIs) and real estate of Beijing. We also deliver several noteworthy features and an instance of charging station placement optimization.

2   System Architecture

Figure 1 shows the architecture of ChargeMap. Initially, the map is divided into geographical grids, each grid could be a candidate address of charging stations. Secondly, various helpful features are generated from the data sets of taxi trajectory, power grid, population, real estate and POIs. Following it, the generated features are integrated to the MF-CFP model. Ultimately, the optimization placement of charging stations is shown by using the visualization model. Considerable modules of the proposed architecture are detailed below.

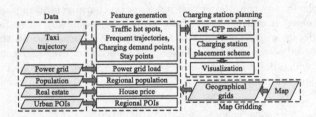

Fig. 1. System architecture

2.1   Feature Generation

The MF-CFP model feature set consists of traffic hot spots, frequent trajectories, charging demand points, stay points, power grid load, population, house price and POIs. All of the features should be normalized to the range [0, 1] through Min-Max Normalization. Three most important features are detailed below.

Traffic hot spots: DBSCAN is adopted to cluster the patterns containing longitude, latitude and time of each journey's end point, thereby attaining the traffic hot spots during a particular time period.

Charging demand points: Using the trajectories of fuel-driven taxis to simulate the trajectories of EVs. It is assumed that the car should charge itself when the energy is 15% left, the location from which the car heads for the charging station is regarded as a charging demand point.

Stay points: The points where the taxi stays for over half an hour (e.g., parking space of company or housing area) are extracted. Each of these points is termed to be the stay point.

2.2   Charging Station Planning

Charging station planning has two main requirements: (1) the charging demand should be fulfilled, which means the number of frequent trajectories, traffic hot spots, stay points and POIs near the charging stations should be maximized. (2) the average transportation cost and construction cost should be minimized. We establish the MF-CFP model to solve this issue, as presented in Eq. (1).

$$min : \alpha \frac{1}{D} \sum_{G_i \in G} \sum_{G_j \in G} D_i X_{ij} C_{ij} + \beta Y P + \gamma Y A + \mu Y O +$$
$$\phi Y S + \eta Y L + \lambda Y M + \varphi Y K \tag{1}$$

Where D, G, P, A, O, S, L, M, and K respectively represent the charging demand points, all of the N grids, house price, population, the number of POIs, the number of stay points, load of power grid, the number of traffic hot spots and the number of frequent trajectories within the grid. If $Y_i = 1$, it means that a charging station should be constructed in G_i. C, X are $N \times N$ matrices, C_{ij} denotes the time needed from G_i to G_j. If $X_{ij=1}$, it means that the vehicles in G_i should go to G_j for charging.

The goal is to minimize the overall cost of station placement (namely the value of Eq. (1)), then the vector Y attained from the solution is the optimization planning scheme of charging stations.

3   Demonstration Scenarios

Using Beijing as a case study, this demo employs a large-scale data set containing 1.5 billion trajectories from 30,270 taxis in one month, 570 thousand POIs and 353 thousand real estate records to illustrate two noteworthy features and an example scheme of charging station planning.

Figure 2(a) shows the feature of traffic hot spots during the evening peak. Each hot spot is represented by an orange circle. From this figure, it can be observed that during the evening peak, the traffic hot spots are concentrated in the traffic hubs, downtown areas and large residential areas. Then the location of large-scale EV charging stations can be considered in these places.

Figure 2(b) highlights the feature of frequent trajectories between Guomao and Capital Airport during the morning peak. The blue line denotes the frequent traffic trajectory, and the red arrow is an indication of the traffic direction. Then new fast charging stations can be constructed along the roads with frequent traffic trajectories.

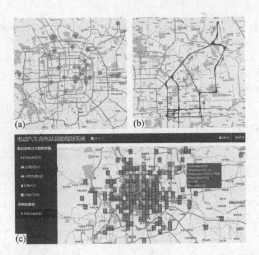

Fig. 2. Demonstration of ChargeMap. (a) The feature of traffic hot spots (b) The feature of frequent trajectories (c) Charging station planning scheme

Figure 2(c) shows the front end of ChargeMap. Different functional pages can be opened by clicking the options at the left. An instance of charging station planning is provided at the right. The blue grids with transverse lines represent existing charging stations, and the red grids with oblique lines indicate the newly planned charging stations. A gray box will pop up after placing the mouse on the grid to bring forth the details of the station.

Acknowledgments. This work was supported by State Grid Basic Research Program (DZ71-15-004).

References

1. Cai, H., Jia, X., Chiu, A.S., Hu, X., Xu, M.: Siting public electric vehicle charging stations in beijing using big-data informed travel patterns of the taxi fleet. Transp. Res. Part D: Transp. Environ. **33**, 39–46 (2014)
2. Han, D., Ahn, Y., Park, S., Yeo, H.: Trajectory-interception based method for electric vehicle taxi charging station problem with real taxi data. Int. J. Sustain. Transp. **10**(8), 671–682 (2016)
3. Li, Y., Luo, J., Chow, C.Y., Chan, K.L., Ding, Y., Zhang, F.: Growing the charging station network for electric vehicles with trajectory data analytics. In: 2015 IEEE 31st International Conference on Data Engineering (ICDE), pp. 1376–1387. IEEE (2015)
4. Vazifeh, M.M., Zhang, H., Santi, P., Ratti, C.: Optimizing the deployment of electric vehicle charging stations using pervasive mobility data. arXiv preprint arXiv:1511.00615 (2015)

Topic Browsing System for Research Papers Based on Hierarchical Latent Tree Analysis

Leonard K.M. Poon[1(✉)], Chun Fai Leung[2], Peixian Chen[2],
and Nevin L. Zhang[2]

[1] Department of Mathematics and Information Technology,
The Education University of Hong Kong, Hong Kong SAR, China
kmpoon@eduhk.hk

[2] Department of Computer Science and Engineering, The Hong Kong University
of Science and Technology, Hong Kong SAR, China
cfleungac@connect.ust.hk, {pchenac,lzhang}@cse.ust.hk

Abstract. New academic papers appear rapidly in the literature nowadays. This poses a challenge for researchers who are trying to keep up with a given field, especially those who are new to a field and may not know where to start from. To address this kind of problems, we have developed a topic browsing system for research papers where the papers have been automatically categorized by a probabilistic topic model. Rather than using Latent Dirichlet Allocation (LDA) for topic modeling, we use a recently proposed method called hierarchical latent tree analysis, which has been shown to perform better than some state-of-the-art LDA-based methods. The resulting topic model contains a hierarchy of topics so that users can browse topics at different levels. The topic model contains a manageable number of general topics at the top level and allows thousands of fine-grained topics at the bottom level.

1 Introduction

New academic papers appear rapidly in the literature nowadays. This makes a good contribution to the acquisition of knowledge but poses a challenge for researchers trying to keep up with a given field. The problem may be even worse for postgraduate students who are new to a field and may not know where to start from. Researchers usually use keywords to search for related papers using a search engine. They may then group related papers together to find out the main topics in the field. This process can be time-consuming.

The above approach can be regarded as a bottom-up approach. In contrast, a top-down approach would start with topic hierarchy. Researchers can then pick a general topic and drill down to find more specific topics. Papers related to any of the topics can be presented to the researchers when requested.

To allow the top-down approach, traditionally a taxonomy has to be defined manually. Papers are then be categorized manually according to the taxonomy. There are two issues with the traditional method. First, it requires much effort. Second, the topics in the taxonomy may not be able to keep up with latest

© Springer International Publishing AG 2017
L. Chen et al. (Eds.): APWeb-WAIM 2017, Part II, LNCS 10367, pp. 341–344, 2017.
DOI: 10.1007/978-3-319-63564-4_32

0.015 mikolov word-representation word-embedding compositionality socher word-vector distribute-representation

0.019 word-representation mikolov

0.012 word-embedding socher word-vector

0.008 compositionality distribute-representation word-phrase

0.048 word2vec dean

Fig. 1. A latent tree model (left) and the topic hierarchy extracted from it (right).

development. To address those issues, we propose a topic browsing system using a recent method for automatically building a topic tree and categorizing papers.

2   Hierarchical Latent Tree Analysis

Topic models are often used to categorize documents automatically. They can detect topics from a collection of documents and classify each document according to the detected topics. We have recently developed a topic modeling method called *hierarchical latent tree analysis* (HLTA) [3]. Unlike Latent Dirichlet Allocation [2], HLTA produces a hierarchy of topics. Our recent work [3] has also shown that HLTA produces topics and topic hierarchies of better quality than two state-of-the-art LDA-based methods, namely hLDA [1] and nHDP [5].

HLTA uses a class of tree-structured probabilistic graphical models called *latent tree models* (LTMs) [4]. Figure 1 shows an example of LTM for topic modeling. In the model, the leaf nodes (unframed nodes) represent observed word variables $\boldsymbol{W}$, whereas its internal nodes (oval nodes) represent unobserved topic variables $\boldsymbol{Z}$. All variables are binary. Each word variable $W \in \boldsymbol{W}$ indicates the presence or absence of the corresponding word in a document. Each topic variable $Z \in \boldsymbol{Z}$ indicates whether a document belongs to the corresponding topic.

An LTM can be regarded as a Bayesian network by rooting at one of its latent nodes. Let $pa(X)$ be the parent of a variable X. Then the LTM defines a joint distribution over all observed and latent variables as follows: $P(\boldsymbol{W}, \boldsymbol{Z}) = \prod_{X \in \boldsymbol{W} \cup \boldsymbol{Z}} P(X|pa(X))$.

Denote a document by $d = (w_1, \ldots, w_M)$, where w_i is the observed value of word variable $W_i \in \boldsymbol{W}$. Whether a document d belongs to a topic $Z \in \boldsymbol{Z}$ can be determined by the probability $P(Z|d)$. The LTM gives a *multi-membership model* since a document can belong to multiple topics. And unlike in LDA, the topic probabilities $P(Z|d)$ in LTM do not necessarily sum to one.

3   System Overview

We have developed a topic browsing system for research papers where the papers can be automatically categorized by a topic model built with HLTA. At the time of writing, the system has categorized 24,307 papers in the field of artificial intelligence published in 7 major conferences and 3 journals from 2000 to 2017.

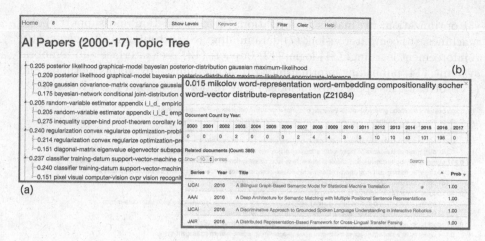

Fig. 2. (a) Home of topic browsing system. (b) Papers for a selected topic.

The resulting topic model contains a hierarchy of topics. The top level contains a manageable number of general topics and the bottom level contains thousands of fine-grained topics.

The home page of our topic browsing system displays a hierarchy of topics (Fig. 2a). When a topic node is clicked, a list of papers belonging to that topic is displayed (Fig. 2b). The system is built as a Node.js web application. Papers information is stored in a MongoDB database and is loaded through Ajax.

To prepare the data, papers are downloaded from related proceedings and journal websites. The text is preprocessed with word lemmatization, word normalization, and removal of stop words. Then, text is converted to data with bag-of-words representation. We consider n-grams, where $1 \leq n \leq 4$. We use the 10,000 n-grams with highest TF-IDF and appearing in less than 25% of papers.

The converted data is used as input to HLTA, resulting in an LTM. The number of levels and the number of latent variables in each level are automatically determined. A topic hierarchy is extracted based on the tree structure of the LTM, with each internal node representing a topic. Each topic is then characterized by those of its descendent words with highest MI with the topic. Figure 1 shows an example of LTM and the hierarchy extracted from it.

The LTM is also used to classify papers according to the topics detected. A paper d is assigned to topic Z if $P(Z = 1|d) > 0.5$. Besides, the size of each topic Z is estimated by the marginal probability $P(Z)$. In the topic hierarchy of Fig. 1, the number on each row indicates the topic size.

The code for processing is available at https://github.com/kmpoon/hlta.

4   Demonstration

The upper part of the hierarchy in our system is shown on the next page. The topics include: (1) graphical models and Bayesian methods; (2) statistical analysis;

(3) optimization; (4) matrix factorization; (5) classification and support vector machines; (6) computer vision; (7) data mining; (8) text mining; (9) agents; (10) reinforcement learning; (11) logic; (12) impact; (13) user interface and temporal systems; (14) publishing information; (15) satisfiability; (16) kernel methods; and (17) solvers. The upper part includes many major topics and looks reasonable.

(1) 0.205 posterior likelihood graphical-model bayesian posterior-distribution gaussian maximum-likelihood
(2) 0.205 random-variable estimator appendix l_l_d_ empirical learning-research-submit sample-size
     0.240 regularization convex regularize optimization-problem gradient diagonal-matrix eigenvalue
(3)   0.214 regularization convex regularize optimization-problem gradient convex-optimization norm
(4)   0.151 diagonal-matrix eigenvalue eigenvector subspace principal-component-analysis diag high_dimensional
     0.237 classifier training-datum support-vector-machine class-label classification-accuracy classification-problem classification-task
(5)   0.240 classifier training-datum support-vector-machine class-label classification-accuracy classify classification-problem
(6)   0.151 pixel visual computer-vision cvpr vision recognition object-recognition
     0.188 document information-retrieval text mining datum-mining baseline corpus
(7)   0.161 mining datum-mining knowledge-discovery sigkdd-international-conference-knowledge sigkdd icdm data-mining
(8)   0.177 document text information-retrieval baseline corpus sigir precision-recall
     0.151 agent multi_agent initial-state agent-agent multiagent assume-agent optimal-policy
(9)   0.124 agent multi_agent agent-agent multiagent assume-agent multi_agent-system multiagent-system
(10)   0.098 initial-state reward optimal-policy transition reinforcement-learning policy markov-decision-process
(11)   0.168 logic logical propositional semantics negation predicate disjunction
     0.227 continue impact earlier effort back initially happen
(12)   0.229 continue impact earlier effort back initially happen
(13)   0.247 interface temporal people dynamic duration cognitive activity
     0.052 intelligence-www_elsevier_com_locate_artint front-matter e_mail-address elsevier-rights-reserve revised-form solver satisfiability
     0.052 intelligence-www_elsevier_com_locate_artint front-matter e_mail-address elsevier-rights-reserve revised-form satisfiability correspond-autho
(14)   0.052 intelligence-www_elsevier_com_locate_artint front-matter e_mail-address revised-form elsevier-rights-reserve correspond-author ailable
(15)   0.117 satisfiability clause assignment satisfiable constraint-satisfaction-problem assignment-variable literal
(16)   0.151 kernel smola kernel-method scho-lkopf vapnik kernel-function statistical-learning-theory
(17)   0.207 solver optimal-solution problem-instance problem-solve search-tree solve-problem search-algorithm

To browse the topics, one can go down the hierarchy (Fig. 2a) and click a topic node to show a list of related papers (Fig. 2b). The papers are sorted in descending order of membership as indicated by $P(Z|d)$. Links to the original papers can be accessed by clicking the titles in the paper list. Numbers of papers for each year are also shown in a table at the top. This shows the trend of the topic.

Our proposed system can be accessed from: https://ltm.eduhk.hk/papers/.

Acknowledgment. The work was supported by the Education University of Hong Kong under project RG90/2014-2015R and Hong Kong Research Grants Council under grants 16202515 and 16212516.

References

1. Blei, D.M., Griffiths, T.L., Jordan, M.I.: The nested Chinese restaurant process and Bayesian nonparametric inference of topic hierarchies. J. ACM **57**(2), 7:1–7:30 (2010)
2. Blei, D.M., Ng, A.Y., Jordan, M.I.: Latent Dirichlet allocation. J. Mach. Learn. Res. **3**, 993–1022 (2003)
3. Chen, P., Zhang, N.L., Poon, L.K.M., Chen, Z.: Progressive EM for latent tree models and hierarchical topic detection. In: Proceedings of the Thirtieth AAAI Conference on Artificial Intelligence (2016)
4. Chen, T., Zhang, N.L., Liu, T., Poon, K.M., Wang, Y.: Model-based multidimensional clustering of categorical data. Artif. Intell. **176**, 2246–2269 (2012)
5. Paisley, J., Wang, C., Blei, D.M., Jordan, M.I.: Nested hierarchical Dirichlet processes. IEEE Trans. Pattern Anal. Mach. Intell. **37**(2), 256–270 (2015)

A Tool of Benchmarking Realtime Analysis for Massive Behavior Data

Mingyan Teng[1], Qiao Sun[2], Buqiao Deng[2], Lei Sun[2], and Xiongpai Qin[3]([✉])

[1] Department of Mathematics, Bohai University, Jinzhou, China
tengmingyan@gmail.com
[2] Beijing Guodiantong Network Technology Co., Ltd., Beijing, China
{sunqiao,dengbuqiao,sunlei3}@sgitg.sgcc.com.cn
[3] Information School, Renmin University of China, Beijing, China
qxp1990@ruc.edu.cn

Abstract. With the increasing development of platforms for massive users, the amount of data generated from these platforms is rapidly increasing. A large number of big data analysis frameworks have been designed to analyze data generated from these platforms. This however requires a specific benchmark to evaluate the system performance. Today, realtime analysis (or streaming analysis) becomes a hot research topic of big data research. However, there is no special benchmark designed for such streaming analysis systems. This paper introduces a tool for evaluating the performance of such streaming analysis systems. Based on the scenario of e-commerce platforms, the benchmark tool is designed using a data generator with certain user models based on the user's habits in e-commerce platforms. A test suite is developed to be responsible for simulated mixed workloads for streaming analysis.

1 Introduction

The fast development of web-scale applications for massive users and its contribution to the economy is becoming increasingly prominent. In the meanwhile, much data has been generated by users, and it allows companies to better analyze and understand customer needs and market dynamics. However, the extracted value is limited by the speed of data processing. As a result, a large number of big data processing systems have emerged. They can analyze and deal with the massive historical data in data warehouse. However, the real-time performance of these systems is relatively poor. They often can not utilize real-time data to timely adjust the model. Therefore, in recent years, streaming analysis is more concerned by industry and academia. Among the top-level Apache projects, there are Apache Storm [1], Apache Flink [2], and Spark Streaming [3], designed for different granularity of streaming analysis.

Streaming analysis has a wide range of applications. For example, in e-commerce applications, the ability to make decisions using real-time sales data is the key to achieve profitability. This is particularly true in many monitoring-based applications, such as meteorological monitoring, geological monitoring,

© Springer International Publishing AG 2017
L. Chen et al. (Eds.): APWeb-WAIM 2017, Part II, LNCS 10367, pp. 345–348, 2017.
DOI: 10.1007/978-3-319-63564-4_33

and environmental monitoring, which require a more accurate and timely response to different situations, based on the analysis of real-time data in conjunction with historical data. Real-time big data analysis is becoming the core competitiveness of many e-commerce applications. Take Taobao as an example, it produces massive log data, including purchase time, purchase quantity, commodity prices and other order information, as well as information of customers such as age, occupation, address. The analysis of these data contribute to item recommendation, shop ranking, and real-time personalized recommendation.

A large number of big data systems have been developed in recent years, and they vary significantly. To select the right system to meet the demand, a more targeted benchmark is desired. At present, the application of real-time analysis in e-commerce is increasing, such as real-time monitoring the effect of promotional activities, real-time monitoring of user behaviors to adjust the strategy and so on. Therefore, a benchmark for streaming analysis over e-commerce data is necessary. This paper introduces a tool for evaluating the performance of streaming analysis systems. Based on the application scenario of e-commerce platforms, the benchmark tool is designed with some workloads. A data generator with user models is designed based on some typical user's habits in e-commerce platforms. A test suite is developed to be responsible for simulated mixed workloads.

2   Data Generator

A general streaming analysis system aims to give fast and accurate decisions by combining the newly generated streaming data with historical data. Therefore, the statistical monitoring of streaming data and the retrieval of the corresponding historical data have become two very important factors affecting the efficiency of streaming analysis. Our test benchmark is designed for the important aspects of data analysis as well as data retrieval in streaming analysis.

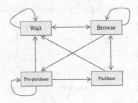

Fig. 1. Stages of user behavior modelling

Users in e-commerce platforms commonly generate the following kinds of behaviors: browse the goods, search for goods, collect goods, add goods to the shopping cart, remove the goods from the shopping cart, remove the goods from the list, buy goods, and provide comments. They can be further categorized into four phases: (1) search for goods and browse the goods for the browse stage;

(2) shopping cart operations for the pre-purchase stage; (3) purchase and comment for the purchase stage; and (4) the waiting (offline) stage. The transfers of these four phases can be roughly represented in Fig. 1.

The Fig. 1 shows examples of the transfers between the various stages of the user's behaviors, simulating the general process of online shopping in e-commerce applications. When a user logs in, he will first browse or search some products. As such, the wait stage can only transfer to the browse stage. The browse stage is the major stage, which has a high probability to transfer to the same stage as users may keep in browsing some products on the website. The browse stage can transfer to pre-purchase or purchase stages. For example, some users tend to compare more products, and they may be in the browse stage and will continue to add other goods to the shopping cart, and then select the goods from the shopping cart for payment. Some users are relatively decisive, and they will make the order after several visits. As the user will not necessarily buy goods, so every stage of operations can transfer to the wait stage. Different types of users can be classified by setting the transfer probability and delays between two stages. For user behavior modeling, we consider three factors: the frequency of visiting the online sites, the number of comparisons for each session, the user operations of each session, and then generate user behavioral data according to the category that a user falls in, based on a Markov chain model.

3   Workloads

The purpose of this benchmark is mainly for the application of realtime analysis in e-commerce platforms. Therefore, when designing the workloads, it is necessary to choose the corresponding workloads according to the application scenarios. The streaming monitoring workloads are mostly based on the data statistics within a time window, rather than the monitoring of discrete events. For performance testing using the workloads, we care more about the query throughput and the query response time, when dealing with mixed workloads.

(1) Real-time monitoring of merchandise sales. Real-time monitoring of the sales of goods within a time window, the query template is as follows: *select itemID from (select itemID, count(*) as time from log where operation = "purchas" group by itemID) as tmp where time ≥ threshold;* To obtain the purchase amount of all the goods in a time window, and retrieve items exceeding a given threshold: *select * from log as l join item as i on l.itemID = i.itemID where itemID = targetItemID and timeStamp ≥ startTime;*

(2) Real-time monitoring of user behaviors. Real-time monitoring of user behaviors within a period of time from the user's clickthrough logs, extract users frequently browsing the same category of goods within a period of time: *select UserID, category from (select UserID, category, count(*) as times from log as l join item as i on l.itemID = i.itemID where operation = "browse" group by UserID, category) tmp as temp where times ≥ threshold;*

(3) Real-time monitoring of the number of online users. It is to aggregate the number of online users in realtime. When the number of online users exceeds

some threshold, return the location name based on the aggregated results: *select province from (select province, count(distinct userID) as time form log as l join user as u on l.userID = u.userID group by province) as tmp where time ≥ threshold;*

(4) Monitor the attention received by goods. Return *itemID* whose access times within a time slot is more than some threshold: *select itemID from (select itemID, count(*) as time from log where operation ="browse" or operation = "search" group by itemID) as temp where time ≥ threshold;*

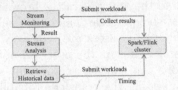

Fig. 2. Framework of performance benchmarking

4 Demonstration

In this demonstration, we will allow the attendees to adjust the percentage of different user groups so that they can generate datasets of different properties. By adjusting the parameters, we are also able to generate user behavior data in different throughputs. This allows us to select particular workload and test the throughputs of the systems. We will use the proposed workloads to test the performance of streaming analysis systems such as Spark streaming and Flink, following the framework shown in Fig. 2. The performance metrics will be shown to the attendees so that they can have a concrete feeling over the performance of the compared streaming systems.

Acknowledgements. This work is supported by Science and Technology Project of the State Grid Corporation of China (SGBJDK00KJJS1500180) and the State Grid Information & Telecommunication Group CO., LTD. (SGITG-KJ-JSKF[2015]0010).

References

1. storm.apache.org
2. Carbone, P., Katsifodimos, A., Ewen, S., Markl, V., Haridi, S., Tzoumas, K.: Apache flink™: stream and batch processing in a single engine. IEEE Data Eng. Bull. **38**(4), 28–38 (2015)
3. Chintapalli, S., Dagit, D., Evans, B., et al.: Benchmarking streaming computation engines: storm, flink and spark streaming. In: IPDPS Workshops 2016, Chicago, 23–27 May 2016, pp. 1789–1792 (2016)

Interactive Entity Centric Analysis of Log Data

Qiao Sun[1], Xiongpai Qin[2(✉)], Buqiao Deng[1], and Wei Cui[1]

[1] Beijing Guodiantong Network Technology Co., Ltd., Beijing, China
{sunqiao,dengbuqiao,cuiwei1}@sgitg.sgcc.com.cn
[2] Information School, Renmin University of China, Beijing, China
qxp1990@ruc.edu.cn

Abstract. Interactive entity centric analysis of log data can help us gain fine granularity insights on business. In this paper, firstly we describe a fiber based partitioning method for log data, which accelerate later entity centric analysis. Secondly, we present our fiber based partitioner which is used by Spark SQL query engine. Fiber based partitioner takes locations of data blocks into account when loading data from HDFS into RDD, and when shuffling data from upstream operators to downstream operators during joining, avoids data interchange between node and speeds up query processing. Finally, we present our experiment results which demonstrates that fiber based partitioner improve entity centric queries.

1 Entity Centric Analysis of Log Data

Log data contains valuable information for decision making. Timely and efficiently analyzing of log data can bring significant business value. For example, by analyzing log data of servers and applications, we can infer the root causes of failures. By analyzing log data of e-commerce sites, we can learn recent changes in browsing and purchasing behaviors of specific customers. Based on that, e-commerce sites can provide more personalized recommendations [1]. In these application scenarios, people need to perform interactive analysis on log data around some specific entities (customers, products, servers, applications etc.).

2 Fiber Based Log Data Partitioning

To facilitate entity centric analysis, we propose a fiber based partitioning method. A tuple of log data contains some information of one event about some entities. For example, in log data of an e-commerce site, each tuple describes an event about some specific customer and some specific product. In this scenario, customer and product are entities. Customer is a primary entity, and product is a secondary entity. Our discussion will center around primary entities. The treatment of secondary entities is similar to the way we process primary entities.

We organized entities into clusters, which are called entity fibers (fiber in short). The mapping from entity to fiber can have some sematic meaning, or we can simply use some Hash or Range function to map entities to fibers. For

© Springer International Publishing AG 2017
L. Chen et al. (Eds.): APWeb-WAIM 2017, Part II, LNCS 10367, pp. 349–352, 2017.
DOI: 10.1007/978-3-319-63564-4_34

example, in mobile communication applications, the call detail records could be partitioned according to calling intensity of different areas to which mobile phones are registered (home location of a the phone). Users of city areas can be split into several fibers, and users of rural areas can be combined into one fiber.

After entities are split into entity fibers, log records are split according to entity fibers. For example, user 1 and user 2 belong to fiber 1, user 3 and user 4 belong to fiber 2 etc. based on that, log records about user 1 and user 2 will go to the same partition - partition 1, and log records about user 3 and user 4 will go to another partition - partition 2 etc. Log data of each partition is organized in blocks. For example, block 11 and block 12, contain log records about user 1 and user 2 but occur at different time. In this paper, we use entity fibers to refer to three things, the fibers, corresponding log partitions, blocks of the partitions.

3   Loading of Log Data into HDFS

We load log data into HDFS for later running entity centric queries over the log data. When loading the log data into HDFS, we firstly partition log records according to above introduced fiber based partitioning method, and stage the data temporarily in a Kafka message queue. Log data of different fiber is written into different partitions of Kafka message queue.

Some log data loaders run on Data Nodes of HDFS. They pull log data from different Kafka partitions. Each loader is responsible for pulling and loading of log data of several fibers. For example, there are three Data Nodes in HDFS, loader 1 is responsible for loading of fiber 1 and fiber 2 log data into HDFS, loader 2 is responsible for loading of fiber 3 and fiber 4 into HDFS.... During loading, primary replica of a data block will be written to local disks, and other replicas are written to different nodes in the cluster.

Each loader launches several threads according to the number of fibers that it is responsible for. Each thread will pull data from one of partitions of the Kafka message queue. Each partition of Kafka contains log records of one fiber. When the total volume of the data accumulated by the threads reach a threshold of one data block (256MB), the loader organizes the temporary data of these threads into a data block. Inside each fiber, the log records are sorted by timestamp; then the fibers are concatenated, and saved into HDFS using the Parquet format.

After a data block is written into HDFS, we record some information about the block into one of meta data tables - the Block table. Several tuples are logged into the table according to number of fibers contained in the data block. Each tuple has the following information: data block id, fiber id, minimum time stamp of the fiber, maximum timestamp of the fiber, record count of the fiber, and the logical file name of the block in HDFS.

The mapping from fiber to Data Nodes is periodically readjusted to guarantee that the log data is dispersed onto the cluster. From the perspective of each fiber, before readjustment of the mapping, the primary replicas of blocks of the fiber are written to some Data Nodes. After readjustment of the mapping, the primary replicas of blocks of the fiber will be written to some new Data Nodes.

4   Loading of Log Data from HDFS into Spark RDD

After loading log data into HDFS using the fiber based partitioning method and registering some data into meta data tables. We run entity centric entity centric queries over the log data using Spark SQL. Spark is a rising tool for big data processing, and Spark SQL is a SQL query engine on Spark. Spark uses RDD (Resilient Distributed Data) to organized data. RDDs are read-only, partitioned data stores, which are distributed across many nodes on a cluster. Partitions of a RDD, scattering on cluster nodes, constitute a data table logically.

To query the log data using Spark SQL, the data should be loaded into in-memory RDDs first. If we rely on the default method of Spark to load the log data into RDDs, Spark will hash log records into different RDD partitions, which will incur much data transmit on the network.

We have design a fiber based partitioner, which can be used by Spark when loading data from HDFS into RDDs. An example is used to tell how the partitioner works. For example, for some specific time period, the loader has loaded some data into three nodes of n1, n2 and n3 according to fiber to node mapping. After a while, there are 10 data blocks containing fiber 1 and fiber 2 data on n1, and 13 data blocks containing fiber 3 and fiber 4 data on n2, and 11 data blocks containing fiber 5 and fiber 6 data n3. Since we readjust the mapping from fibers to nodes periodically, so there are also 2 data blocks containing fiber 3 and fiber 4 data on n1, 2 data blocks containing fiber 5 and fiber 6 data on n2, and 2 data blocks containing fiber 1 and fiber 2 on n3 etc.

During loading data by Spark from HDFS to RDDs, fiber based partitioner works as follows. Firstly, it uses the Block table to filter out some data blocks containing unrelated data, and partitions the data according to the information derived from statistics. For example, it finds that, for fiber 1, most of its data blocks resides on n1, and only a small fraction resides on n3, so it partitions the fiber 1 data to the RDD on n1. By the same principle, it partitions the fiber 2 data to the RDD on n1 etc. if Spark use the default partition method which does not considers the information, it may partition fiber 1 to RDD on n2, in such situation, the RDD should have loaded much data from other nodes (n1 and n3), which incur much network transmission.

5   Shuffling Data from Upstream to Downstream Operators in Join

In Spark SQL, there are three joining algorithms, i.e. Broadcast Join, Hash Join, and Sort Merge Join. When the data volume of both side of the joining are large Spark SQL resort to Hash Join or Sort Merge Join. The fiber based partitioner can also be used to replace the default data shuffling partitioner used between upstream and down-stream operators during join. We use Hash Join as an example to go into more details. For example, when joining some table to the log data using the hash join algorithm, Spark SQL will blindly hash tuples of both tables onto a set of nodes. On each node, joining can then be conducted locally. In our system, the log data, and the table (for example a User table and so on) to join to the log data, have been partitioned by a fiber based partitioning method and loaded into HDFS. Joining can be done locally without network transmission. However, Spark SQL does not leverage such information, the fiber based partitioner using in loading data from HDFS into RDDs, can be used when shuffling data from the User table RDDs and log data RDDs, to do later join operator.

6   Experiment Results

We have used TPC-H data to test our system. The volume of Lineitem table is 25 GB, and the size of Customer table is around 700 MB. We have run a single table query and a join query to test our method. The single table query first filter some data from Lineitem table using a date range and do some aggregation. The join query joins the Customer table and the Lineitem table on the cust_key, which is the key used in fiber based data partitioning.

The experiment result for the single table query is as follows, when the selectivity changes from 5%, to 10%, to 20% and to 50%, the run time of the query on our system changes from 7 s, to 10 s, to 15.5 s and 28.8 s. Since Spark SQL does not use query condition to filter out some data blocks, and blind loads data into in-memory RDDs, the run times are as high as around 55 s, no matter how low the selectivity is.

Runtimes of join query on our scheme compared to Spark SQL default setting is consistent with the result of the single table query as follows. When the selectivity changes from 5%, to 10%, to 20% and to 50%, the run time of the query on our scheme changes from 27 s, to 36 s, to 48 s and 69 s. On the other hand, the run time of the query on Spark SQL (default setting) is as high as around 128 s.

Acknowledgements. This work is supported by Science and Technology Project of the State Grid Corporation of China (SGBJDK00KJJS1500180) and the State Grid Information & Telecommunication Group CO., LTD. (SGITG-KJ-JSKF[2015]0010).

References

1. Xiongpai, Q.I.N., Guodong, J.I.N., Yang, L.I.U., Yiming, C., Xiaoyong, D.U.: Entity fiber based partitioning, no loss staging and fast loading of log data. In: PDCAT, pp. 199–203. IEEE Press, New York (2016)

A Tool for 3D Visualizing Moving Objects

Weiwei Wang and Jianqiu Xu[(✉)]

Nanjing University of Aeronautics and Astronautics, Nanjing, China
{wei,jianqiu}@nuaa.edu.cn

Abstract. Visually representing query results and complex data structures in a database system provides a convenient way for users to understand and analyze the data. In this demo, we will present a tool for 3D visualizing moving objects, i.e., spatial objects continuously changing locations over time. Instead of simply reporting numbers and lines, moving objects as well as dynamic attributes are animated and graphically displayed in an unified way. The tool benefits comprehensively understanding the spatio-temporal movement. Furthermore, we introduce how the tool provides powerful visual metaphors to explore the index structure such that one can fast determine whether the structure has a good shape, i.e., well preserving the spatio-temporal proximity. This is not a standalone software but a tool embedded in a database system.

1 Introduction

Data visualization in a database system aims to visually and graphically represent the query result for users to facilitate understanding and analyzing the data [5]. Some tools have been developed to visualize queries and the optimizer such as SILURIAN [3] and Picasso [8]. Due to a wide diversity of datasets such as spatial, graph and web, different visualization tools are essentially needed.

Recently, due to the widespread use of GPS-enabled devices such as mobile phones and car navigation systems, the recording of location data is convenient and a large amount of mobile data is collected [4]. Such data has been widely used in various applications such as geographic information system and traffic management. Although a lot of work has been conducted on indexing and querying moving objects [7], few efforts are made on visualizing the data and indexes, which are not easily animated and displayed by a simple query interface.

One common method is to assume that the objects move in a 2D space. Then, the data will be displayed in a 2D viewer by projecting the movement into spatial curves. The time dimension is missing or separately shown from spatial curves. The method is able to graphically display the spatial data but cannot well illustrate the relationship between spatial and temporal dimensions, and simultaneously visualize the data in different dimensions. Furthermore, the index built on moving objects such as 3D R-tree cannot be well displayed in a 2D viewer. In addition to location and time, moving objects can be associated with dynamic attributes such as *speed* and *direction*. Displaying dynamic attribute values as well as location and time benefits users because the viewer provides a comprehensive picture of moving objects.

© Springer International Publishing AG 2017
L. Chen et al. (Eds.): APWeb-WAIM 2017, Part II, LNCS 10367, pp. 353–357, 2017.
DOI: 10.1007/978-3-319-63564-4_35

This motivates us to develop a visualization tool for moving objects. The tool should have the functionalities: (i) comprehensively animating moving objects in both 2D and 3D spaces; (ii) graphically visualizing the 3D R-tree; (iii) showing the query path in the index. This benefits understanding the query procedure and enhancing the analysis for the optimization. Based on [10], we generalize 3D visualizing and animating moving objects to support not only indoor moving objects but also outdoor moving objects. The tool is developed in Java3D [1] and embedded in an extensible database system SECONDO [6]. The user manual is provided at [2] such that the tool will be available for other researchers.

The developed tool enhances the user interface and the visualization capability of moving objects databases because query results are accompanied by an appropriate representation of the content. Users can quickly justify the correctness of the result (e.g., similarity, nearest neighbors) and the goodness of the index structure rather than running some programs to analyze numeric values and texts. In the demonstration, we show how to use the tool to 3D visualize moving objects, dynamic attributes and index structures. The animation is provided such that one can clearly figure out how the data changes over time, which is an important functionality for displaying moving objects.

2   Overview

When a user sends a query in the system, the execution procedure goes to the kernel level to look for the data. Query results are sent from the kernel to the interface in which the proposed tool automatically transforms the data into a certain format accepted by Java3D. Figure 1 outlines the data flow in the tool.

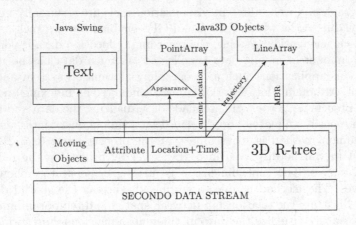

Fig. 1. The framework

In Java3D universe, two data types are used: PointArray and LineArray. PointArray, defined by 3D points, is used to transform location and time to

Java3D objects. LineArray is used to transform spatial lines and bounding boxes. We support animating moving objects in the 3D viewer in which the 3D axes correspond to time, latitude and longitude, respectively. Dynamic attributes such as speed and direction are displayed in the text form. The tool supports visualizing the 3D R-tree that is one of the most popular index structures for moving objects. One can clearly observe the distribution of tree nodes and analyze the locality and shape of the structure. This helps users determine whether the index well preserves spatial and temporal proximities.

3 Data Representation

In principle, moving objects are spatial objects continuously changing their locations over time, but they are also associated with characteristic attributes that should be represented in order to have comprehensive knowledge about real entities and further be queried by users. To enrich the data representation, we define that each moving object contains a set of dynamic attributes, represented by available temporal data types such as moving bool and moving int. We use the taxi trips as sample data, as listed in Table 1. Each taxi contains a sequence of time-stamped locations and three attributes: *Speed*, *Direction* and *Occupancy*.

Table 1. Taxi trips

Id	Trip:mpoint	Speed:mreal	Direction:mint	Occupancy:mbool
1	{ ([07:00:36, 07:01:41) (37.76, -122.43)(37.76, -122.43)); ([07:01:41, 07:02:36) (37.76, -122.43)(37.76, -122.43)); ...... ([07:49:45, 07:50:08) (37.75, -122.39)(37.75, -122.39))};	{ ([07:00:36, 07:01:41) 4.81); ([07:01:41, 07:02:36) 6.43); ...... ([07:49:45, 07:50:08) 1.28)};	{ ([07:00:36, 07:01:41) 180); ([07:01:41, 07:02:36) 258); ...... ([07:49:45, 07:50:08) 339)};	{ ([07:00:36, 07:24:52) 0); ([07:24:52, 07:35:43) 1); ...... ([07:35:43, 07:50:08) 0)};

A relation is defined in which the composite data type *mpoint* is provided and embedded in the schema. Dynamic attributes are also integrated into the relational interface. The advantage of employing an relational interface is that we can leverage relational operators to formulate queries. A comprehensive set of operators on spatio-temporal data types is provided in the system.

4 Demonstration

We use the real dataset cab mobility traces from [9] in the demonstration.

Displaying moving objects and dynamic attributes. The tool visually represents moving objects in a 3D space and provides the animation. Different colors are used to mark attribute values changing over time. For example, a taxi is occupied for a while and then free, as illustrated in Fig. 2a.

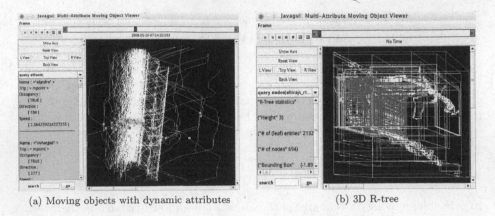

(a) Moving objects with dynamic attributes   (b) 3D R-tree

Fig. 2. 3D visualization

Visualizing R-tree. Due to the data distribution and the way of creating the index (e.g., bulk load), the structure will be in different shapes that significantly affect the query performance. Making a quick judgment about how well the structure is calls for not only analyzing the index statistics but also graphically viewing the structure. The tool distinguishes between leaf and non-leaf nodes, as shown in Fig. 2b The traversal path in the tree for a query can also be displayed, benefiting comparing different algorithms and queries.

A hybrid viewer. The tool is also able to 3D and 2D visualize moving objects in an unified viewer, as illustrated in Fig. 3. This helps users understand the data from different aspects as one can simultaneously see how the objects move in 2D and 3D spaces and make explicit comparisons. The user can flexibly adjust the view angle to highlight the primary information and ignore others. For example, if we only concern about long/lat, the time axis can be made to point to users such that the time will not be observed in the viewer.

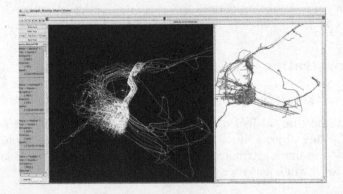

Fig. 3. A hybrid viewer

Acknowledgement. The work is funded by NSFC under grant number 61300052, Fundation of Graduate Innovation Center in NUAA under grant number KFJJ20161603, the Fundamental Research Funds for the Central Universities, and the Funding of Security Ability Construction of Civil Aviation Administration of China (AS-SA2015/21).

References

1. https://java3d.java.net
2. http://dbgroup.nuaa.edu.cn/jianqiu/
3. Castillo, S., Palma, G., Vidal, M.: SILURIAN: a SPARQL visualizer for understanding queries and federations. In: ISWC, pp. 137–140 (2013)
4. Cuzzocrea, A.: Advanced query answering techniques over big mobile data. In: IEEE MDM, pp. 4–7 (2016)
5. Gatterbauer, W.: Databases will visualize queries too. PVLDB **4**(12), 1498–1501 (2011)
6. Güting, R.H., Behr, T., Düntgen, C.: SECONDO: a platform for moving objects database research and for publishing and integrating research implementations. IEEE Data Eng. Bull. **33**(2), 56–63 (2010)
7. Güting, R.H., Behr, T., Xu, J.: Efficient k-nearest neighbor search on moving object trajectories. VLDB J. **19**(5), 687–714 (2010)
8. Haritsa, J.R.: The Picasso database query optimizer visualizer. PVLDB **3**(2), 1517–1520 (2010)
9. Piorkowski, M., Sarafijanovic-Djukic, N., Grossglauser, M.: CRAWDAD dataset epfl/mobility (v. 2009-02-24). http://crawdad.org/epfl/mobility/20090224
10. Xu, J., Güting, R.H.: Infrastructures for research on multimodal moving objects. In: IEEE MDM, pp. 329–332 (2011)

Author Index

Bao, Xuguang II-329
Bao, Yuqing II-324
Bao, Zhifeng I-74

Cai, Peng I-311, II-245
Cai, Yang I-642
Cao, Zhao I-362
Chao, Han-Chieh I-239
Chen, Ben I-527
Chen, Enhong I-575
Chen, Guidan II-294
Chen, Haibao I-411
Chen, Hong I-484, I-495
Chen, Ming I-251
Chen, Peixian II-341
Chen, Wei II-64, II-337
Chen, Yunfeng I-575
Chen, Zitong I-346
Cheng, Hong I-132
Cheng, Reynold I-3
Cui, Bin II-313
Cui, Wei II-349

Deng, Buqiao II-345, II-349
Ding, Zhiyuan I-626
Dong, Lei I-495
Du, He I-158
Du, Xiaoyong II-169
Duan, Lei II-185

Fan, Shiliang II-201
Fang, Zepeng I-565
Fei, Chaoqun I-266
Feng, Chong I-610
Feng, Zhiyong I-149, I-297, I-427
Fournier-Viger, Philippe I-215, I-239
Fu, Ada Wai-Chee I-346

Gan, Wensheng I-239
Gao, Dawei I-41
Gao, Yang I-610
Gao, Zihao II-276
Guo, Jinwei I-311
Guo, Xianjun II-219

He, Ben II-153
He, Yukun I-320
Hong, Shenda II-33
Hong, Xiaoguang I-399
Hou, Shengluan I-266
Hu, Huiqi I-320, II-210
Hu, Xiaoyi I-57
Huang, Heyan I-610
Huang, Jinjing I-100
Huang, Min II-124
Huang, Ming II-313
Huang, Ruizhang I-230, I-626
Huang, Ting I-230, I-626
Huang, Weijing II-64
Huang, Yu I-266
Huang, Zhipeng I-3

Ishikawa, Yoshiharu I-511
Iwaihara, Mizuho II-276

Ji, Yudian I-41
Jia, Mengdi II-83
Jiang, Jiawei II-313
Jiang, Jie II-313
Jiang, Jing II-245
Jiang, Shouxu II-114
Jiang, Yong II-73
Jin, Cheqing I-11
Jin, Hai I-411
Jin, Peiquan I-331, I-556, II-319
Jin, Yaohui I-27
Jin, Zhongxiao I-116

Keyaki, Atsushi II-133
Kishida, Shuhei II-133
Kito, Naoki I-391

Leung, Chun Fai II-341
Li, Cuiping I-484
Li, Gang II-3
Li, Guohua I-116
Li, Hongyan II-33
Li, Huanhuan I-57
Li, Kaixia I-362

Li, Li   I-650, II-124
Li, Lin   I-251
Li, Minglan   I-460
Li, Ning   I-642
Li, Qiong   I-427
Li, Xiang   II-48
Li, Xiaoming   II-285
Li, Xilian   II-64
Li, Xin   I-575
Li, Xuhui   I-541
Li, Zhijun   II-114
Li, Zhixu   I-100, I-200
Liang, Yun   I-565
Liao, Qun   I-158
Lin, Chen   I-565
Lin, Jerry Chun-Wei   I-215, I-239
Lin, Wutao   II-337
Liu, An   I-100, I-200
Liu, Bowei   I-230, I-626
Liu, Chuang   I-85, I-475
Liu, Guanfeng   I-200
Liu, Guiquan   I-575
Liu, Hao   II-219
Liu, Mengzhan   II-210
Liu, Qizhi   I-74
Liu, Shijun   I-444
Liu, Shushu   I-200
Liu, Tong   II-18
Liu, Yangming   I-484
Liu, Yixuan   II-276
Liu, Yuan   II-333
Long, Cheng   I-346
Lu, Ruqian   I-266
Lu, Wenyang   II-201
Lu, Yanmin   I-484
Luo, Qiong   II-219
Lv, Xia   I-331
Lv, Zhijin   I-527

Ma, Can   I-626
Ma, Ningning   I-185
Ma, Qiang   I-66
Mao, Jiali   I-11
Min, Zhang   I-591
Miyazaki, Jun   II-133
Mu, Lin   I-331

Ng, Eddie   I-3
Ni, Lionel M.   II-219
Ni, Weijian   II-18

Niu, Junyu   II-142
Niu, Zhendong   I-282
Nummenmaa, Jyrki   II-185

Pan, Li   I-444
Pang, Tianze   II-245
Pang, Tinghai   II-185
Peng, Hui   I-495
Peng, Zhaohui   I-399
Poon, Leonard K.M.   II-341

Qian, Tieyun   I-541
Qian, Weining   I-311, I-320, II-210, II-245
Qiang, Siwei   I-27
Qiao, Yu   I-377
Qin, Dong   I-391
Qin, Xiongpai   II-345, II-349
Qu, Dacheng   I-362
Qu, Wenwen   II-229

Rao, Guozheng   I-297
Rao, Weixiong   I-116
Ren, Yongli   I-391

Satoh, Shin'ichi   I-169
Sha, Chaofeng   II-142
Shen, Yizhu   I-66
Shi, Wei   I-132
Shi, Xiutao   I-444
Song, Guangxuan   II-229
Song, Ping   II-201
Sugiura, Kento   I-511
Sun, Haohao   II-18
Sun, Hui   I-495
Sun, Lei   I-158, II-345
Sun, Qiao   II-345, II-349

Tan, Haoyu   II-219
Tang, Suhua   I-169
Teng, Mingyan   II-345
Thom, James   I-391
Tian, Yuan   I-556
Tong, Yongxin   I-41

Ueda, Seiji   II-133

Wan, Shouhong   I-331, I-556
Wang, Bai   II-260
Wang, Chongjun   I-377, I-642

Wang, Dengbao   II-124
Wang, Donghui   II-245
Wang, Feng   II-3
Wang, Hao   II-98
Wang, Jiahao   I-311
Wang, Jing   I-650
Wang, Jingyuan   II-124
Wang, Junhu   I-149
Wang, Liqiang   I-444
Wang, Lizhen   II-329
Wang, Ning   II-324
Wang, Qinyong   II-98
Wang, Rui   I-626
Wang, Sen   I-282
Wang, Shan   II-169
Wang, Tengjiao   II-64, II-337
Wang, Weiwei   II-353
Wang, Xiaoliang   II-319
Wang, Xiaoling   II-229
Wang, Xiaorong   II-337
Wang, Xin   I-149, II-333
Wang, Xinghua   I-399
Wang, Xingjun   I-411
Wang, Yafang   I-444
Wang, Yilin   II-229
Wang, Ying   I-541
Wang, Yong   II-124
Wang, Yongheng   II-294
Wang, Yongkun   I-27
Wang, Zengwang   II-294
Wang, Zhuren   II-303
Wei, Linjing   I-610
Wei, Xiaochi   I-610
Wei, Zhensheng   II-18
Wen, Hui   I-460
Wong, Ka Yu   I-3
Wu, Bin   II-260
Wu, Hesheng   I-642
Wu, Jianguo   I-57
Wu, Jiayu   I-282
Wu, Jun   I-377
Wu, Lei   I-444
Wu, Meng   II-33
Wu, Song   I-411
Wu, Xiaoyu   II-324
Wu, Yang   I-346
Wu, Zhengwu   II-33

Xi, Yihai   II-324
Xia, Shu-Tao   II-73

Xiang, Jianwen   I-57
Xiao, Jiang   II-219
Xiao, Lin   I-591
Xiao, Qing   II-329
Xiao, Xi   I-185, II-73, II-303
Xie, Qing   I-57, I-251
Xie, Tao   II-260
Xie, Zizhe   I-74
Xing, Kai   I-169
Xing, Xiaolu   II-142
Xu, Jiajie   I-200
Xu, Jianqiu   II-353
Xu, Jungang   II-153
Xu, Ke   I-41
Xu, Liyang   I-626
Xu, Longlong   II-337
Xu, Ming   I-642
Xu, Qiang   I-149, II-333
Xu, Yanxia   I-100
Xu, Zhenhui   II-337

Yan, Jing   I-3
Yan, Rui   II-48
Yan, Yingying   I-230, I-626
Yang, Chengcheng   II-319
Yang, Chenhao   II-153
Yang, Wenyan   II-83
Yang, Xiaofei   II-114
Yang, Xiaolin   I-11
Yang, Yajun   I-149
Yang, Yubin   II-201
Yang, Yulu   I-158
Yang, Zhanbo   II-124
Yao, Xin   II-73
Ye, Qi   II-3
Ye, Zhili   I-460
Yin, Hongzhi   I-100, II-98
Yin, Jun   II-285
Yongfeng, Zhang   I-591
Yu, Fei   II-114
Yu, Jeffrey Xu   I-132
Yu, Lu   I-85, I-475
Yu, Wenli   II-124
Yu, Xiaohui   I-527
Yu, Yi   I-169
Yue, Lihua   I-331, I-556, II-319

Zhang, Chunxia   I-282
Zhang, Chuxu   I-85, I-475

Zhang, Dezhi II-319
Zhang, Jiexiong I-215
Zhang, Jinjing I-650
Zhang, Lei I-377, I-575
Zhang, Ming II-48
Zhang, Nevin L. II-341
Zhang, Peng II-185
Zhang, Shuhan I-266
Zhang, Xiao II-169
Zhang, Xiaowang I-427
Zhang, Xiaoying I-495
Zhang, Xiuzhen I-391
Zhang, Zhigang I-11
Zhang, Zi-Ke I-85, I-475
Zhao, Dongdong I-57
Zhao, Lei I-100, I-200
Zhao, Qiong II-210
Zhao, Suyun I-484
Zhao, Yan II-83
Zhao, Zhenyu I-297
Zheng, Hai-Tao I-185, II-73, II-303
Zheng, Kai I-200, II-83

Zheng, Weiguo I-132
Zheng, Yudian I-3
Zhong, Chunlin I-169
Zhong, Ming I-541
Zhou, Aoying I-11, I-311, I-320, II-210, II-245
Zhou, Huan I-320
Zhou, Ningnan II-169
Zhou, Tao I-85, I-475
Zhou, Wutong II-324
Zhou, Xiangmin I-391
Zhou, Xuan II-169
Zhu, Changlei II-3
Zhu, Jia II-83
Zhu, Tao I-320, II-210
Zhu, Tianchen I-399
Zhu, Xiaohang I-311
Zhu, Yuanyuan I-541
Zhuang, Chenyi I-66
Zong, Yu I-575
Zou, Lei I-132
Zuo, Jie II-185

Printed in the United States
By Bookmasters